FUZZY SET THEORY AND ADVANCED MATHEMATICAL APPLICATIONS

INTERNATIONAL SERIES IN INTELLIGENT TECHNOLOGIES

Other books in the series:

Applied Research in Fuzzy Technology
by Anca L. Ralescu

Analysis and Evaluation of Fuzzy Systems
by Akira Ishikawa and Terry L. Wilson

Fuzzy Logic and Intelligent Systems
by Hua Li and Madan Gupta

FUZZY SET THEORY AND ADVANCED MATHEMATICAL APPLICATIONS

EDITED BY

Da Ruan
Belgian Nuclear Research Centre (SCK • CEN)
Mol, Belgium

KLUWER ACADEMIC PUBLISHERS
Boston/London/Dordrecht

Distributors for North America:
Kluwer Academic Publishers
101 Philip Drive
Assinippi Park
Norwell, Massachusetts 02061 USA

Distributors for all other countries:
Kluwer Academic Publishers Group
Distribution Centre
Post Office Box 322
3300 AH Dordrecht, THE NETHERLANDS

Library of Congress Cataloging-in-Publication Data

A C.I.P. Catalogue record for this book is available
from the Library of Congress.

Printed on acid-free paper.

Printed in the United States of America

to Etienne Kerre

for his 50th birthday

CONTENTS

CONTRIBUTORS xiii

FOREWORD
Hans-Jürgen Zimmermann xv

EDITOR'S PREFACE
Da Ruan xvii

Part 1: FOUNDATIONS OF FUZZY SET THEORY

1 ON THE EVOLUTION OF THE MATHEMATICS OF FUZZINESS
Etienne Kerre 3
1 Mathematics of fuzziness during the seventies 3
2 Mathematics of fuzziness during the eighties 9
3 Current mathematics of fuzziness 18
REFERENCES 23

2 CLOSEDNESS AND UPPER SEMI-CONTINUITY OF FUZZY MULTIVALUED MAPPINGS
Elena Tsiporkova-Hristoskova 35
1 Introduction 35
2 Preliminaries 37
3 Closedness of fuzzy multivalued mappings 42
4 Upper semi-continuity of fuzzy multivalued mappings 46
5 Relationship between closedness and upper semi-continuity 50

6 Properties derived from the interaction of closedness and upper semi-continuity 55
7 Examples of families of implication operators 63
REFERENCES 65

3 AN ORDER-THEORETHIC APPROACH TO SOLVING SUP-T EQUATIONS
Bernard De Baets 67
1 Some order-theoretic notions 68
2 The equation $\mathcal{T}(\alpha, x) = \beta$ 71
3 Sup-$\mathcal{T}$ equations 75
4 The real unit interval 80
REFERENCES 86

4 TOWARDS A POSSIBILISTIC LOGIC
Gert de Cooman 89
1 Introduction 89
2 Preliminary definitions 91
3 Possibility theory and the extension principle 97
4 Possibilistic truth values and their combinations 105
5 Possibilistic extension logics 115
6 An interesting special case 122
7 Conclusion 130
REFERENCES 131

Part 2: DECISION MAKING AND SYSTEMS SCIENCE

5 FUZZY DYNAMIC PROGRAMMING: A NEW QUALITY THROUGH FUZZY SETS
Janusz Kacprzyk 137
1 Introduction 137
2 Decision making and control under fuzziness 138
3 FDP formulation for a fixed and specified termination time 139
4 FDP formulation with a fuzzy termination time 144

5 Remarks on the cases of an implicitly given and infinite termination times 147
6 Remarks on some further extensions 149
7 Applications of fuzzy dynamic programming 149
8 Concluding remarks 150
REFERENCES 150

6 ON TRANSITIVITY OF FUZZY PREFERENCE RELATIONS IN RANKING FUZZY NUMBERS
Xuzhu Wang and Da Ruan 155
1 Introduction 155
2 Further investigations into relations between some transitivity-related notions 157
3 The transitivity of indices introduced by Dubois and Prade 162
4 Ordering method under $W3$-Transitivity 166
5 Concluding remarks 170
REFERENCES 171

7 INFORMATION AGGREGATION: ETHICAL AND COMPUTATIONAL ISSUES
Vincenzo Cutello and Javier Montero 175
1 Introduction and preliminaries 175
2 Truth values aggregation 177
3 Aggregation of preference intensities 179
4 Cover-based hierarchical aggregation rules 184
5 OWA operators and ordered hierarchies 187
6 Maximizing and minimizing dispersion and orness 193
7 Final Comments 197
REFERENCES 198

8 ON THE NOTION STATE IN MULTISTATE STRUCTURE FUNCTION THEORY
Bart Cappelle 201
1 Introduction 201
2 From binary to multistate models: a summary 203

3 Some seminal ideas about the notion state 205
4 Systems, components and their possible states 207
5 On the notion state space 212
6 Conclusion 217
REFERENCES 218

Part 3: FUZZY LOGIC, FUZZY CONTROL, AND NEURO-FUZZY CONTROL

9 FUZZY LOGIC AND NEURAL SYSTEMS
Madan M. Gupta 225
1 Introduction 225
2 Cognition, uncertainty, and intelligence 227
3 Cognition and relative grades 229
4 Introduction to fuzzy logic 230
5 Perspectives 230
6 Virtual intelligence 231
7 Human Cognition 232
8 The neuron 233
9 Nonlinear systems 235
10 Neural-fuzzy logic 236
11 Neuronal learning 238
12 Virtual cognition 239
13 Some remarks 241
REFERENCES 242

10 SOME MATHEMATICAL PROBLEMS OF FUZZY CONTROLLERS
László T. Kóczy 245
1 Introduction 245
2 The linguistic approach to fuzzy control 246
3 Mathematical formulae of fuzzy controllers 248
4 Fuzzy controllers as universal approximators 253
5 Rule interpolation controllers 254
6 Control with structured rule bases 259

	7 Conclusions	261
	REFERENCES	262
11	**KEEP FUZZY CONTROLLERS FUZZY**	
	Ann Nowé	267
	1 introduction	267
	2 fuzzy controller laws are fuzzy valued mappings rather than crisp valued mappings	269
	3 Fuzzy controllers as fuzzy graphs	272
	4 An inference mechanism based on fuzzy piecewise interpolation	276
	5 conclusion	288
	REFERENCES	288
12	**COMPARATIVE STUDY OF FUZZY CONTROL, NEURAL NETWORK CONTROL AND NEURO-FUZZY CONTROL**	
	Jelena Godjevac	291
	1 Introduction	291
	2 Fuzzy control	293
	3 Khepera mobile robot	301
	4 Neural networks	305
	5 Fuzzy neurons	310
	6 Neuro-fuzzy controller	313
	7 Conclusion	318
	REFERENCES	319
SUBJECT INDEX		323

CONTRIBUTORS

Bart Cappelle
Dept. of Applied Mathematics and CS
University of Gent
Gent, Belgium

Vincenzo Cutello
Dept. of Mathematics
University of Catania
Catania, Italy

Bernard De Baets
Dept. of Applied Mathematics and CS
University of Gent
Gent, Belgium

Gert de Cooman
Vakgroep Elektrische Energietechniek
University of Gent
Gent, Belgium

Jelena Godjevac
Microcomputing Laboratory
Swiss Federal Institute of Technology
Lausanne, Switzerland

Madan M. Gupta
Intelligent Systems Research Laboratory
University of Saskatchewan
Saskatoon, Sask., Canada

Janusz Kacprzyk
Systems Research Institute
Polish Academy of Sciences
Warsaw, Poland

László T. Kóczy
Dept. of Telecomm. and Telematics
Technical University of Budapest
Budapest, Hungary

Etienne Kerre
Fuzziness and Uncertainty Modeling
University of Gent,
Gent, Belgium

Javier Montero
Dept. of Statistics and OR
Complutense University
Madrid, Spain

Ann Nowé
Department Info
University of Brussel
Brussels, Belgium

Da Ruan
Fuel Research Unit
Nuclear Research Centre (SCK•CEN)
Mol, Belgium

Elena Tsiporkova-Hristoskova
Dept. of Applied Mathematics and CS
University of Gent
Gent, Belgium

Xuzhu Wang
Fuel Research Unit
Nuclear Research Centre (SCK•CEN)
Mol, Belgium

FOREWORD

Professor Etienne Kerre is one of the earliest pioneers in the area of Fuzzy Set Theory. It would by far exceed the scope of this foreword to describe his opus in this area in detail. Nevertheless it might be appropriate to mention a few of his earlier contributions in a volume dedicated to his 50th birthday and anniversary of almost two decades of extremely creative and productive scientific work in the theory and application of fuzzy set theory.

Already in 1978 he presented a contribution on "Fuzzy topologizing with preassigned operations" at the International Congress for Mathematicians in Helsinki. In 1980 he spoke about "The usefulness of fuzzy sets in medical diagnosis" in Gent. In 1983 he already talked about "Application of fuzzy programming to ship steering" and about "An axiomatic approach to fuzzy connectives" at the NAFIPS conference in the USA and already in 1984 he started his work in fuzzy document retrieval with a contribution at the Founding Conference of IFSA in Hawaii.

This was followed by his presentation "An expert system development tool using fuzzy logic: an application to speech recognition" at the NAFIPS conference in 1986 in New Orleans.

Simultaneous with these almost missionary activities Etienne started publishing in various international scientific journals, such as the *Quarterly Journal of Pure and Applied Mathematics* in 1979, in the *Journal of Fuzzy Sets and Systems.* His numerous publications culminated with his book *Introduction to the Basic Principles of Fuzzy Set theory and some of its Applications* in 1991.

What is still more impressive than the wide scope of his publications in the early times of fuzzy set theory is that he did not only contribute scientifically, but that already in the academic year 1979–80 he started offering classes on fuzzy set theory to his students at the University of Gent and he proceeded with this academic service continuously until today.

These are only a few of Etienne Kerre's contributions to the area of fuzzy set theory. They might, however, already hint at the wide scope of his contributions, his worldwide missionary activity in the area and strong and creative productivity in research, teaching and supervision of PhD and Master students, and applications. It is probably not surprising, even though highly worth of recognition, that Dr Ruan untertook the challenging and certainly not easy task to celebrate Etienne's 50th birthday by dedicating to him a book, written only by his disciples and friends. In it three areas are covered, which have always been of particularly interest to Professor Kerre: Mathematical Foundations of Fuzzy Set Theory, Decision Making and Systems Science, and the newest development in this area, namely, Fuzzy Neural Control. The book will certainly please Professor Etienne Kerre, but it will also be extremely useful to all those working in these areas of fuzzy set theory. Dr Ruan deserves highest appreciation for this excellent book and Prof. Kerre congratulations to such disciples and friends that go out of their way to show him their gratefulness and appreciation. I am joining them from all my heart and hope that the book will be of benefit to many readers around the world.

Hans-Jürgen Zimmermann

EDITOR'S PREFACE

This book is a congratulatory volume on the occasion of Professor Etienne Kerre's 50th birthday, for his almost twenty years of teaching and research on fuzzy set theory and its applications at the University of Gent, Belgium.

As one of Prof. Kerre's former PhD students (as well as one of Etienne's friends), I have benefited from a close collaboration with him for more than seven years. In the last two years in particular, I enjoyed a fruitful cooperation between the Belgian Nuclear Research Centre and the University of Gent, in the framework of the new international forum FLINS (Fuzzy Logic and Intelligent technologies in Nuclear Science).

I am very pleased to edit, on behalf of FLINS, this issue on the Stimulating Role of Fuzzy Set Theory in Mathematics in the series of FLINS volumes, in honour of Prof. Kerre (honorary chair of FLINS). This issue consists of a collection of chapters on fuzzy set theory, written by researchers who received a PhD no more than five years ago in Fuzzy Set Theory and Application (most of them under supervision of Prof. Kerre) and by leading experts, as well as Prof. Kerre's friends. Due to the limitation of time and space, unfortunately, some of the contributions submitted by both Prof. Kerre's PhD students and friends could not be included in this volume.

The issue is to provide a readable and practical toolbook with a clear introduction to fuzzy set theory and its evolution in mathematics (Chapter 1) and new results on foundations of fuzzy set theory (Chapters 2-4), on decision making and systems science (Chapters 5-8), and on fuzzy control and neural systems (Chapters 9-12) for university students, researchers, and engineers to better understand the role of fuzzy set theory. Each chapter of the issue is self-contained and I hope this volume will not only be a major reference about Fuzzy Logic for FLINS, but also be an important textbook on new achievements in Fuzzy Logic and related intelligent systems for both academic and industrial research.

Special thanks are due to Prof. Zimmermann (Chairman of the international scientific advisory committee of FLINS) for his encouragements and advice regarding the FLINS activities, and for his foreword to this issue; to all contributors for their kind cooperation to this congratulatory FLINS volume; to Alexander Greene (Publisher, Kluwer Academic Publishers) for his kind advice and help to publish this volume, to Dr J.-L. Doumont (JL Consulting, Brussels) for his proofreading of parts of the manuscript and his final typesetting of this book, and to the Belgian Nuclear Research Centre for its support.

Da Ruan

Part 1

FOUNDATIONS OF FUZZY SET THEORY

1

ON THE EVOLUTION OF THE MATHEMATICS OF FUZZINESS

Etienne Kerre

Fuzziness and Uncertainty Modeling
Universiteit Gent
Krijgslaan 281-S9
B-9000 Gent, Belgium

ABSTRACT

Since the publication in 1965 of the seminal paper "Fuzzy Sets" by L.A. Zadeh, mathematicians started to fuzzify the classical mathematical structures. Three stages can be considered: firstly the straightforward fuzzification during the seventies, secondly the explosion of the possible choices in the generalization process during the eighties and currently the standardization, axiomatization and L-fuzzification. In this contribution we will illustrate these three stages by many examples and we will outline the stimulating role of fuzzy set theory in the expansion of mathematics. Most of the illustrative examples in the different stages in the development of the mathematics of fuzziness are taken from the results of our research group during the past twenty years.

1 MATHEMATICS OF FUZZINESS DURING THE SEVENTIES

The first stage (1965-1980) in the development of the mathematics of fuzziness can be described as the period of straightforward fuzzifications. During that period almost every domain of pure as well as of applied mathematics that relies on classical binary logic and classical set theory has been fuzzified: fuzzy topology, fuzzy algebraic structures such as fuzzy groups and fuzzy vector spaces, fuzzy arithmetic, fuzzy measure theory, fuzzy relational calculus ... were born. The key issues during this first stage were:

(i) the definition of a suitable fuzzification of the existing classical notions; i.e., suitable in the sense that when the extension is applied to classical sets, the classical notion is obtained;

(ii) the determination of the properties of the classical concept that remain valid for the generalization given in (i).

In the following we will give three examples of this straightforward fuzzification process; each example comes from a different domain of mathematics: set theory, algebra and topology.

1.1 The fuzzification of the cartesian product

Undoubtedly the cartesian product of sets may be considered as one of the most fundamental notions in set theory. Indeed as soon as the classifier $\{\ |\ \}$ has been defined, either intuitively or based on the well-known axiom of abstraction that states that for an arbitrary formula P there exists precisely one set $\{x|P\}$ whose elements satisfy P, one obtains successively the concepts of a singleton, a doubleton and an ordered pair:

$$\begin{aligned}\{x\} &= \{z|z = x\}\\ \{x,y\} &= \{z|z = x \text{ or } z = y\}\\ (x,y) &= \{\{x\},\{x,y\}\}.\end{aligned}$$

Already now we are able to define the cartesian product $A \times B$ of two sets A and B as the set of all ordered pairs (a,b) such that $a \in A$ and $b \in B$. It is well known that a relation R from A to B is defined as a subset of $A \times B$. Recall that functions have been defined as special relations. The importance of the concept of the cartesian product and hence of its generalization may be clear now.

Let A_1 and A_2 be two fuzzy sets in a universe X. Zadeh [1] defined the cartesian product of A_1 and A_2 as the fuzzy set in $X \times X$ given by $A_1 \times A_2(x,y) = \min(A_1(x), A_2(y)), \forall (x,y) \in X \times X$.

It is easily verified that this definition clearly extends the cartesian product of two crisp (sub)sets in X. In the following we have listed the main properties of the extended cartesian product; A_1, A_2 and A_3 are three fuzzy sets in a universe X,

(1) $A_1 \times A_2 = \phi \Longleftrightarrow A_1 = \phi$ or $A_2 = \phi$

(2) $A_1 = A_2 \Longrightarrow A_1 \times A_2 = A_2 \times A_1$

(2') $A_1 \times A_2 = A_2 \times A_1 \not\Longrightarrow A_1 = A_2$

(3) $A_1 \times (A_2 \cup A_3) = (A_1 \times A_2) \cup (A_1 \times A_3)$

(3') $A_1 \times (A_2 \cap A_3) = (A_1 \times A_2) \cap (A_1 \times A_3)$

(4) $A_1 \times (A_2 \hat{+} A_3) \subseteq (A_1 \times A_2) \hat{+} (A_1 \times A_3)$

(4') $(A_1 \times A_2).(A_1 \times A_3) \subseteq A_1 \times (A_2.A_3)$

(5) $A_1 \times (A_2 +_b A_3) \subseteq (A_1 \times A_2) +_b (A_1 \times A_3)$

(5') $(A_1 \times A_2) \bigcap (A_1 \times A_3) \subseteq A_1 \times (A_2 \bigcap A_3)$

(6) $A_1 \times (A_2 \setminus A_3) \subseteq (A_1 \times A_2) \setminus (A_1 \times A_3)$

(6') $(A_1 \times A_2) \setminus (A_1 \times A_3) \nsubseteq A_1 \times (A_2 \setminus A_3)$

(7) $A_2 \subseteq A_3 \Longrightarrow A_1 \times A_2 \subseteq A_1 \times A_3$

(7') $A_1 \times A_2 \subseteq A_1 \times A_3 \not\Longrightarrow A_2 \subseteq A_3$, if $A_1 \neq \phi$

(7") $A_1 \times A_2 \subseteq A_1' \times A_2' \not\Longrightarrow A_1 \subseteq A_1'$ and $A_2 \subseteq A_2'$, if $A_1 \times A_2 \neq \phi$

(8) $A_1 \times (A_2 \cup_S A_3) \subseteq (A_1 \times A_2) \cup_S (A_1 \times A_3)$

(8') $(A_1 \times A_2) \cap_T (A_1 \times A_3) \subseteq A_1 \times (A_2 \cap_T A_3)$

These properties reveal some deviations with respect to the cartesian product of crisp sets. Property (2) states that contrastedly to classical set theory the cartesian product of two fuzzy sets may be commutative without the two components being equal. Property (6) expresses that the cartesian product of two fuzzy sets no longer remains distributive with respect to the extended difference operation. Properties (7') and (7") indicate that the cartesian product of fuzzy sets no longer obeys the cancellation law. Properties (8) and (8') reveal a kind of semi-distributivity of the cartesian product with respect to an arbitrary S-union and an arbitrary T-intersection where S is a triangular conorm and T a triangular norm, properties (3) and (3') show that we get the distributivity if S = max and T = min, i.e., if Zadeh's original extensions for union and intersection are taken. For a proof of the properties in table 1 we refer to [1].

1.2 Elementary fuzzy algebraic structures

The fuzzification of the algebraic structure of a group and an ideal has been initiated by Rosenfeld [3] in 1971. Negoita and Ralescu [4] introduced the concept of a fuzzy module, while Katsaras was the first to investigate fuzzy vector spaces in 1977. As an example of this fuzzification process we recall the following concepts from [3].

Let $(X, .)$ be a groupoid and A be a fuzzy set in the underlying universe X.

The structure $(A, .)$ is called:

- a fuzzy groupoid iff $(\forall(x, y) \in X^2)(A(x.y) \geq \min(A(x), A(y)))$
- a fuzzy left ideal iff $(\forall(x, y) \in X^2)(A(x.y) \geq A(y))$
- a fuzzy right ideal iff $(\forall(x, y) \in X^2)(A(x.y) \geq A(x))$
- a fuzzy ideal iff $(\forall(x, y) \in X^2)(A(x.y) \geq \max(A(x), A(y))$.

It is easily seen that these concepts are real extensions of their classical counterparts, i.e., for A a classical subset of X holds:

$$(A, .) \text{ is a left ideal } \Longleftrightarrow (\chi_A, .) \text{ is a fuzzy left ideal}$$

where χ_A stands for the characteristic mapping of A in X.

Secondly one easily proves the following characterization in terms of the weak α-levels, i.e., for $A \in \mathcal{F}(X)$ holds:

$$(A, .) \text{ is a fuzzy groupoid } \Longleftrightarrow (\forall \alpha \in]0, 1])((A_\alpha, .) \text{ is a subgroupoid})$$

Furthermore one has proved the following generalizations:

- the Zadeh intersection of an arbitrary family of fuzzy groupoids of a groupoid is again a fuzzy groupoid;
- the inverse image of a fuzzy groupoid under a homomorphism is a fuzzy groupoid.

Finally, with respect to the direct image one could only prove the following restricted generalization of the well-known classical property: the direct image of a fuzzy groupoid with the sup-property, under a homomorphism is a fuzzy groupoid.

1.3 Fuzzy topology

The concept of a fuzzy topology has been introduced by Chang [5] in 1968: A fuzzy topology on a universe X is a class τ of fuzzy sets in X satisfying the conditions:

(O.1) $(\forall k \in \{0,1\})(\underline{k} \in \tau)$

(O.2) $O_1 \in \tau$ and $O_2 \in \tau \Longrightarrow O_1 \cap O_2 \in \tau$

(O.3) $(\forall j \in J)(O_j \in \tau) \Longrightarrow \bigcup_{j \in J} O_j \in \tau$

The definition of a fuzzy topology has been modified by Lowen [6] in 1976. He has given several reasons to justify the replacement of condition (O.1) by the stronger condition:

(O.1') $(\forall k \in [0,1])(\underline{k} \in \tau)$

where $\underline{k}$ denotes the constant $X - \{k\}$ mapping on the given X.

We will restrict ourselves here to a Chang fuzzy topology. As in classical topology one has defined:

- the class τ' of all closed fuzzy sets in (X, τ) as

$$\tau' = \{F \mid \operatorname{co}F \in \tau\};$$

- the interior of a fuzzy set A in X as:

$$\operatorname{int}_\tau(A) = \cup\{O \mid O \in \tau \text{ and } O \subseteq A\};$$

- the closure of a fuzzy set A in X as:

$$\operatorname{cl}_\tau(A) = \cap\{F \mid F \in \tau' \text{ and } A \subseteq F\};$$

- the neighbourhood of a fuzzy singleton p in X as:

$$A \in \mathcal{U}_p \Longleftrightarrow (\exists O \in \tau)(p \subseteq O \subseteq A)$$

where a fuzzy singleton p in X is defined as a fuzzy set in X whose support is an ordinary singleton in X; the content of this support is called the support of p and the unique nonzero degree of membership is called the value of the fuzzy singleton. Usually a fuzzy singleton with support x and value ε is denoted as x_ε.

It is well known that a classical topology can be characterized by means of preassigned operations such as the closed sets, the interior operator, the closure operator and by means of neighbourhood systems. My first contribution to fuzzy set theory concerned the generalization of these classical characterizations into the framework of fuzzy topology [7]. We could easily prove that a fuzzy topology can be characterized by means of the class of closed fuzzy sets, by means of an interior operator and by means of a closure operator. As an example we mention the characterization in terms of an interior operator. Let i be a $\mathcal{F}(X) - \mathcal{F}(X)$ mapping satisfying the conditions:

(i.1) $i(X) = X$

(i.2) $(\forall A \in \mathcal{F}(X))(i(A) \subseteq A)$

(i.3) $(\forall A \in \mathcal{F}(X))(i(i(A)) = i(A))$

(i.4) $(\forall (A, B) \in \mathcal{F}(X)^2)(i(A \cap B) = i(A) \cap i(B))$

and let τ be the subset of $\mathcal{F}(X)$ given by

$$\tau = \{A | A \in \mathcal{F}(X) \text{ and } i(A) = A\}$$

Then we could prove that τ constitutes a fuzzy topology on X such that $\text{int}_\tau = i$, i.e., the interior operator corresponding to the fuzzy topology τ equals the given operator i.

We also tried the characterization by means of neighbourhood systems. As in the previous characterizations we started from a straightforward fuzzification of the classical neighbourhood axioms. More explicitly let X be a universe and for every fuzzy singleton p in X a class $\mathcal{U}_p$ of fuzzy sets in X is given. Suppose that this class $\mathcal{U}_p$ satisfies the following conditions:

(N.1) $\mathcal{U}_p \neq \phi$

(N.2) $A_1 \in \mathcal{U}_p$ and $A_2 \in \mathcal{U}_p \Longrightarrow A_1 \cap A_2 \in \mathcal{U}_p$

(N.3) $A_1 \in \mathcal{U}_p$ and $A_1 \subseteq A_2 \Longrightarrow A_2 \in \mathcal{U}_p$

(N.4) $(\forall A_1 \in \mathcal{U}_p)(\exists A_2 \in \mathcal{U}_p)(A_2 \subseteq A_1 \text{ and } (\forall q \subseteq A_2)(A_1 \in \mathcal{U}_q))$.

Now we define the class $\tau_{\mathcal{U}}$ as:

$$\tau_{\mathcal{U}} = \{O | O \in \mathcal{F}(X) \text{ and } (\forall p \subseteq O)(O \in \mathcal{U}_p)\}.$$

i.e. as the class of all fuzzy sets that are a neighbourhood of all the fuzzy singletons contained in them. In [7] we were not able to prove that $\tau_{\mathcal{U}}$ constitutes a fuzzy topology on X. The main reason for this failure was the following deviation from the crisp case. For an arbitrary class $(O_i)_{i \in I}$ of fuzzy sets in a universe X and p a fuzzy singleton in X one has:

$$(\exists i \in I)(p \subseteq O_i) \Longrightarrow p \subseteq \bigcup_{i \in I} O_i$$

but not the reverse implication, i.e., a fuzzy singleton can be contained in an arbitrary union of fuzzy sets without being contained in one of them.

At the end of the seventies Pu and Liu [8] claimed the characterization mentioning a straightforward proof! We became very curious and we started a very detailed research on this topic. We will return on this in the next section.

Other examples of the straightforward fuzzification of the classical structures are the calculus of fuzzy relations, the introduction of fuzzy metric spaces and the initiation of fuzzy measure theory and fuzzy integrals [9].

2 MATHEMATICS OF FUZZINESS DURING THE EIGHTIES

The second stage can be described as the stage of explosion of the possible choices in the process of generalization. The discovery of triangular norms and triangular conorms that have been introduced in the framework of probabilistic metric spaces [10] has considerably influenced the process of generalization. During the second stage we have seen a deep study of the alternative operations on a logical level, i.e., negation, disjunction, conjunction, implication, as well as on a fuzzy-set-theoretic level, i.e., complementation, union, intersection, inclusion. Many of pioneers in fuzzy set theory have contributed to this domain, among them Bellman, Zadeh, Giertz [11], Dubois and Prade [12], Yager [13], Zimmermann, Zysno [14], Bandler, Kohout [15].

This second period is also characterized by diversification due to the non-equivalence of the generalizations. Important examples of this diversification process can be found in the disjointness of fuzzy sets, the neighbourhood concept in a fuzzy topology, the transitivity of a fuzzy relation and the concept of a fuzzy number; we will now outline some of them.

2.1 On the extension of normality in fuzzy topological spaces

Let us first repeat the definitions in a classical topological space. Let (X, T) be topological space, then we define:

$$(X, T) \text{ is normal } \Longleftrightarrow (\forall (F_1, F_2) \in (T')^2)(F_1 \cap F_2 = \phi \Longrightarrow$$
$$(\exists (O_1, O_2) \in T^2)(O_1 \cap O_2 = \phi \text{ and } F_1 \subseteq O_1 \text{ and } F_2 \subseteq O_2)$$

where T' denotes the class of all closed sets. It is well known that Urysohn has proven the following characterization of normality:

$$(X, T) \text{ is normal}$$
$$\Updownarrow$$
$$(\forall O \in T)(\forall F, F \in T' \text{ and } F \subseteq O)(\exists V \in \mathcal{P}(X))(F \subseteq \text{Int}V \text{ and } \text{cl}V \subseteq O)$$

where Int$V$ (cl$V$) denotes the interior (the closure) of V.

Now, let (X, τ) be a fuzzy topological space. In 1975 Hutton [16] introduced the following definition of normality:

$$(X, \tau) \text{ is normal}$$
$$\Updownarrow$$
$$(\forall O \in \tau)(\forall F, F \in \tau' \text{ and } F \subseteq O)(\exists V \in \mathcal{F}(X))(F \subseteq \text{Int}V \text{ and } \text{cl}V \subseteq O)$$

i.e., a straightforward fuzzification of Urysohn's characterization. An important question remained unanswered: why did Hutton take Urysohn's form to generalize normality in fuzzy topological spaces? The answer has been given in [17, 18]. To fuzzify the classical concept of normality we need a suitable fuzzification of disjointness. For crisp sets the following equivalence holds:

$$F_1 \cap F_2 = \phi \Longleftrightarrow F_1 \subseteq \text{co}F_2.$$

For fuzzy sets however only the implication

$$F_1 \cap F_2 = \phi \Longrightarrow F_1 \subseteq \text{co}F_2.$$

holds. Based on this observation we introduced the following concepts for a fuzzy topological space (X, τ):

(X, τ) is normal $\iff (\forall (F_1, F_2) \in \tau'^2)(F_1 \subseteq \text{co } F_2 \implies (\exists (O_1, O_2) \in \tau^2)(O_1 \subseteq \text{co } O_2$ and $F_1 \subseteq O_1$ and $F_2 \subseteq O_2))$

(X, τ) is weakly normal $\iff (\forall (F_1, F_2) \in \tau'^2)(F_1 \cap F_2 = \phi \implies (\exists (O_1, O_2) \in \tau^2)(O_1 \subseteq \text{co } O_2$ and $F_1 \subseteq O_1$ and $F_2 \subseteq O_2))$

(X, τ) is completely normal $\iff (\forall (A_1, A_2) \in \mathcal{F}(X)^2))(A_1 \subseteq \text{co cl}(A_2)$ and $A_2 \subseteq \text{co cl}(A_1) \implies (\exists (O_1, O_2) \in \tau^2)(A_1 \subseteq O_1$ and $A_2 \subseteq O_2$ and $O_1 \subseteq \text{co } O_2)$.

It is easily seen that for crisp topological spaces both concepts normality and weakly normality coincide with the classical normality concept. The following properties have been proven [17, 18]:

(P.1) (X, τ) is completely normal
$\implies (X, \tau)$ is normal.

(P.2) (X, τ) is normal
$\implies (X, \tau)$ is weakly normal.

(P.3) (X, τ) is normal
$\iff (X, \tau)$ is normal in the sense of Hutton.

(P.4) (X, τ) is completely normal
$\implies$ every subspace of (X, τ) is normal.

The converse of property (P.2) is not valid, since the fuzzy Sierpinski space introduced in [19] constitutes a counterexample. From property (P.3) it can be seen that our definition of normality, i.e., interpreting the disjointness of F_1 and F_2 as $F_1 \subseteq \text{co } F_2$, leads to the fuzzy analogue of Urysohn's theorem. On the other hand we couldn't prove the fuzzy extension of Tietze's characterization theorem of complete normality, i.e., the converse of property (P.4) is not true.

2.2 On the different notions of neighbourhood in fuzzy topological spaces.

The concept of a neighbourhood in a fuzzy topological space relies on the fuzzification of a point and the corresponding membership relation. We already have introduced the concept of a fuzzy singleton in a universe X as a fuzzy set in X with support a crisp singleton, or more explicitly:

$$\begin{aligned} x_\varepsilon \;:\; & X \to [0,1] \\ & x \longmapsto \varepsilon \\ & y \longmapsto 0, \text{ if } y \in X \setminus \{x\} \end{aligned}$$

where $\varepsilon \in]0,1]$.

A fuzzy singleton x_ε will be called a fuzzy point iff $\varepsilon \in]0,1[$. Note that every point x in X can be considered as the fuzzy singleton x_1 but not as a fuzzy point. The following membership relations have been introduced [8, 20, 21] for $A \in \mathcal{F}(X)$ and x_ε a fuzzy singleton in X:

- x_ε is called a subset of A (denoted as $x_\varepsilon \subseteq A$) iff $\varepsilon \leq A(x)$;
- x_ε is called quasi-coincident with A (denoted as $x_\varepsilon q A$) iff $\neg(x_\varepsilon \subseteq \text{co} A)$, or equivalently, $\varepsilon + A(x) > 1$.

and for $A \in \mathcal{F}(X)$ and x_ε a fuzzy point in X:

- x_ε is said to belong to A (denoted as $x_\varepsilon \underset{\sim}{\in} A$), iff $\varepsilon < A(x)$.

In [22] we have proven that fuzzy singletons and fuzzy points reveal a complementary attitude towards the operations of Zadeh. Let us give some examples of this complementary behaviour. Let $(A_j)_{j \in J}$ be an arbitrary family of fuzzy sets in X, s a fuzzy singleton in X and p a fuzzy point in X. Then the following properties hold:

(P.1) $s \subseteq \bigcap_{j \in J} A_j \Longleftrightarrow (\forall j \in J)(s \subseteq A_j)$

(P.2) $s q \bigcap_{j \in J} A_j \Longrightarrow (\forall j \in J)(s q A_j)$

(P.3) $p \underset{\sim}{\in} \bigcap_{j \in J} A_j \Longrightarrow (\forall j \in J)(p \underset{\sim}{\in} A_j)$

(P.4) $(\exists j \in J)(s \subseteq A_j) \Longrightarrow s \subseteq \bigcup_{j\in J} A_j$

(P.5) $(\exists j \in J)(sqA_j) \Longleftrightarrow sq \bigcup_{j\in J} A_j$

(P.6) $(\exists j \in J)(p \underset{\sim}{\in} A_j) \Longleftrightarrow p \underset{\sim}{\in} \bigcup_{j\in J} A_j$.

Note that the converse implication in (P.2), (P.3) and (P.4) is true for a finite index set J, but not for an arbitrary index set J. This means that contrastedly to the crisp case, a fuzzy point p may belong to every element of an arbitrary class of fuzzy sets, yet not belong to its intersection.

Now we are ready to state the different neighbourhood concepts that have been introduced by Ludescher and Roventa [23], Kerre [7], Warren [24], Pu and Liu [8], Mashhour, Ghanim, Abd. El-Monsef and Kerre [25,26].

Let (X, τ) be a fuzzy topological space, A a fuzzy set in X, x an element of X, s a fuzzy singleton in X and p a fuzzy point in X. Then we define:

- A is a Ludescher neighbourhood of x
  $\Longleftrightarrow (\exists O \in \tau)(x \in \text{supp } O \text{ and } O \subseteq A)$.
- A is a Kerre neighbourhood of s
  $\Longleftrightarrow (\exists O \in \tau)(s \subseteq O \subseteq A)$.
- A is a Warren neighbourhood of x
  $\Longleftrightarrow (\exists O \in \tau)(x \in \text{supp } O \text{ and } O \subseteq A \text{ and } O(x) = A(x))$.
- A is a Pu neighbourhood of s
  $\Longleftrightarrow (\exists O \in \tau)(sqO \text{ and } O \subseteq A)$.
- A is a Mashhour neighbourhood of p
  $\Longleftrightarrow (\exists O \in \tau)(p \underset{\sim}{\in} O \text{ and } O \subseteq A)$.

The class of all Ludescher neighbourhoods of x is denoted as: $\mathcal{L}_x(\tau)$. Similar notations are used to denote the other neighbourhood classes: $\mathcal{K}_x(\tau)$, $\mathcal{W}_x(\tau)$, $\mathcal{P}_s(\tau)$, $\mathcal{M}_p(\tau)$.

In a series of papers [22,27-31] we have reported about a detailed research concerning the different notions of neighbourhood and the relations between them. Let us summarize the main results of this research:

(i) All neighbourhood systems, except that of Warren, satisfy the purely formal translated properties of classical neighbourhood systems $N_1 - N_4$, as listed in section 1.3. Warren's neighbourhood concept exhibits the departure from ordinary topology that enlarging a neighbourhood not necessarily produces a new neighbourhood.

(ii) We completely solved the characterization problem, i.e., given a class of fuzzy sets satisfying some conditions is there a unique fuzzy topology having the given fuzzy sets as neighbourhoods?

- A Ludescher system defines a fuzzy topology that is not necessarily unique i.e., different fuzzy topologies can lead to the same Ludescher neighbourhood system. So there is no characterization of a fuzzy topology by means of a Ludesher neigbourhood system.
- A Kerre system defines no fuzzy topology but only a base for a fuzzy topology and hence it does not provide a characterization of a fuzzy topology, as was nevertheless claimed by Pu and Liu in [8]. A counterexample for the conjecture of Pu and Liu can be found in [22].
- A Pu-Liu and Mashhour-Ghanim-Kerre system both define a unique fuzzy topology and hence a characterization of a fuzzy topology. More explicitly consider a mapping $\mathcal{M}$ from the class $P(X)$ of all fuzzy points in X into the power class of $\mathcal{F}(X)$ satisfying the conditions:

  (P.1) $(\forall p \in P(X))(\mathcal{M}_p \neq \phi)$

  (P.2) $(\forall p \in P(X))(\forall A \in \mathcal{M}_p)(p \underset{\sim}{\in} A)$

  (P.3) $(\forall p \in P(X))(\forall (A,B) \in \mathcal{M}_p^2)(A \cap B \in \mathcal{M}_p)$

  (P.4) $(\forall p \in P(X))(\forall A \in \mathcal{M}_p)(\forall B \in \mathcal{F}(X))(A \subseteq B \Longrightarrow B \in \mathcal{M}_p)$

  (P.5) $(\forall p \in P(X))(\forall A \in \mathcal{M}_p)(\exists B \in \mathcal{M}_p)(B \subseteq A$ and $(\forall q \in P(X))(q \underset{\sim}{\in} B \Longrightarrow B \in \mathcal{M}_q))$

  then the subset τ of $\mathcal{F}(X)$ defined by

$$\tau = \{O | O \in \mathcal{F}(X) \text{ and } (\forall p \in P(X))(p \underset{\sim}{\in} O \Longrightarrow O \in \mathcal{M}_p)\}$$

  is the unique fuzzy topology on X having the specified fuzzy sets as Mashhour neighbourhoods.

From these observations it may be concluded that only the approaches presented by Pu-Liu and Ghanim-Kerre-Mashhour totally induce a complete formal parallelism with ordinary topology. From this point of view they can be considered as the most suitable extensions of the classical local description of a topology by means of neighbourhood systems.

(iii) Using the concept of complemented fuzzy point, we were able to link the different approaches mentioned above. The Mashhour neighbourhoods of a fuzzy singleton could be expressed in terms of the Pu neighbourhoods and in terms of the Kerre neighbourhoods in the following way:

$$\mathcal{M}_{x_\varepsilon}(\tau) = \mathcal{P}_{x_{1-\varepsilon}}(\tau)$$
$$\mathcal{M}_{x_\varepsilon}(\tau) = \bigcup_{\varepsilon' \in]\varepsilon,1]} \mathcal{K}_{x_{\varepsilon'}}(\tau).$$

(iv) A deep lattice theoretic study of the different approaches to the neighbourhood concept has lead to the introduction of several subclasses of fuzzy topological spaces such as the surjective fuzzy topological spaces, the conditionally closed Kerre spaces, the conditionally closed Mashhour spaces, the conditionally closed Pu spaces. This again constitutes a substantial enrichment of the concept of a classical topology.

2.3 On the concept of a fuzzy number

The concept of a fuzzy number has been introduced by Mizumoto and Tanaka in 1979 [32]. Since that time a great number of papers on that topic appeared, unfortunately many of them using a different definition for the concept of a fuzzy number. Mizumoto and Tanaka simply imposed a convexity condition on the fuzzy sets in **R**. Afterwards Dubois and Prade [33] introduced some monotonicity conditions as well as several forms of continuity. Another category of authors follow the line of Rodabough [34] who defined a fuzzy number as an equivalent class of decreasing fuzzy sets in **R** satisfying some supplementary boundary conditions. In this approach a close link between fuzzy numbers and one-dimensional probability distributions has been established. The fact that there exist a lot of different definitions for an important concept such as a fuzzy number is not quite comfortable the more so as the properties of the extended operations such as addition and multiplication heavily depend upon the definition used. The importance of this fuzzy number concept is still growing since its application in the framework of expert systems: roughly speaking a fuzzy number may be considered as a representation for a ill-known quantity. Besides the explosion in the possible definitions for a fuzzy number, many methods to tackle the problem of ranking fuzzy numbers were introduced.

At this time already more than 30 methods are available. For an excellent overview of the ranking methods we refer to Degani and Bortolan [35]. In a series of papers [36-45] we have contributed to several topics in the theory of fuzzy numbers. Our main results concern:

(i) The structuring of the existing conditions, starting from the most general notion of a fuzzy quantity and moving to the most specific concept of a fuzzy real number by successively adding supplementary conditions of monotonicity, continuity, convexity and normalization.

(ii) The introduction of a practical ranking method based on the generalized maximum and the Hamming distance.

(iii) The introduction of a class of fuzzy quantities that is very easy to implement, i.e., the class of piecewise linear fuzzy quantities.

(iv) A deep study of the continuity properties of the extended arithmetic and order-theoretic operations on fuzzy quantities. We could prove very remarkable results such as for example: the sum of two fuzzy quantities yields a continuous (respectively uniform continuous, respectively infra semicontinuous, respectively supra semicontinuous) fuzzy quantity as soon as only one of them is continuous (respectively uniform continuous, respectively infra semicontinuous, respectively supra semicontinuous).

2.4 The enlargement of the classical toolkit of the relational calculus

In section 1.1 we have already mentioned that the concept of a relation is very fundamental in pure as well as in applied mathematics. Remember the use of relations in the development of relational databases, in information retrieval and in (medical) diagnosis problems. Unfortunately classical mathematics has not lavished her gifts upon relational calculus. Most of the time functional relations or functions are treated. The above mentioned applications have made it worth to study relations. Moreover during the eighties new images and products have been introduced due to the pioneering work of Bandler and Kohout [48]. Through numerous papers we have been contributing to the theory and the application of crisp relations as well as fuzzy relations [49-67]. As an example of the enrichment of classical relational calculus inspired by fuzzy set theory, let us briefly recall the new images and the new compositions as they have been introduced by Bandler and Kohout [48]. Let R be a relation from a universe X to a universe Y, S a relation from the universe Y to a universe Z and A a subset of X. The classical direct image of A under R is defined as:

$$\begin{aligned} R(A) &= \{y \mid (\exists x \in A)((x,y) \in R)\} \\ &= \{y \mid A \cap Ry \neq \phi\} \end{aligned}$$

where Ry is the so-called ($R-$) foreset of y that is defined as the set of all x that are R-related to y, i.e., $Ry = \{x \mid (x,y) \in R\}$.

This classical image can be refined by means of the subdirect image $R^{\lhd}(A)$, the superdirect image $R^{\rhd}(A)$ and the squaredirect image $R^{\square}(A)$, defined by:

$$\begin{aligned} y \in R^{\lhd}(A) &\iff A \subseteq Ry \\ y \in R^{\rhd}(A) &\iff Ry \subseteq A \\ y \in R^{\square}(A) &\iff Ry = A. \end{aligned}$$

The classical composition $R \circ S$ (R before S) of two relations R and S, as well as its refinements: the subcomposition $R \lhd S$, the supercomposition $R \rhd S$ and the squarecomposition $R \square S$ are defined by:

$$\begin{aligned} (x,z) \in R \circ S &\iff xR \cap Sz \neq \phi \\ (x,z) \in R \lhd S &\iff xR \subseteq Sz \\ (x,z) \in R \rhd S &\iff xR \supseteq Sz \\ (x,z) \in R \square S &\iff xR = Sz \end{aligned}$$

where xR is the so-called ($R-$) afterset of x that is defined as the set of all y that are R-related to x, i.e., $xR = \{y \mid (x,y) \in R\}$.

The fuzzification of a relation i.e., instead of dealing with related or nonrelated objects one considers objects that are related in some intermediate degree, has thoroughly enriched the applicability of this basic notion. As already mentioned, our work on (fuzzy) relations has been twofold: purely theoretical issues as well as contributions to the solution of practical problems. Our first theoretical contribution concerned the new forms of composition for crisp as well as for fuzzy relations. We have identified some undesirable shortcomings of these definitions and we were able to set up acceptable modifications. A second theoretical realization concerns the fuzzification of a multivalued mapping, so that we have laid down the basis for the fuzzification of C. Berge's work [68]. Current research on fuzzy multivalued mappings has shown that this domain is an unexploited goldmine for researchers! Regarding the practical applications of fuzzy relations we mention our contributions to information retrieval [49, 50], to medical diagnosis problems [36, 37, 38, 56] and the relational database systems [52, 57, 63-67].

3 CURRENT MATHEMATICS OF FUZZINESS

The main themes of research in the mathematics of fuzziness during the nineties are: standardization, axiomatization, L-fuzzification and a critical comparison of the fuzzy model to other models for the representation and manipulation of imprecision and uncertainty. In section 2 we have seen the explosion of the possible generalizations for classical notions during the eighties. The purpose of the standardization process that started in the beginning of the nineties, is the motivated selection of the most suitable definition among the many possibilities. We have to agree upon the most suitable fuzzified version of a point or an atom in set theory: fuzzy point or fuzzy singleton and which membership relation has to be chosen? Which definition will be used for the concept of neighbourhood in a fuzzy topological space? Which triangular norm and conorm will be used to represent the extension of the classical intersection and union? During the current period may papers appear that are entitled: fuzzy this or that revisited. Undoubtedly the selection of the most suitable definition is very important and has to be based upon a good balancing of the pro's and contra's. Once the choice has been made then every deviation from the standard should be clearly specified.

Secondly more and more axiomatic systems appear in fuzzy set theory during the nineties. An axiomatic system represents a consensus with respect to the desirable properties of fuzzy concepts such as fuzzy union, fuzzy intersection, fuzzy implication, measures of fuzziness, fuzzy measures.... Thirdly, the current period can also be seen as the period of L-fuzzification where L represents a complete lattice. Already in 1967 Goguen [20] has introduced the L-fuzzy sets in order to take into account the incomparability of elements, states and systems. During the current period papers entitled L-fuzzy groups, L-fuzzy ideals, L-fuzzy topology ... are coming into the picture.

Lastly, in this period researchers have started the critical comparison between the fuzzy model and other models such as rough sets, subdefinite sets, intuitionistic sets, to represent imprecision and uncertainty.

Let us give some examples of the research in this period. Again most of these examples are taken from the results of our research group.

3.1 Interrelation between the numerous extensions of continuity of mappings between fuzzy topological spaces

Let f be a mapping between two crisp topological spaces (X_1, T_1) and (X_2, T_2). Then the following equivalent formulations of the continuity of f are well known:

$$
\begin{aligned}
f \text{ is continuous} \quad &\Longleftrightarrow \quad (\forall O_2 \in T_2)(f^{-1}(O_2) \in T_1) \\
&\Longleftrightarrow \quad (\forall F_2 \in T_2')(f^{-1}(F_2) \in T_1') \\
&\Longleftrightarrow \quad (\forall A_1 \subseteq X_1)(f(\,\mathrm{cl}_1(A_1)) \subseteq \mathrm{cl}_2 f(A_1)) \\
&\Longleftrightarrow \quad (\forall A_2 \subseteq X_2)(f^{-1}(\,\mathrm{int}_2(A_2)) \subseteq \mathrm{int}_1 f^{-1}(A_2)).
\end{aligned}
$$

Now f be a mapping between two fuzzy topological spaces (X_1, τ_1) and (X_2, τ_2). In his seminal paper [5] Chang has introduced the concept of fuzzy continuity of f as:

$$f \text{ is fuzzy continuous} \quad \Longleftrightarrow \quad (\forall O_2 \in \tau_2)(f^{-1}(O_2) \in \tau_1)$$

where $f^{-1}(O_2)$ is the fuzzy set in X_1 defined by:

$$f^{-1}(O_2)(x) = O_2(f(x)), \forall x \in X_1.$$

During the eighties already 26 different forms of fuzzy continuity have been introduced: fuzzy weak continuous [69], fuzzy almost continuous [69], fuzzy semicontinuous [69], fuzzy weak semicontinuous [70], fuzzy θ semicontinuous [70], fuzzy almost semicontinuous [70], fuzzy δ-continuous [71], fuzzy semi irresolute continuous [72], fuzzy strongly irresolute continuous [72], fuzzy irresolute continuous [73], fuzzy θ-continuous [74], fuzzy strongly θ-continuous [75], fuzzy almost strongly θ-continuous [75], fuzzy super continuous [76], fuzzy weak θ-continuous [76], fuzzy weak precontinuous [77], fuzzy weak λ-continuous [77], fuzzy semi strongly θ-continuous [78], fuzzy (θ, S) continuous [78], fuzzy quasi irresolute continuous, fuzzy λ-irresolute continuous and fuzzy semi weak continuity [79]. Through the introduction in 1991 of the notion of an operation on a fuzzy topological space, we could obtain all these notions of fuzzy continuity as particular instances of the central notion of fuzzy $\varphi\psi$ continuity where φ and ψ denote operations on (X, τ) i.e., $\mathcal{F}(X) - \mathcal{F}(X)$ mappings satisfying the condition int $(A) \subseteq \varphi(A), \forall A \in \mathcal{F}(X)$. For a detailed report upon this remarkable unification process we refer to [80-82]. In a similar way we used the concept of an operation to unify several existing notions of separation, compactness, open and closed mappings [74, 83-86]. Our most recent research on fuzzy topology concerns two additional topics: smooth topological spaces and modifications of crisp as well as fuzzy continuity of mappings. The original definition of a fuzzy topology by Chang [5] as well as the modifications: L-fuzzy topology by Goguen [20] and stratified fuzzy topology by Lowen [6], are not completely in the spirit of fuzzy set theory: in these spaces a fuzzy set may be either open or not open. Part of our research concerns a real fuzzification of topology where a fuzzy set may assume a partial or intermediate degree of openness. These structures are called smooth topologies. For some of our results on smooth topological spaces we refer to [87, 88]. Using the new images of sets under mappings mentioned in section 2.4, we have introduced new forms of crisp as well as fuzzy continuity of mappings between (fuzzy) topological spaces. For the first results on this promising new topic we refer to [89-90].

3.2 Axiomatization of fuzzy implication operators

As an example of an axiomatic system in the development of fuzzy set theory we give the axioms of Smets and Magrez [91] with respect to a fuzzy implication operator i.e. a suitable extension of the binary implication to the unit square $[0,1]^2$. A fuzzy implication operator is a $[0,1]^2 - [0,1]$ mapping $\mathcal{I}$ satisfying the following axioms:

(A.1) Law of contraposition

$$(\forall (x,y) \in [0,1]^2)(\mathcal{I}(x,y) = \mathcal{I}(1-y, 1-x))$$

(A.2) Exchange principle

$$(\forall (x,y,z) \in [0,1]^3)(\mathcal{I}(x, \mathcal{I}(y,z)) = \mathcal{I}(y, \mathcal{I}(x,z)))$$

(A.3) Hybrid monotonicity

$$(\forall x \in [0,1])(\mathcal{I}(\cdot,x) \text{ is decreasing and } \mathcal{I}(x,\cdot) \text{ is increasing})$$

(A.4) Boundary conditions

$$(\forall (x,y) \in [0,1]^2)(x \le y \Longrightarrow \mathcal{I}(x,y) = 1)$$

(A.5) Neutrality principle

$$(\forall x \in [0,1])(\mathcal{I}(1,x) = x)$$

(A.6) Continuity condition

$$\mathcal{I} \text{ is a continuous } [0,1]^2 - [0,1] \text{ mapping.}$$

We have been contributing a lot to the study of the behaviour of fuzzy implication operators. In [92] the generalized modus ponens inference rule has been examined in a very abstract way and a characterization of the $[0,1]^2 - [0,1]$ mappings, especially the fuzzy implication operators, was given according to their behaviour with respect to the sup-T inference rule. The analogies between triangular norms and conorms on the one hand and fuzzy implication operators on the other hand were described. It was also shown that a fuzzy implication operator that is an extension of the classical NOT P OR Q definition in the sense that the negation NOT is replaced by a strict complement operator and the disjunction OR by a triangular conorm, never satisfies the sup-min inference rule. In [93] we gave a detailed study of some popular fuzzy implication operators such as the Luckasiewicz, the Gödel-Brouwer and the Kleene-Dienes operator. In [94] we presented a natural continuation of the work of Mizumoto [95] on the influence of the choice of the fuzzy implication operator on the validity of the generalized modus ponens, modus tollens and syllogism. In that paper we gave a detailed investigation of the validity of a fuzzified version of the well-known method-of-cases inference rule under seventeen different fuzzy implication operators. In [96, 97] this work has been extended to the generalized method of cases based upon the extension of a triangular norm to n arguments and a general uniformization of Zadeh's compositional rule of inference; moreover we have tested 18 widely used fuzzy implication operators w.r.t. the axioms of Smets and Magrez mentioned above.

Another topic of interest in our research group concerned the fuzzification of classical bivalued logic. In [98-100] a theoretical model based on Zadeh's possibility theory, for approximate reasoning in rule-based systems has been presented. This work has been extended using a generalized sup-T extension principle [101]. In the latter paper we have introduced a family of propositional logics the so-called sup-T extension logics; a fuzzy truth value has been defined as a normalized fuzzy set on the doubleton {true, false} and by means of the sup-T extension principle we extended all the classical logical operations.

3.3 Other topics of current research

During the last 5 years several new topics of research have been tackled by our research group: measures of fuzziness, fuzzy reliability theory, fuzzy morphology, fuzzy measures and fuzzy multivalued mappings. In [102,103] we have treated the problem of measuring the fuzziness of a fuzzy set as an evaluation problem. A general framework has been created in which most existing measures of fuzziness such as entropy and energy measures could be incorporated and hence it constitutes a new example of standardization and uniformization. With respect to reliability, since the thirties the reliability of a system has been exclusively defined in a probabilistic way where the system and its components assume either one out of two possible states. In 1978 [104] the concept of a multistate structure function has been introduced to overcome some of the shortcomings of the binary approach. In a series of papers [105-112] we have shown the power of fuzzy sets and possibility theory to design the reliability of complex systems. Perhaps the most important of our recent contributions were given on the theory of fuzzy measures and fuzzy integrals. We were able to develop a general theory of fuzziness and uncertainty based on confidence relations and that incorporate Sugeno's fuzzy measures [9], the Kolmogorov probability measures, the credibility and plausibility measures, the possibility and necessity measures. Furthermore the semi-norm and semiconorm integral of Suarez Garcia and Gil Alvarez [113] has been generalized. For more details we refer to [114-119].

Finally let us refer to our contributions to miscellaneous topics among them medical diagnosis, fuzzy mathematical programming, intelligent data reduction methods, fuzzy mathematical morphology and fuzzy multivalued mappings [120-136].

REFERENCES

[1] L. Zadeh, "Fuzzy Sets", Information Control 8 (1965), 338-353.

[2] E.E. Kerre, "Basic principles of fuzzy set theory for the representation and manipulation of imprecision and uncertainty", in: Introduction to the Basic Principles of Fuzzy Set Theory and some of its Applications (E.E. Kerre, ed.) Communication and Cognition, Gent, 1993, second revised edition, 1-158.

[3] A. Rosenfeld, "Fuzzy groups", J. Math. Anal. Appl. 35 (1971), 512-517.

[4] C. Negoita, D. Ralescu, "Application of Fuzzy sets to System Analysis", Birkhäuser Verlag, Basel, 1975.

[5] C.L. Chang, "Fuzzy topological spaces", J. Math. Anal. Appl. 24 (1988), 182-190.

[6] R. Lowen, "Fuzzy topological spaces and fuzzy compactness", J. Math. Anal. Appl. 56 (1976), 621-633.

[7] E.E. Kerre, "Fuzzy topologizing with preassigned operations", International Congress for Mathematicians, Helsinki, 1978, 61-62.

[8] P.M. Pu, Y.M. Liu, "Fuzzy topology I", J. Math. Anal. Appl. 76 (1980), 571-599.

[9] M. Sugeno, Theory of fuzzy integrals and its applications, PhD Thesis, Tokyo Institute of Technology, Tokyo, 1974.

[10] B. Schweizer and A. Sklar, "Associative functions and statistical triangle inequalities", Publ.Math. Debrecen 8 (1961), 169-186.

[11] R. Bellman, L. Zadeh, M. Giertz, "On the analytic formalism of the theory of fuzzy sets", Information Sciences 5 (1973), 149-156.

[12] D. Dubois, H. Prade, "New results about properties and semantics of fuzzy-set-theoretic operators", in: Fuzzy Sets, Theory and Application to Policy Analysis and Information Systems (P.P. Wang, S.K. Chang, eds.) Plenum Press, New York, 1980, 59-75.

[13] R. Yager, "On the implication operator in fuzzy logic", Information Sciences 31 (1983), 141-164.

[14] H. Zimmermann, P. Zysno, "Latent connectives in human decision making", Fuzzy Sets and Systems 4 (1980), 37-51.

[15] W. Bandler, L. Kohout, "Fuzzy power sets and fuzzy implication operators", Fuzzy Sets and Systems 4 (1980), 13-30.

[16] B. Hutton, "Normality in fuzzy topological spaces", J. Math. Anal. Appl. 50 (1975), 74-79.

[17] E.E. Kerre, "Characterization of normality in fuzzy topological spaces", Simon Stevin, 53 (1979), 239-248.

[18] E.E. Kerre, P. Ottoy, "Fuzzy subspace of a fuzzy topological space", in: Proceedings NAFIPS'88 (S. Ovchinnikov, ed.) San Francisco, 1988, 141-146.

[19] E.E. Kerre, "Fuzzy Sierpinski space and its generalizations", J. Math. Anal. Appl. 74 (1980), 318-324.

[20] J. Goguen, "L-fuzzy sets", J. Math. Anal. Appl. 18 (1967), 145-174.

[21] C.K. Wong, "Fuzzy points and local properties of fuzzy topology", J. Math. Anal. Appl. 46 (1974), 316-318.

[22] E.E. Kerre, P. Ottoy, "On the different notions of neighbourhood in Chang-Goguen fuzzy topological spaces", Simon Stevin 61 (1987), 131-146.

[23] H. Ludescher, E. Roventa, "Sur les topologies floues définies à l'aide des voisinages", C. R. Acad. Sc. Paris t. 283 (1976) Série A, 575-577.

[24] R.H. Warren, "Fuzzy topologies characterized by neighbourhood systems", Rocky Mt. J. Math. 9 (1979), 761-764.

[25] M.E. Abd. El-Monsef, M.H. Ghanim, E.E. Kerre, A.S. Mashhour, "Fuzzy topological results", in: Proceedings Fifth Prague Topological Symposium (J. Novak, ed.) Helderman Verlag, Berlin, 1983, 1-5.

[26] M. Ghanim, E.E. Kerre, A. Mashhour, "Separation axioms, subspaces and sums in fuzzy topology", J. Math. Anal. Appl. 102 (1984), 189-202.

[27] E.E. Kerre, P. Ottoy, "On the characterization of a Chang fuzzy topology by means of a Kerre neighbourhood system", in: Proceedings NAFIPS 87 (J.L. Chameau, J. Yao, eds.), Purdue University Press, Purdue, 1987, 302-307.

[28] E.E. Kerre, P. Ottoy, "Lattice properties of neighbourhood systems in Chang fuzzy topological spaces", Fuzzy Sets and Systems 30 (1989), 205-213.

[29] E.E. Kerre, P. Ottoy, "A comparison of the different notions of neighbourhood systems for Chang topologies", in: Proceedings of the First Joint IFSA-EC EURO-WG Workshop on Progress in Fuzzy Sets in Europe, (J. Kacprzyck, A. Straszak, eds.), Prace ibs pan, Warsaw, vol. 169, 1989, 241–251.

[30] E.E. Kerre, P. Ottoy, "On α-generated fuzzy topologies", Fasciculi Mathematici 19 (1990), 127–134.

[31] E.E. Kerre, P. Ottoy, "Counterexamples in fuzzy topology", to appear in: Proceedings of the Third Polish Symposium on Fuzzy Sets and Interval Analysis in Pure and Applied Mathematics (J. Albrycht, ed.) Poznan, 1989.

[32] M. Mizumoto, K. Tanaka, "Some properties of fuzzy numbers", in: Advances in Fuzzy Set Theory and Applications (Gupta, Ragade, Yager, eds.) North Holland, Amsterdam, 1979, 153–164.

[33] D. Dubois, H. Prade, Fuzzy Sets and Systems, Theory and Applications, Academic Press, New York, 1980.

[34] S. Rodabaugh, "Fuzzy addition in the L-fuzzy real line", Fuzzy Sets and Systems 8 (1982), 39–52.

[35] G. Bortolan, R. Degani, "A review of some methods for ranking fuzzy subsets", Fuzzy Sets and Systems 15 (1985), 1–19.

[36] E.E. Kerre, "The use of fuzzy set theory in electrocardiological diagnostics", in: Approximate Reasoning in Decision Analysis (M. Gupta, E. Sanchez, eds.) North-Holland, Amsterdam, 1982, 277–282.

[37] E.E. Kerre, "Fuzzy-set-theoretic approach to medical diagnosis problems", in: Proceedings Second World Conference on Mathematics at the Service of Man (A. Ballester, D. Cardus, E. Trillas, eds.) Universidad Politechnica de Las Palmas, Las Palmas, 1982, 388–393.

[38] E.E. Kerre, "Fuzzy approach to EGG diagnosis", in: Encyclopedia of Systems and Control (M. Singh, ed.) Pergamon Press, Oxford, 1987, 1405–1407.

[39] E.E. Kerre, A. Van Schooten, "A deeper look on fuzzy numbers from a theoretical as well as a practical point of view", in: Fuzzy Logic in Knowledge-Based Systems, Decision and Control (M. Gupta, T. Yamakawa, eds.) North-Holland, Amsterdam, 1988, 173–196.

[40] R. Baekeland, E.E. Kerre, "Remarkable continuity-preserving properties of the extended operations on fuzzy numbers", in : Proceedings 2nd joint IFSA-EC EURO-WG Workshop on Progress in Fuzzy Sets in Europe (H. Hansen, W. Janko, eds.) Wirtschäftuniversität, Wien, 1988, 28–31.

[41] R. Baekeland, E.E. Kerre, "Piecewise linear fuzzy quantities : a way to implement fuzzy information into expert systems and fuzzy databases", in: Uncertainty and Intelligent Systems, Lecture Notes in Computer Science n° 313, Springer-Verlag, Berlin, 1988, 119–126.

[42] H. Steyaert, F. Van Parys, R. Baekeland, E.E. Kerre, "A *C*-implementation of the algebraic operations on piecewise linear fuzzy quantities", in: Advances in Fuzzy Sets and Applications (H. Teodorescu, I. Tofan, J.G. Aluya, O. Costinescu, eds.) Editura Universitati, Iasi, 1992, 87–93.

[43] H. Steyaert, F. Van Parys, R. Baekeland, E.E. Kerre, "The implementation of piecewise linear fuzzy quantities", submitted to International Journal of Intelligent Systems, paper n° 9249 (1992).

[44] R. Baekeland, E.E. Kerre, "Operations on piecewise linear fuzzy quantities : a theoretical approach", submitted to Fuzzy Sets and Systems (1992).

[45] B. De Baets, E.E. Kerre, "Analytical inference results for triangular fuzzy data", in: Proceedings of the Second International Conference on Fuzzy Logic and Neural Networks, (T. Yamakawa, ed.), FLSI, Tokyo, 1992, 295–300.

[46] B. De Baets, E.E. Kerre, M. Gupta, "Expert knowledge representation by means of piecewise linear fuzzy quantities", in: Proceedings Third IFSA Congress, Seattle, 1989, 618–621.

[47] X. Wang, E.E. Kerre, B. Cappelle, D. Ruan, "Transitivity of fuzzy orderings based on pairwise comparisons", in: Fuzzy Logic and Intelligent Technologies in Nuclear Science (D. Ruan, a.o., eds.) World Scientific Publishing Company, Singapore, 1994, 38–43.

[48] W. Bandler, L. Kohout, "Fuzzy relational products as a tool for analysis and synthesis of the behaviour of complex natural and artificial systems", in: Fuzzy Sets Theory and Application to Policy Analysis and Information Systems (P.P. Wang, S.K. Chang, eds.) Plenum Press, New York, 1980, 341–367.

[49] R. Zenner, E.E. Kerre, R. De Caluwe, "Practical determination of document description relations in document retrieval systems", in: Proceedings Workshop on the Membership Function, (C. Carlsson, ed.) EIASM, Brussels, 1984, 127–138.

[50] R. Zenner, E.E. Kerre, R. De Caluwe, "A new approach to information retrieval systems using fuzzy expressions", Fuzzy Sets and Systems 17 (1985), 9–22.

[51] E.E. Kerre, R. Zenner, R. De Caluwe, "The use of fuzzy set theory in information retrieval and databases: a survey", JASIS 37 (1986), 341–345.

[52] R. Vandenberge, E.E. Kerre, A. Van Schooten, R. De Caluwe, "A practical application of fuzzy database techniques to criminal investigation", in: Proceedings Second IFSA Congress, Tokyo, 1987, 661–664.

[53] E.E. Kerre, P. Ottoy, "On the fuzzification of multivalued mappings", in: Proceedings of the Workshop on Knowledge-Based Systems and Models of Logical Reasoning, Scomi-Scientific Center, F.A. El-Cherif, Caïro (1988).

[54] R. Vandenberge, A. Van Schooten, R. De Caluwe, E.E. Kerre, "Some practical aspects of fuzzy database techniques: an example", Information Sciences 14, n° 6 (1989), 465–472.

[55] E.E. Kerre, "Fuzzy relational calculus in terms of after- and foresets and its application to information retrieval", in: Proceedings of Assuit First International Conference on Mathematics and Statistics, Part VIII (Al-Hussaini, Abou El-Ela, eds.) Assuit, 1990, pp. 133–157.

[56] E.E. Kerre, "A walk through fuzzy relations and their application to information retrieval, medical diagnosis and expert systems", in: Proceedings of ISUMA '90 (Bilal M. Ayyub, ed.), IEEE Computer Society Press, Los Alamitos, CA, 1990, 84–89.

[57] G. Chen, E.E. Kerre, J. Vandenbulcke, "A step towards the theory of fuzzy relational database design", in: Proceedings IFSA 91, (B. Lowen, M. Roubens, eds.) Brussels, 1991, 44–47.

[58] E.E. Kerre, "A walk through fuzzy relations and their applications", in: Analysis and Management of Uncertainty (B. Ayyub, M. Gupta, L. Kanal, eds.) North-Holland, Amsterdam, 1992, 141–151.

[59] B. De Baets, E.E. Kerre, "Some reflections on fuzzy relational compositions", in: Proceedings IPMU 92 (B. Bouchon, R. Yager, eds.) Palma de Mallorca, 1992, 251–254.

[60] B. De Baets, E.E. Kerre, "Triangular fuzzy relational compositions revisited", in: Uncertainty in Intelligent Systems (B. Bouchon, L. Valverde, R. Yager, eds.) North-Holland, Amsterdam, 1993, 257–268.

[61] B. De Baets, E.E. Kerre, "A revision of Bandler-Kohout compositions of relations", Mathematica Pannonica 4/1 (1993), 1–21.

[62] B. De Baets, E.E. Kerre, "FRESH, A Pascal library for solving fuzzy relational equations on the unit interval", in: Proceedings EUFIT 93, Aachen, 1993, 537–541.

[63] G. Chen, E.E. Kerre, J. Vandenbulcke, "A general treatment of data redundancy in a fuzzy relational data model", JASIS 43, n° 4 (1992), 304-311.

[64] G. Chen, E.E. Kerre, J. Vandenbulcke, "Fuzzy functional dependency and axiomatic system in a fuzzy relational data model", in: Proceedings IPMU 92 (B. Bouchon, R. Yager, eds.) Palma de Mallorca, 1992, 313-316.

[65] G. Chen, E.E. Kerre, J. Vandenbulcke, "The axiomatic system of fuzzy functional dependency in a fuzzy relational data model", in: Information and Systems 92 (Zhang Shengkai, Zou Kaiqi, eds.) Dalian Maritime University Publishing House, Dalian, 1992, 876–880.

[66] G. Chen, E.E. Kerre, J. Vandenbulcke, "On the lossless-join decomposition of relation scheme(s) in a fuzzy relational data model", in: Proceedings of ISUMA 93 (Bilal M. Ayyub, ed.) IEEE Computer Society Press, Maryland, 1993, 440–446.

[67] G. Chen, E.E. Kerre, J. Vandenbulcke, "A computational algorithm for the FFD transitive closure and a complete axiomatization of fuzzy functional dependency", Int. J. Intelligent Systems, 9 (1994) 421–439.

[68] C. Berge, Espaces Topologiques et Fonctions Multivoques, T. Dunod, Paris, 1966.

[69] K.K. Azad, "On fuzzy semicontinuity, fuzzy almost continuity and fuzzy weakly continuity", J. Math. Anal. Appl. 82 (1981), 14–32.

[70] S.P. Arya, M.P. Bhamini, "A note on semi *US* spaces", Ranchi Univ. Math. J. 13 (1982), 60–67.

[71] S. Saha, “Fuzzy δ-continuous mappings”, J. Math. Anal. Appl. 126 (1987), 130–142.

[72] S. Malakar, “On fuzzy semi-irresolute and strongly irresolute mappings”. Private communication.

[73] M.N. Mukherjee, S.P. Sinha, “Irresolute and almost open functions between fuzzy topological spaces”, Fuzzy Sets and Systems 29 (1989), 381–388.

[74] A. Kandil, E.E. Kerre, M. El-Shaffi, A. Nouh, “Fuzzy strongly θ-continuous mappings”, in: Proceedings of Assiut First International Conference on Mathematics and Statistics, Part VIII (E.K. Al-Hussaini, A.M. Abou-El-Ela, eds.), Assiut, Egypt, 1990, 97–113.

[75] M.N. Mukherjee, B. Ghosh, “Some stronger forms of fuzzy continuous mappings on fuzzy topological spaces”, Fuzzy Sets and Systems 38 (1990), 375–387.

[76] S.P. Sinha, “On S^*-closedness in fuzzy setting”. Private communication.

[77] I. Hanafy, “A study on non-continuous functions in fuzzy topological spaces and good extensions of R_1-ness and paracompactness”, PhD thesis, Tanta, Egypt, 1990.

[78] M.E. Abd El-Monsef, I.M. Hanafy, S.N. El-Deeb, “Semi strongly θ-continuous functions”, Indian J. Pure Appl. Math. 28 (8) (1990), 721–728.

[79] T. Noiri, “On θ-semicontinuous functions”, Indian J. Pure Appl. Math. 21 (5) (1990), 410–415.

[80] E.E. Kerre, N. Morsi, “On co-optimal lifts in the category of fuzzy neighbourhood spaces”, in: Proceedings IFSA 91 (B. Lowen, M. Roubens, eds.) Brussels, 1991, 105–108.

[81] A. Kandil. E.E. Kerre. A. Nouh, “Operations and mappings on fuzzy topological spaces”, Annales de la Societé Scientifique de Bruxelles, T.105, 4 (1991), 167–188.

[82] E.E. Kerre, A. Nouh, A. Kandil, “Operations on the class of fuzzy sets on a universe endowed with a fuzzy topology”, J. Math. Anal. Appl. 180 (2) (1993), 325–341.

[83] A. Kandil, E.E. Kerre, M.E. El-Shafee, A. Nouh, Quasi "θ-spaces and pairwise θ-perfect irreducible mappings", J. Austral. Math. Soc. (Series A) 52 (1992), 322–333.

[84] A. Kandil, E.E. Kerre, A. Nouh, M El-Shafee, "Generalized mappings between fuzzy topological spaces", Mathematica Pannonica 3/2 (1992), 59–71.

[85] A. Kandil, E.E. Kerre, A. Nouh, M. El-Shafee, "Fuzzy θ-perfect irreducible mappings and fuzzy θ-proximity spaces", Fuzzy Sets and Systems 45 (1992), 93–102.

[86] E.E. Kerre, A. Nouh, A. Kandil, "Generalized compactness in fuzzy topological spaces", Matematicki Vesnik 43 (1991), 29–40.

[87] M.K. El Gayyar, E.E. Kerre, A.A. Ramadan, "Almost compactness and near compactness in smooth topological spaces", Fuzzy Sets and Systems 62 (1994), 193–202.

[88] M.K. El Gayyar, E.E. Kerre, A.A. Ramadan, "Separation axioms in smooth topological spaces", submitted to Far-East Journal of Mathematical Sciences (1993).

[89] B. De Baets, E.E. Kerre, "Fuzzy subcontinuity", in: Proceedings of NAFIPS '93, (M. McAllister, ed.) Allentown, PA, USA, 1993, 232–236.

[90] B. De Baets, E.E. Kerre, "On the supercomposition of subcontinuous and fuzzy subcontinuous mappings and related topics", Journal of Fuzzy Mathematics 2 (1994), 17–34.

[91] P. Smets, P. Magrez, "Implication in fuzzy logic", Int. J. Approximate Reasoning 1 (1987), 327–347.

[92] B. Cappelle, E.E. Kerre, D. Ruan, F. Vanmassenhove, "Characterization of the binary operations on the unit interval satisfying the generalized modus ponens inference rule", Mathematica Pannonica 2/1 (1991), 105–121.

[93] E.E. Kerre, "A comparative study of the behavior of some popular fuzzy implication operators on the generalized modus ponens", in: Fuzzy Logic for the Management of Uncertainty (L. Zadeh, J. Kacprzyck, eds.) J. Wiley and Sons, New York, 1992, 281–296.

[94] D. Ruan, E.E. Kerre, G. De Cooman, B. Cappelle, F. Vanmassenhove, "Influence of the fuzzy implication operator on the method-of-cases inference rule", in: Int. J. Approx. Reas. vol.4, n° 4 (1990), 307–318.

[95] M. Mizumoto, H. Zimmerman, "Comparison of fuzzy reasoning", Fuzzy Sets and Systems 8 (1982), 253-283.

[96] D. Ruan, E.E. Kerre, "On the extension of the compositional rule of inference", Int. J. Intell. Syst. 8 (1993), 807-817.

[97] D. Ruan, E.E. Kerre, "Fuzzy implication operators and generalized fuzzy method of cases", Fuzzy Sets and Systems 54 (1993), 23-37.

[98] A. Van Schooten, E. Kerre, R. De Caluwe, "An expert system development tool using fuzzy logic" in: Proceedings Intelligent Autonomous Systems (L. Hertzberger, ed.) Elsevier, Amsterdam, 1986, pp. 668-677.

[99] A. Van Schooten, E. Kerre, R. De Caluwe, "Fuzzy expert systems", in: Proceedings The Intelligent Access to Information (R. De Caluwe, ed.) FBVI, Brussels, 1987, 245-257.

[100] A. Van Schooten, E. Kerre, R. De Caluwe, "Approximate reasoning based on possibility theory", in: Cybernetics and Systems (R. Trappl, ed.) Kluwer Academic Publishers, Dordrecht, 1988, 643-650.

[101] G. De Cooman, E.E. Kerre, B. Cappelle, D. Ruan, F. Vanmassenhove, "On the extension of classical propositional logic by means of a triangular norm", Int. J. Intell. Syst. 5, n°3 (1990), 307-322.

[102] B. De Baets, E.E. Kerre, S. Nasuto, "Measuring fuzziness: an evaluation problem", in Proceedings ISUMA 93 (B. Ayyub, ed.) Maryland, 1993, 65-72.

[103] B. De Baets, S. Nasuto, E.E. Kerre, "An order - theoretic approach to the measuring of fuzziness", in: Uncertainty Modeling and Analysis: Theory and Applications (B. Ayyub, M.M. Gupta, eds.), Elsevier Science Publishers, Amsterdam, 1994, 173-191.

[104] R. Barlow, S. Wu, "Coherent systems with multistate components", Mathematics of Operations Research, 3 (1978), 275-281.

[105] B. Cappelle, E.E. Kerre, F. Vanmassenhove, "A possibilistic uncertainty model applied to the Barlow-Wu class of multistate structures", in: Proceedings IFSA 91 (B. Lowen, M. Roubens, eds.) Brussels, 1991, 25-28.

[106] B. Cappelle, E.E. Kerre, "On a possibilistic approach to reliability theory", in: Proceedings ISUMA 93 (B. Ayyub, ed.) Maryland, 1993, 415-418.

[107] J. Montero, B. Cappelle, E.E. Kerre, "The usefulness of complete lattices in reliability theory", in: Fuzzy Sets and Possibilistic Theory in Reliability and Safety Analysis (J. Kacprzyck, T. Onisawa, eds.), Omnitech Press, Warsaw and Physica-Verlag, Heidelberg, 1993, in press.

[108] B. Cappelle, E.E. Kerre, "Issues on possibilistic reliability theory", in: Fuzzy Sets and Possibilistic Theory in Reliability and Safety Analysis (J. Kacprzyck, T. Onisawa, eds.), Omnitech Press, Warsaw and Physica-Verlag, Heidelberg, 1993, in press.

[109] B. Cappelle, E.E. Kerre, "Computer assisted reliability analysis: an application of possibilistic reliability theory to a subsystem of a nuclear power plant", in: Fuzzy Logic and Intelligent Technologies in Nuclear Science (D. Ruan, P. D'hondt, P. Govaerts, E.E. Kerre, eds.) World Scientific Publishing Company, Singapore, 1994, 260–265.

[110] B. Cappelle, E.E. Kerre, "The CARA system: an intelligent tool to support qualitative reliability analysis", in: Proceedings International AMSE Conference on Systems, Analysis, Control & Design, 1 (1994), 25–35.

[111] B. Cappelle, E.E. Kerre, "A general possibilistic framework for reliability theory", in: Proceedings 5th International Conference Information Processing and Managament of Uncertainty in Knowledge-Based Systems 1 (1994), 832–837.

[112] B. Cappelle, E.E. Kerre, "Possibilistic and necessistic reliability functions: fundamental concepts and theorems to represent non-probabilistic uncertainty in reliability theory", in: Uncertainty Modeling and Analysis: Theory and Applications (B. Ayyub, M.M. Gupta, eds.) Elsevier Science Publishers, Amsterdam, 1994, 131–144.

[113] F. Suarez Garcia, P. Gil Alvarez, "Two families of fuzzy integrals", Fuzzy Sets and Systems 18 (1986), 67–81.

[114] G. De Cooman, E.E. Kerre, B. Cappelle, D. Ruan, F. Vanmassenhove, "On the extension of classical propositional logic by means of a triangular norm", in:Proceedings Third IFSA Congress, Seattle, 1989, 821–824.

[115] G. De Cooman, E.E. Kerre, F. Vanmassenhove, "Possibility theory: an integral theoretic approach", Fuzzy Sets and Systems 46 (1992), 287–299.

[116] G. De Cooman, E.E. Kerre, "Order norms on partially ordered sets", in: Proceedings Conference Intellectual Systems, Moscow, 1993.

[117] G. De Cooman, E.E. Kerre, "Ample fields", Simon Stevin 67 (1993), 235–244.

[118] G. De Cooman, E.E. Kerre, "Order norms on bounded partially ordered sets", Journal of Fuzzy Mathematics 2 (1994), 281–310.

[119] G. De Cooman, E.E. Kerre, "A new approach to possibilistic independence", in: Proceedings 3rd. IEEE International Conference on Fuzzy Systems, 1994, 1446–1451.

[120] E.E. Kerre, B. Raman, "Application of fuzzy programming to ship steering", in: Approximate Reasoning in Expert Systems, (M. Gupta, A. Kandel, W. Bandler, J. Kiszka, eds.) North-Holland, Amsterdam, 1985, 719–730.

[121] E.E. Kerre, "Outline of an expert system for ECG diagnosis using fuzzy sets", Artificial Intelligence in Medicine 1 (1989), 139–144.

[122] E.E. Kerre, "A call for crispness in fuzzy set theory", Fuzzy Sets and Systems 29 (1989), 57–65.

[123] E.E. Kerre, "Fuzzy set theory in information retrieval and databases", in: Encyclopedia of Computer Science and Technology (A. Kent, J. Williams, eds.) Marcel Dekker Inc., New York, vol 21, Suppl.6, 1990, 117–138.

[124] P. Poulosky, D. Holliday, J. Niemann, E.E. Kerre, "Calculus of α-level fuzzy sets", Report Series #115, University of Nebraska, Lincoln, U.S.A. (1990).

[125] D. Guinan, K. Streicher, E.E. Kerre, "Set-theoretic properties of the class of fuzzy sets endowed with the bounded sum and the bold intersection", Report Series #114, University of Nebraska, Lincoln, U.S.A. (1990).

[126] E.E. Kerre, "The suitability of fuzzy set theory for the representation and manipulation of imprecision and uncertainty in knowledge based systems", Communication & Cognition, vol.8, n°1 (1991), 49–79.

[127] E.E. Kerre, "Fuzzy logic and medical diagnosis", in: Proceedings of the Symposium Clear Applications of Fuzzy Logic, Delft, 1991, 160–174.

[128] B. Cappelle, G. De Cooman, E.E. Kerre, F. Vanmassenhove, "Intelligent reductions methods based upon fuzzy set theory as a helpful tool for controlling industrial plants", Fuzzy Sets and Systems 50, (1992), 119-126.

[129] E.E. Kerre, "Operations on fuzzy sets", in: Vage verzamelingen en vage logica (J. De Kerf, ed.) Technologisch Instituut KVIV, Antwerpen, 1992, 1-27.

[130] B. De Baets, E.E. Kerre, "Kite-tail lattices and their characterization", Far-East Journal of Mathematical Sciences, vol. 1 (1993), 1-11.

[131] B. De Baets, E.E. Kerre, "The generalized modus ponens and the triangular fuzzy data model", Fuzzy Sets and Systems, Special Issue on Fuzzy Data Analysis (K. Hirota, ed.), 59 (1993), 305-317.

[132] B. De Baets, E.E. Kerre, "An introduction to fuzzy mathematical morphology", in: Proceedings NAFIPS 93 (Mc-Allister, ed.) Allentown, U.S.A., 1993, 129-133.

[133] B. De Baets, E.E. Kerre, "Fuzzy relational compositions", Fuzzy Sets and Systems 60 (1993), 109-120.

[134] B. De Baets, E.E. Kerre, "A primer on solving fuzzy relational equations on the unit interval", International Journal of Uncertainty, Fuzziness and Knowledge-Based Systems 2 (1994), 205-225.

[135] B. Van de Walle, D. Ruan, E.E. Kerre, "Applications of fuzzy reasoning in nuclear decision aiding systems", Fuzzy Systems and Artificial Intelligence, vol. II, n°2 (1993).

[136] E. Tsiporkova-Hristoskova, E.E. Kerre, "Pure and stable fuzzy sets", in: Proceedings International AMSE Conference on Systems, Analysis, Control & Design, vol. 1, 1994, 3-12.

2

CLOSEDNESS AND UPPER SEMI-CONTINUITY OF FUZZY MULTIVALUED MAPPINGS

Elena Tsiporkova-Hristoskova

Visiting Scholar from the Technical University of Plovdiv (Bulgaria)
Department of Applied Mathematics and Computer Science
Universiteit Gent
Krijgslaan 281-S9
B-9000 Gent, Belgium

ABSTRACT

In this paper, the definitions of three different types of closedness (direct, lower and upper) and two different types of upper semi-continuity (upper semi-continuity^{d} and upper semi-continuity^{u}) of fuzzy multivalued mappings are recalled, and some of their properties are presented. Very interesting relationships between direct closedness and upper semi-continuity^{d}, and upper closedness and upper semi-continuity^{u} are established with respect to regular and strongly compact fuzzy topological spaces. Using these relationships, a lot of new properties of closed and upper semi-continuous fuzzy multivalued mappings are derived. They include mixed intersection, mixed union, and mixed composition of closed and upper semi-continuous fuzzy multivalued mappings, intersection of upper semi-continuous fuzzy multivalued mappings, and composition of closed fuzzy multivalued mappings. Examples of families of implication operators, which satisfy some of the conditions required in the studied relationships and properties are presented.

1 INTRODUCTION

The continuity of multivalued mappings is characterized by two types of semi-continuity: lower and upper semi-continuity. We have discussed lower semi-continuity of fuzzy multivalued mappings in [10] and have shown that this type of continuity is closely related to level lower semi-continuity and also to the classical lower semi-continuity of the cuts of the mapping. Classically, upper semi-continuity is related to the closedness of the mapping. In [14] we have introduced three different types of closedness for fuzzy multivalued mappings: direct, lower and upper closedness. Their properties and the relationships between them have been completely studied. Using a similar approach, we have defined in [15] two different types of upper semi-continuity for fuzzy multivalued mappings: upper semi-continuity^{d} and upper semi-continuity^{u}. Their properties and the relationships between them have been studied as well. In this paper, our main task is to investigate under which conditions classical relationships between closedness and upper semi-continuity can be extended to the fuzzy case, for instance by imposing additional conditions on the fuzzy topological spaces involved, or by imposing restrictions on the mappings themselves. Let us therefore first recall the definitions of closedness and upper semi-continuity of a multivalued mapping (MM), the relationships between them, and the properties derived from these relationships [1, 2].

Consider a set A in X, a set B in Y and two MMs F and G from X into Y.

Throughout this paper, a classical topology on a universe X is denoted by π_X.

Definition 1.1. (Closedness). [1, 2]
F is closed at $x \in \mathrm{dom}(F)$ with respect to (X, π_X) and (Y, π_Y) iff for all $y \in Y$ such that $y \notin F(x)$, there exist two open neighbourhoods $\mathcal{U}(x)$ and $\mathcal{V}(y)$ such that

$$(\forall x' \in \mathcal{U}(x))(F(x') \cap \mathcal{V}(y) = \emptyset).$$

F is closed iff it is closed at all $x \in \mathrm{dom}(F)$.

Definition 1.2. (Upper semi-continuity). [1, 2]
F is upper semi-continuous (USC) at $x \in \mathrm{dom}(F)$ with respect to (X, π_X) and (Y, π_Y) iff for each open set $O \in \pi_Y$ such that $F(x) \subseteq O$, there exists an open neighbourhood $\mathcal{U}(x)$ such that

$$(\forall x' \in \mathcal{U}(x))(F(x') \subseteq O).$$

F is USC iff it is USC at all $x \in \mathrm{dom}(F)$.

If $(\forall x \in X)(F(x)$ is closed or compact) then F is called closed-valued or compact-valued, respectively.

Proposition 1.1. [1, 2] *The following relationships hold, with respect to (X, π_X) and (Y, π_Y):*

(i) If (Y, π_Y) is separated then closed-valued and USC F is closed.

(ii) If (Y, π_Y) is compact then closed F is USC.

Proposition 1.2. [1, 2] *The following properties hold, with respect to (X, π_X) and separated (Y, π_Y):*

(i) The intersection $F \cap G$ of closed F and compact-valued and USC G is USC.

(ii) The intersection of a family of compact-valued and USC MMs from X into Y is USC.

In the fuzzy case, due to some lack of properties available in ordinary topological spaces, separated and compact fuzzy topological spaces are not sufficient to obtain relationships and properties similar to the above ones. Instead, in the interaction between upper semi-continuity and closedness of fuzzy multivalued mappings regular and strongly compact fuzzy topological spaces play a more crucial role.

2 PRELIMINARIES

2.1 Fuzzy multivalued mappings

In this subsection, we recall the definition of a fuzzy multivalued mapping (FMM) and the expressions for the images of fuzzy sets under FMMs [11, 12, 13].

If with each element x of a universe X we associate a fuzzy set $F(x)$ in a universe Y then we say that the correspondence $x \mapsto F(x)$ establishes an FMM F from X into Y. The degree of membership of y in $F(x)$ is denoted by $F(x).y$. The domain and the range of F are defined by:

$$\begin{aligned} \mathrm{dom}(F)(x) &= \sup_{y \in Y} F(x).y \\ \mathrm{rng}(F)(y) &= \sup_{x \in X} F(x).y. \end{aligned}$$

The inverse mapping of F is the FMM F^- from Y into X defined by $F^-(y).x = F(x).y$. One easily verifies that $\mathrm{dom}(F^-) = \mathrm{rng}(F)$ and $\mathrm{rng}(F^-) = \mathrm{dom}(F)$.

A FMM F from X into Y is called [13]:

(i) non-void iff $(\forall x \in X)\ (F(x) \neq \varnothing)$;

(ii) surjective iff $\text{rng}(F) = Y$;

(iii) normalized iff $(\forall x \in X)(\exists y \in Y)(F(x).y = 1)$.

The following relationships hold:

$$F \text{ is normalized } \Rightarrow F^{-} \text{ is surjective } \Rightarrow F \text{ is non-void.}$$

Notice that for an MM the above implications become equivalences. An interesting property is that the union of an arbitrary family of non-void (resp. surjective, normalized) fuzzy multivalued mappings is also non-void (resp. surjective, normalized).

Consider a fuzzy set A in X, a fuzzy set B in Y, an FMM F from X into Y, and an FMM G from Y into Z.

A $[0,1]^2 \to [0,1]$ mapping $\mathcal{T}$ is called a triangular norm (t-norm) iff it satisfies the conditions, for all $(x, y, z) \in [0,1]^3$ [9]:

(i) boundary condition: $\mathcal{T}(1, x) = x$;

(ii) monotonicity: $x \leq y \Rightarrow \mathcal{T}(x, z) \leq \mathcal{T}(y, z)$;

(iii) associativity: $\mathcal{T}(x, \mathcal{T}(y, z)) = \mathcal{T}(\mathcal{T}(x, y), z)$;

(iv) commutativity: $\mathcal{T}(x, y) = \mathcal{T}(y, x)$.

Given a t-norm $\mathcal{T}$, we can define the degree of compatibility $\text{Com}_{\mathcal{T}}(C, D)$ of two fuzzy sets C and D in X as follows [12]:

$$\text{Com}_{\mathcal{T}}(C, D) = \sup_{x \in X} \mathcal{T}(C(x), D(x)).$$

The composition of F and G is the FMM $G \circ F$ from X into Z, defined by [11]:

$$(G \circ F)(x).z = \text{Com}_{\mathcal{T}}(F(x), G^{-}(z)).$$

If F and G are non-void (resp. normalized) then $G \circ F$ is also non-void (resp. normalized). Moreover, if $(\forall x \in [0,1])(\mathcal{T}(x, \cdot)$ is left-continuous) and G^{-} is surjective then $\text{dom}(G \circ F) = \text{dom}(F)$.

The direct image of A under F and the lower inverse image of B under F are the fuzzy sets $F_{\mathcal{T}}(A)$ in Y and $F^-_{\mathcal{T}}(B)$ in X, defined by [11]:

$$\begin{aligned} F_{\mathcal{T}}(A)(y) &= \mathrm{Com}_{\mathcal{T}}\left(F^-(y), A\right) \\ F^-_{\mathcal{T}}(B)(x) &= \mathrm{Com}_{\mathcal{T}}\left(F(x), B\right). \end{aligned}$$

The upper inverse image has been introduced for FMMs [12, 13] as based on the concept of a fuzzy inclusion, defined from an implication operator.

A $[0,1]^2 \to [0,1]$ mapping $\mathcal{I}$ is called an implication operator iff it satisfies the boundary conditions $\mathcal{I}(0,0) = \mathcal{I}(0,1) = \mathcal{I}(1,1) = 1$ and $\mathcal{I}(1,0) = 0$. Some useful potential properties of implication operators are, for all $(x,y) \in [0,1]^2$ [4]:

(i) hybrid monotonicity: $\mathcal{I}(x,\cdot)$ is increasing and $\mathcal{I}(\cdot,y)$ is decreasing;

(ii) neutrality principle: $\mathcal{I}(1,x) = x$.

Given an implication operator $\mathcal{I}$, we can define the degree of inclusion $\mathrm{Inc}_{\mathcal{I}}(C,D)$ of a fuzzy set C in X in a fuzzy set D in X as follows [4]:

$$\mathrm{Inc}_{\mathcal{I}}(C,D) = \inf_{x\in X} \mathcal{I}(C(x), D(x)).$$

The upper inverse images of B under F are the fuzzy sets $F^+_{\mathcal{I},\mathrm{e}}(B)$, $F^+_{\mathcal{I},}(B)$ and $F^+_{\mathcal{I},\mathcal{T}}(B)$ in X, defined by [12, 13]:

$$\begin{aligned} F^+_{\mathcal{I},\mathrm{e}}(B)(x) &= \begin{cases} \mathrm{Inc}_{\mathcal{I}}(F(x),B), & \text{if } F(x) \neq \emptyset \\ 0, & \text{otherwise} \end{cases} \\ F^+_{\mathcal{I},}(B)(x) &= \min\left(\mathrm{Inc}_{\mathcal{I}}\left(F(x),B\right), \mathrm{dom}(F)(x)\right) \\ F^+_{\mathcal{I},\mathcal{T}}(B)(x) &= \min\left(\mathrm{Inc}_{\mathcal{I}}\left(F(x),B\right), F^-_{\mathcal{T}}(B)(x)\right). \end{aligned}$$

Further on, we will use the notation $F^+_{\mathcal{I}}(B)$ whenever the particular definition is irrelevant to the usage of the upper inverse image.

2.2 Fuzzy topological spaces

In this subsection, we recall the definition of a fuzzy topological space and some related concepts [6, 7, 8]. The definitions of strongly quasi-separated, regular, locally closed, normal, and strongly compact fuzzy topological spaces and the relationships established between them are presented as well [16].

Consider a universe X. A subset $\varphi_X \subseteq I^X$ is a fuzzy topology on X iff [6]:

(i) $\emptyset \in \varphi_X \ \wedge \ X \in \varphi_X$,

(ii) $O_1 \in \varphi_X \ \wedge \ O_2 \in \varphi_X \Rightarrow O_1 \cap O_2 \in \varphi_X$,

(iii) $(\forall i \in I)(O_i \in \varphi_X) \Rightarrow \bigcup_{i \in I} O_i \in \varphi_X$.

The fuzzy sets in φ_X are called open fuzzy sets.

Throughout this paper, a cotopology of the fuzzy topology φ_X is denoted by φ'_X, i.e.:

$$\varphi'_X = \{\mathrm{co}\, O \mid O \in \varphi_X\}.$$

The closure $\mathrm{Cl}\, A$ of a fuzzy set A in a fuzzy topological space (X, φ_X) is the smallest closed fuzzy set which contains A, i.e. [7]:

$$\mathrm{Cl}\, A = \cap\{C \mid C \in \varphi'_X \ \wedge \ A \subseteq C\}.$$

A fuzzy singleton x_δ in X, with $\delta \in]0, 1]$, is the fuzzy set in X, defined by [8]:

$$x_\delta(z) = \begin{cases} \delta, & \text{if } z = x \\ 0, & \text{otherwise} \end{cases},$$

where x and δ are called support and value of x_δ, respectively.

A fuzzy set A in a fuzzy topological space (X, φ_X) is a fuzzy neighbourhood of a fuzzy set B in X iff there exists an open fuzzy set $O \in \varphi_X$ such that $B \subseteq O \subseteq A$ [6].

A fuzzy set is open iff it is a fuzzy neighbourhood of all of its fuzzy singletons.

Throughout this paper, an arbitrary open fuzzy neighbourhood of a fuzzy set A in X or a fuzzy singleton x_δ in X is denoted by $\mathcal{U}(A)$ and $\mathcal{V}(A)$ or $\mathcal{U}(x_\delta)$ and $\mathcal{V}(x_\delta)$, respectively.

A family $\mathcal{O} = (O_i \mid i \in I)$ of open fuzzy sets in a fuzzy topological space (X, φ_X) is an open fuzzy covering of a fuzzy set A in X iff [6]:

$$A \subseteq \bigcup_{i \in I} O_i.$$

A subcovering of an open fuzzy covering O of a fuzzy set A in a fuzzy topological space (X, φ_X) is a subfamily of O, which is still a covering of A [6].

A fuzzy set A in a fuzzy topological space (X, φ_X) is compact iff each open fuzzy covering of A has a finite subcovering [6].

Definition 2.1. (Strongly quasi-separation). [16]
A fuzzy topological space (X, φ_X) is strongly quasi-separated iff every fuzzy singleton in X is closed.

Definition 2.2. (Regularity). [16]
A fuzzy topological space (X, φ_X) is regular iff for any fuzzy singleton x_δ in X and any closed set F in X such that $x_\delta \subseteq \operatorname{co} F$, there exist two open fuzzy neighbourhoods $\mathcal{U}(x_\delta)$ and $\mathcal{V}(F)$ such that

$$\mathcal{U}(x_\delta) \subseteq \operatorname{co} \mathcal{V}(F).$$

Definition 2.3. (Locally closedness). [16]
A fuzzy topological space (X, φ_X) is locally closed iff for any fuzzy singleton x_δ in X and any open fuzzy neighbourhood $\mathcal{U}(x_\delta)$ of x_δ, there exists an open fuzzy neighbourhood $\mathcal{V}(x_\delta)$ such that

$$\operatorname{Cl} \mathcal{V}(x_\delta) \subseteq \mathcal{U}(x_\delta).$$

Definition 2.4. (Normality). [16]
A fuzzy topological space (X, φ_X) is normal iff for any two closed sets F_1 and F_2 in X such that $F_1 \subseteq \operatorname{co} F_2$, there exist two open fuzzy neighbourhoods $\mathcal{U}(F_1)$ and $\mathcal{V}(F_2)$ such that

$$\mathcal{U}(F_1) \subseteq \operatorname{co} \mathcal{V}(F_2).$$

Definition 2.5. (Strongly compactness). [16]
A fuzzy topological space (X, φ_X) is strongly compact iff every closed fuzzy set in X is compact.

Proposition 2.1. [16] *The following relationships hold:*

(i) *A fuzzy topological is regular iff it is locally closed.*

(ii) *A strongly quasi-separated and normal fuzzy topological space is regular.*

(iii) *A regular and strongly compact fuzzy topological space is normal.*

3 CLOSEDNESS OF FUZZY MULTIVALUED MAPPINGS

In this section, we recall the definitions of three different types of closedness of FMMs, their properties, and the relationships established between them [14].

From here on, we assume that the implication operator involved is hybrid monotonous and satisfies the neutrality principle.

3.1 Direct, lower and upper closedness

Consider an FMM F from X into Y.

Definition 3.1. (Direct closedness). [14]
F is direct-closed ($DC_{\mathcal{T}}$) at a fuzzy singleton $x_\delta \subseteq \mathrm{dom}(F)$ with respect to (X, φ_X) and (Y, φ_Y) iff for all y_ϵ in Y such that $y_\epsilon \subseteq \mathrm{co}\, F_{\mathcal{T}}(x_\delta)$, there exist two open fuzzy neighbourhoods $\mathcal{U}(x_\delta)$ and $\mathcal{V}(y_\epsilon)$ such that

$$F_{\mathcal{T}}(\mathcal{U}(x_\delta)) \subseteq \mathrm{co}\, \mathcal{V}(y_\epsilon).$$

F is $DC_{\mathcal{T}}$ iff it is $DC_{\mathcal{T}}$ at all $x_\delta \subseteq \mathrm{dom}(F)$.

Definition 3.2. (Lower closedness). [14]
F is lower-closed ($LC_{\mathcal{T}}$) at a fuzzy singleton $x_\delta \subseteq \mathrm{dom}(F)$ with respect to (X, φ_X) and (Y, φ_Y) iff for all y_ϵ in Y such that $x_\delta \subseteq \mathrm{co}\, F_{\mathcal{T}}^{-}(y_\epsilon)$, there exist two open fuzzy neighbourhoods $\mathcal{U}(x_\delta)$ and $\mathcal{V}(y_\epsilon)$ such that

$$\mathcal{U}(x_\delta) \subseteq \mathrm{co}\, F_{\mathcal{T}}^{-}(\mathcal{V}(y_\epsilon)).$$

F is $LC_{\mathcal{T}}$ iff it is $LC_{\mathcal{T}}$ at all $x_\delta \subseteq \mathrm{dom}(F)$.

Definition 3.3. (Upper closedness). [14]
F is upper-closed (UC_1) at a fuzzy singleton $x_\delta \subseteq \mathrm{dom}(F)$ with respect to (X, φ_X) and (Y, φ_Y) iff for all y_ϵ in Y such that $x_\delta \subseteq F_1^+(\mathrm{co}\, y_\epsilon)$, there exist two open fuzzy neighbourhoods $\mathcal{U}(x_\delta)$ and $\mathcal{V}(y_\epsilon)$ such that

$$\mathcal{U}(x_\delta) \cap \mathrm{dom}(F) \subseteq F_1^+(\mathrm{co}\, \mathcal{V}(y_\epsilon)).$$

F is UC_1 iff it is UC_1 at all $x_\delta \subseteq \mathrm{dom}(F)$.

Proposition 3.1. [14] *If F^- is surjective then F is UC_1 at a fuzzy singleton x_δ in X with respect to (X, φ_X) and (Y, φ_Y) iff for all y_ϵ in Y such that $x_\delta \subseteq F_1^+(\mathrm{co}\, y_\epsilon)$, there exist two open fuzzy neighbourhoods $\mathcal{U}(x_\delta)$ and $\mathcal{V}(y_\epsilon)$ such that*

$$\mathcal{U}(x_\delta) \subseteq F_1^+(\mathrm{co}\, \mathcal{V}(y_\epsilon)).$$

Since upper inverse images under FMMs have been introduced in three different ways, it is logical that three different types of upper closedness should also be considered. These will be denoted by the obvious notations $UC_{1,\mathrm{e}}$, $UC_{1,}$, and $UC_{1,\mathcal{T}}$.

3.2 Relationships

Consider an FMM F from X into Y and an MM G from X into Y. The relationships between the different types of closedness of F and G, presented in this subsection, are with respect to (X, φ_X) and (Y, φ_Y) [14].

Proposition 3.2. [14] *For the upper closedness of F the following equivalences hold:*

(i) If F^- is surjective then: F is $UC_{1,\mathrm{e}} \Leftrightarrow F$ is $UC_{1,}$.

(ii) If F is normalized then: F is $UC_{1,\mathrm{e}} \Leftrightarrow F$ is $UC_{1,} \Leftrightarrow F$ is $UC_{1,\mathcal{T}}$.

Proposition 3.3. [14] *For the upper closedness of G the following equivalences hold:*

$$G \text{ is } UC_{1,\mathrm{e}} \Leftrightarrow G \text{ is } UC_{1,} \Leftrightarrow G \text{ is } UC_{1,\mathcal{T}}.$$

From Propositions 3.2 and 3.3 we can conclude that for normalized FMMs or MMs, with respect to fuzzy topological spaces, there exists only one type of upper closedness. In the case of normalized FMMs we will apply the notation $UC_{\mathcal{I}}$.

Moreover, since we have shown in [11, 12] that the expressions of the direct and lower inverse images of fuzzy sets under an MM are independent of the t-norm involved and respectively, the expressions of the upper inverse images are independent of the implication operator involved, in that case, i.e. for MMs, we can apply the notations DC, LC and UC.

Proposition 3.4. [14] *If G is non-void then for the direct, lower, and upper closedness of G the following equivalences hold:*

$$G \text{ is } DC \Leftrightarrow G \text{ is } LC \Leftrightarrow G \text{ is } UC.$$

From the foregoing proposition we can conclude that for a non-void MM with respect to fuzzy topological spaces there exists only one type of closedness. In that case we can simply talk about closedness.

Proposition 3.5. [14] *If $\mathcal{I}$ is defined from $\mathcal{T}$ by:*

$$(\forall (x,y) \in [0,1]^2)\ (\mathcal{I}(x,y) = 1 - \mathcal{T}(x, 1-y))$$

then for the lower and upper closedness of F the following relationships hold:

(i) If F is non-void then: F is $LC_{\mathcal{T}} \Rightarrow$ F is $UC_{\mathcal{I},\mathrm{e}}$.

(ii) If F^- is surjective then: F is $LC_{\mathcal{T}} \Leftrightarrow$ F is $UC_{\mathcal{I},\cdot}$.

(iii) If F is normalized then: F is $LC_{\mathcal{T}} \Leftrightarrow$ F is $UC_{\mathcal{I}}$.

Proposition 3.6. [14] *If $\mathcal{I}$ satisfies the conditions, for all $(x,y) \in [0,1]^2$:*

(i) $\mathcal{I}(\cdot, y)$ is left-continuous

(ii) $x \leq y \Leftrightarrow \mathcal{I}(x,y) = 1$

and is connected with $\mathcal{T}$ by the relationship:

$$(\forall (x,y,z) \in [0,1]^3)\ (\mathcal{I}(\mathcal{T}(x,y),z) = \mathcal{I}(x, \mathcal{I}(y,z)))$$

then for the direct and upper closedness of F the following relationships hold:

(i) If F is non-void then: F is $DC_{\mathcal{T}} \Rightarrow$ F is $UC_{\mathcal{I},\mathrm{e}}$.

(ii) If F^- is surjective then: F is $DC_{\mathcal{T}} \Leftrightarrow$ F is $UC_{\mathcal{I},\cdot}$.

(iii) If F is normalized then: F is $DC_{\mathcal{T}} \Leftrightarrow$ F is $UC_{\mathcal{I}}$.

3.3 Properties

Definition 3.4. (Closed-valued fuzzy multivalued mapping).
A FMM F from X into Y is closed-valued with respect to (X, φ_X) *and* (Y, φ_Y) *iff*

$$(\forall x_\delta \subseteq \mathrm{dom}(F))(F_{\mathcal{T}}(x_\delta) \in \varphi'_Y).$$

Theorem 3.1. [14] *A* $DC_{\mathcal{T}}$ *FMM F from X into Y with respect to* (X, φ_X) *and* (Y, φ_Y) *is closed-valued.*

The FMMs considered in the following six theorems are from X into Y and the different types of closedness which they exhibit are with respect to (X, φ_X) and (Y, φ_Y).

Theorem 3.2. [14] *If* $\mathcal{T}$ *is continuous then the following properties hold:*

(i) The intersection of a finite family of $DC_{\mathcal{T}}$ *FMMs is* $DC_{\mathcal{T}}$.

(ii) The intersection of a finite family of $LC_{\mathcal{T}}$ *FMMs is* $LC_{\mathcal{T}}$.

Theorem 3.3. [14] *If* $\mathcal{I}$ *satisfies the condition:*

$$(\forall (x,y) \in [0,1]^2)(\mathcal{I}(x,\cdot) \text{ and } \mathcal{I}(\cdot,y) \text{ are right-continuous})$$

then the following properties hold:

(i) The non-void intersection of a finite family of non-void and $UC_{\mathcal{I},\mathrm{e}}$ *FMMs is* $UC_{\mathcal{I},\mathrm{e}}$.

(ii) The intersection of a finite family of $UC_{\mathcal{I},}$ *FMMs with surjective inverse mappings, which has also a surjective inverse mapping, is* $UC_{\mathcal{I},}$.

(iii) The normalized intersection of a finite family of normalized and $UC_{\mathcal{I}}$ *FMMs is* $UC_{\mathcal{I}}$.

Theorem 3.4. *The non-void intersection of a finite family of non-void and closed MMs is closed.*

Theorem 3.5. *If* $\mathcal{T}$ *satisfies the condition:*

$$(\forall x \in [0,1])(\mathcal{T}(x,\cdot) \text{ is left-continuous})$$

then the following properties hold:

(i) *The union of a finite family of* $DC_{\mathcal{T}}$ *FMMs with surjective inverse mappings is* $DC_{\mathcal{T}}$.

(ii) *The union of a finite family of* $LC_{\mathcal{T}}$ *FMMs with surjective inverse mappings is* $LC_{\mathcal{T}}$.

Theorem 3.6. *If* $\mathcal{I}$ *satisfies the condition:*

$$(\forall y \in [0,1])(\mathcal{I}(\cdot, y) \text{ is left-continuous})$$

then the following properties hold:

(i) *The union of a finite family of* $UC_{\mathcal{I},\mathrm{e}}$ *(resp.* $UC_{\mathcal{I},}$*) FMMs with surjective inverse mappings is* $UC_{\mathcal{I},\mathrm{e}}$ *(resp.* $UC_{\mathcal{I},}$*).*

(ii) *The union of a finite family of normalized and* $UC_{\mathcal{I}}$ *FMMs is* $UC_{\mathcal{I}}$.

Theorem 3.7. *The union of a finite family of non-void and closed MMs is closed.*

4 UPPER SEMI-CONTINUITY OF FUZZY MULTIVALUED MAPPINGS

4.1 Upper semi-continuity^{d} and upper semi-continuity^{u}

Consider an FMM F from X into Y.

Definition 4.1. (Upper semi-continuity^{d}). [15]
F is upper semi-continuous in terms of direct images ($USC^{\mathrm{d}}_{\mathcal{T}}$) at a fuzzy singleton $x_\delta \subseteq \mathrm{dom}(F)$ with respect to (X, φ_X) and (Y, φ_Y) iff for each open fuzzy set $O \in \varphi_Y$ such that $F_{\mathcal{T}}(x_\delta) \subseteq O$, there exists an open fuzzy neighbourhood $\mathcal{U}(x_\delta)$ such that

$$F_{\mathcal{T}}(\mathcal{U}(x_\delta)) \subseteq O.$$

F is $USC^{\mathrm{d}}_{\mathcal{T}}$ iff it is $USC^{\mathrm{d}}_{\mathcal{T}}$ at all $x_\delta \in \mathrm{dom}(F)$.

Definition 4.2. (Upper semi-continuity^{u}). [15]
F is upper semi-continuous in terms of upper inverse images (USC^{u}_{1}) at a fuzzy singleton $x_\delta \subseteq \mathrm{dom}(F)$ with respect to (X, φ_X) and (Y, φ_Y) iff for each open fuzzy set $O \in \varphi_Y$ such that $x_\delta \subseteq F^{+}_{1}(O)$, there exists an open fuzzy neighbourhood $\mathcal{U}(x_\delta)$ such that

$$\mathcal{U}(x_\delta) \cap \mathrm{dom}(F) \subseteq F^{+}_{1}(O).$$

F is USC^{u}_{1} iff it is USC^{u}_{1} at all $x_\delta \in \mathrm{dom}(F)$.

Proposition 4.1. [15] *If F^{-} is surjective then F is USC^{u}_{1} at a fuzzy singleton x_δ in X with respect to (X, φ_X) and (Y, φ_Y) iff for each open fuzzy set $O \in \varphi_Y$ such that $x_\delta \subseteq F^{+}_{1}(O)$, there exists an open fuzzy neighbourhood $\mathcal{U}(x_\delta)$ such that*

$$\mathcal{U}(x_\delta) \subseteq F^{+}_{1}(O).$$

Three different types of upper semi-continuity^{u} can be considered corresponding to the three different definitions of upper inverse images. These will be denoted by the obvious notations $USC^{\mathrm{u}}_{1,\mathrm{e}}$, $USC^{\mathrm{u}}_{1,}$, and $USC^{\mathrm{u}}_{1,\mathcal{T}}$.

4.2 Relationships

Consider an FMM F from X into Y and an MM G from X into Y. The relationships between the different types of upper semi-continuity of F and G, presented in this subsection, are with respect to (X, φ_X) and (Y, φ_Y) [15].

Proposition 4.2. [15] *For the upper semi-continuity^{u} of F the following equivalences hold:*

(i) *If F^{-} is surjective then: F is $USC^{\mathrm{u}}_{1,\mathrm{e}}$ $\Leftrightarrow$ F is $USC^{\mathrm{u}}_{1,}$.*

(ii) *If F is normalized then: F is $USC^{\mathrm{u}}_{1,\mathrm{e}}$ $\Leftrightarrow$ F is $USC^{\mathrm{u}}_{1,}$ $\Leftrightarrow$ F is $USC^{\mathrm{u}}_{1,\mathcal{T}}$.*

Proposition 4.3. [15] *For the upper semi-continuity^{u} of G the following equivalences hold:*

$$G \text{ is } USC^{\mathrm{u}}_{1,\mathrm{e}} \Leftrightarrow G \text{ is } USC^{\mathrm{u}}_{1,} \Leftrightarrow G \text{ is } USC^{\mathrm{u}}_{1,\mathcal{T}}.$$

As for upper closedness, from the two foregoing propositions we obtain that for normalized FMMs or MMs, with respect to fuzzy topological spaces, there exists only one type of upper semi-continuity^{u}. In the case of normalized FMMs we will apply the notation $USC_{\mathcal{I}}^{\mathrm{u}}$.

Reasoning in the same way as in subsection 3.2, we can conclude that upper semi-continuity^{d} and upper semi-continuity^{u} of MMs are independent of the t-norm and respectively, of the implication operator involved. Therefore in that case, i.e. for MMs, we will apply the notations USC^{d} and USC^{u}.

Proposition 4.4. [15] *If G is non-void then for the upper semi-continuity*^{u} *and upper semi-continuity*^{d} *of G the following equivalence holds:*

$$G \text{ is } USC^{\mathrm{d}} \Leftrightarrow G \text{ is } USC^{\mathrm{u}}.$$

This proposition implies the existence of only one type of upper semi-continuity for a non-void MM with respect to fuzzy topological spaces. In that case we will use the notation USC.

Proposition 4.5. [15] *If $\mathcal{I}$ satisfies the conditions, for all $(x,y) \in [0,1]^2$:*

(i) $\mathcal{I}(\cdot, y)$ is left-continuous

(ii) $x \leq y \Leftrightarrow \mathcal{I}(x,y) = 1$

and is connected with $\mathcal{T}$ by the relationship:

$$(\forall (x,y,z) \in [0,1]^3)\ (\mathcal{I}(\mathcal{T}(x,y),z) = \mathcal{I}(x,\mathcal{I}(y,z)))$$

then for the upper semi-continuity^{u} *and upper semi-continuity*^{d} *of F the following relationships hold:*

(i) If F is non-void then: F is $USC_{\mathcal{T}}^{\mathrm{d}} \Rightarrow$ F is $USC_{\mathcal{I},\mathrm{e}}^{\mathrm{u}}$.

(ii) If F^- is surjective then: F is $USC_{\mathcal{T}}^{\mathrm{d}} \Leftrightarrow$ F is $USC_{\mathcal{I},}^{\mathrm{u}}$.

(iii) If F is normalized then: F is $USC_{\mathcal{T}}^{\mathrm{d}} \Leftrightarrow$ F is $USC_{\mathcal{I}}^{\mathrm{u}}$.

4.3 Properties

Consider an FMM F_1 from X into Y, an FMM F_2 from Y into Z, an MM G_1 from X into Y, and an MM G_2 from Y into Z. The different types of upper semi-continuity considered in the following two theorems are with respect to (X, φ_X) and (Y, φ_Y) when F_1 and G_1 are involved, with respect to (Y, φ_Y) and (Z, φ_Z) when F_2 and G_2 are involved, and with respect to (X, φ_X) and (Z, φ_Z) when $F_2 \circ F_1$ and $G_2 \circ G_1$ are involved.

Theorem 4.1. [15] *If $\mathcal{I}$ satisfies the conditions, for all $(x, y) \in [0,1]^2$:*

(i) $\mathcal{I}(x,\cdot)$ is right-continuous

(ii) $\mathcal{I}(\cdot,y)$ is left-continuous

and is connected with $\mathcal{T}$ by the relationship:

$$(\forall (x,y,z) \in [0,1]^3)\ (\mathcal{I}(\mathcal{T}(x,y),z) = \mathcal{I}(x,\mathcal{I}(y,z)))$$

then the following properties hold:

(i) If $\mathcal{T}$ satisfies the condition:

$$(\forall x \in [0,1])(\mathcal{T}(x,\cdot) \text{ is left-continuous}),$$

and F_1^- and F_2^- are surjective then

(a) If F_1 and F_2 are $USC^{\mathrm{u}}_{\mathcal{I},\mathrm{e}}$ (resp. $USC^{\mathrm{u}}_{\mathcal{I},}$) then $F_2 \circ F_1$ is $USC^{\mathrm{u}}_{\mathcal{I},\mathrm{e}}$ (resp. $USC^{\mathrm{u}}_{\mathcal{I},}$).

(b) Moreover, if $\mathcal{I}$ satisfies the condition:

$$(\forall (x,y) \in [0,1]^2)(x \le y \Leftrightarrow \mathcal{I}(x,y) = 1),$$

and F_1 and F_2 are $USC^{\mathrm{d}}_{\mathcal{T}}$ then $F_2 \circ F_1$ is $USC^{\mathrm{d}}_{\mathcal{T}}$.

(ii) If F_1 and F_2 are normalized then

(a) If F_1 and F_2 are $USC^{\mathrm{u}}_{\mathcal{I}}$ then $F_2 \circ F_1$ is $USC^{\mathrm{u}}_{\mathcal{I}}$.

(b) Moreover, if $\mathcal{I}$ satisfies the condition:

$$(\forall (x,y) \in [0,1]^2)(x \le y \Leftrightarrow \mathcal{I}(x,y) = 1),$$

and F_1 and F_2 are $USC^{\mathrm{d}}_{\mathcal{T}}$ then $F_2 \circ F_1$ is $USC^{\mathrm{d}}_{\mathcal{T}}$.

Theorem 4.2. [15] *If G_1 and G_2 are non-void and USC then $G_2 \circ G_1$ is USC.*

The FMMs considered in the following three theorems are from X into Y and the different types of upper semi-continuity, which they exhibit are with respect to (X, φ_X) and (Y, φ_Y).

Theorem 4.3. [15] *If $\mathcal{T}$ satisfies the condition:*

$$(\forall x \in [0,1])(\mathcal{T}(x,\cdot) \text{ is left-continuous})$$

then the union of a finite family of $USC^{\mathrm{d}}_{\mathcal{T}}$ FMMs with surjective inverse mappings is $USC^{\mathrm{d}}_{\mathcal{T}}$.

Theorem 4.4. [15] *If $\mathcal{I}$ satisfies the condition:*

$$(\forall y \in [0,1])(\mathcal{I}(\cdot, y) \text{ is left-continuous})$$

then the following properties hold:

(i) *The union of a finite family of $USC^{u}_{\mathcal{I},e}$ (resp. $USC^{u}_{\mathcal{I},}$) FMMs with surjective inverse mappings is $USC^{u}_{\mathcal{I},e}$ (resp. $USC^{u}_{\mathcal{I},}$).*

(ii) *The union of a finite family of normalized and $USC^{u}_{\mathcal{I}}$ FMMs is $USC^{u}_{\mathcal{I}}$.*

Theorem 4.5. *The union of a finite family of non-void and USC MMs is USC.*

5 RELATIONSHIP BETWEEN CLOSEDNESS AND UPPER SEMI-CONTINUITY

In this section, we study the relationships between direct closedness and upper semi-continuity^{d}, and upper closedness and upper semi-continuity^{u} of a fuzzy multivalued mapping. Lower closedness is not separately considered because there is no direct relationship between it and any of the two types of upper semi-continuity.

5.1 Direct closedness and upper semi-continuity^{d}

Theorem 5.1. *A closed-valued and $USC^{d}_{\mathcal{T}}$ FMM F from X into Y with respect to (X, φ_X) and regular (Y, φ_Y) is $DC_{\mathcal{T}}$.*

Proof. Consider $x_\delta \subseteq \mathrm{dom}(F)$ and y_ϵ in Y such that $y_\epsilon \subseteq \mathrm{co}\, F_{\mathcal{T}}(x_\delta)$. Since $F_{\mathcal{T}}(x_\delta)$ is closed and (Y, φ_Y) is regular, it follows that there exist two open fuzzy neighbourhoods $\mathcal{V}_1(y_\epsilon)$ and $\mathcal{V}_2(F_{\mathcal{T}}(x_\delta))$ such that

$$\mathcal{V}_1(y_\epsilon) \subseteq \mathrm{co}\, \mathcal{V}_2(F_{\mathcal{T}}(x_\delta)).$$

Since F is $USC^{d}_{\mathcal{T}}$ and $F_{\mathcal{T}}(x_\delta) \subseteq \mathcal{V}_2(F_{\mathcal{T}}(x_\delta))$, it follows that there exists an open fuzzy neighbourhood $\mathcal{U}(x_\delta)$ such that

$$F_{\mathcal{T}}(\mathcal{U}(x_\delta)) \subseteq \mathcal{V}_2(F_{\mathcal{T}}(x_\delta)) \subseteq \mathrm{co}\, \mathcal{V}_1(y_\epsilon).$$

Hence F is $DC_{\mathcal{T}}$ at x_δ. □

Corollary 5.1. *A non-void, closed-valued and USC MM G from X into Y with respect to* (X, φ_X) *and regular* (Y, φ_Y) *is closed.*

Proof. Follows from Propositions 3.4 and 4.4 and Theorem 5.1. □

Corollary 5.2. *A closed-valued and* $USC^{\mathrm{d}}_{\mathcal{T}}$ *FMM F from X into Y with respect to* (X, φ_X) *and locally closed* (Y, φ_Y) *is* $DC_{\mathcal{T}}$.

Proof. Follows from Proposition 2.1 (i) and Theorem 5.1. □

Corollary 5.3. *A closed-valued and* $USC^{\mathrm{d}}_{\mathcal{T}}$ *FMM F from X into Y with respect to* (X, φ_X) *and normal and strongly quasi-separated* (Y, φ_Y) *is* $DC_{\mathcal{T}}$.

Proof. Follows from Proposition 2.1 (ii) and Theorem 5.1. □

Theorem 5.2. *A* $DC_{\mathcal{T}}$ *FMM F from X into Y with respect to* (X, φ_X) *and strongly compact* (Y, φ_Y) *is* $USC^{\mathrm{d}}_{\mathcal{T}}$.

Proof. Consider $x_\delta \subseteq \mathrm{dom}(F)$ and $O \in \varphi_Y$ such that $F_{\mathcal{T}}(x_\delta) \subseteq O$. For any y_ϵ in Y such that $y_\epsilon \subseteq \mathrm{co}\, O$, we have that

$$y_\epsilon \subseteq \mathrm{co}\, O \subseteq \mathrm{co}\, F_{\mathcal{T}}(x_\delta).$$

Since F is $DC_{\mathcal{T}}$, it follows that there exist two open fuzzy neighbourhoods $\mathcal{U}^{y_\epsilon}(x_\delta)$ and $\mathcal{V}(y_\epsilon)$ such that

$$F_{\mathcal{T}}\left(\mathcal{U}^{y_\epsilon}(x_\delta)\right) \subseteq \mathrm{co}\, \mathcal{V}(y_\epsilon).$$

Taking into account that $\mathrm{co}\, O$ can be decomposed in terms of all the fuzzy singletons contained in it, we obtain that

$$\mathrm{co}\, O = \bigcup_{y_\epsilon \subseteq \mathrm{co}\, O} y_\epsilon \subseteq \bigcup_{y_\epsilon \subseteq \mathrm{co}\, O} \mathcal{V}(y_\epsilon).$$

Since $\mathrm{co}\, O$ is compact, it follows that there exists $\{y_{1\epsilon_1}, y_{2\epsilon_2}, \ldots, y_{n\epsilon_n}\}$ such that

$$\mathrm{co}\, O \subseteq \bigcup_{i=1}^{n} \mathcal{V}(y_{i\epsilon_i}).$$

Consider the following neighbourhoods

$$\mathcal{V}(\operatorname{co} O) = \bigcup_{i=1}^{n} \mathcal{V}(y_{i\epsilon_i}) \text{ and } \mathcal{U}(x_\delta) = \bigcap_{i=1}^{n} \mathcal{U}^{y_{i\epsilon_i}}(x_\delta).$$

Taking into account the monotonicity of the direct images of fuzzy sets under an FMM [11], we obtain that

$$\begin{aligned} F_{\mathcal{T}}(\mathcal{U}(x_\delta)) &= F_{\mathcal{T}}\left(\bigcap_{i=1}^{n} \mathcal{U}^{y_{i\epsilon_i}}(x_\delta)\right) \subseteq \bigcap_{i=1}^{n} F_{\mathcal{T}}(\mathcal{U}^{y_{i\epsilon_i}}(x_\delta)) \\ &\subseteq \bigcap_{i=1}^{n} \operatorname{co}\mathcal{V}(y_{i\epsilon_i}) = \operatorname{co}\bigcup_{i=1}^{n} \mathcal{V}(y_{i\epsilon_i}) = \operatorname{co}\mathcal{V}(\operatorname{co} O) \subseteq O. \end{aligned}$$

Hence F is $USC_{\mathcal{T}}^{\mathrm{d}}$ at x_δ. □

Corollary 5.4. *A non-void and closed MM G from X into Y with respect to (X, φ_X) and strongly compact (Y, φ_Y) is USC.*

Proof. Follows from Propositions 3.4 and 4.4 and Theorem 5.2. □

Theorem 5.3. *A FMM F from X into Y is $DC_{\mathcal{T}}$ with respect to (X, φ_X) and regular and strongly compact (Y, φ_Y) iff it is closed-valued and $USC_{\mathcal{T}}^{\mathrm{d}}$.*

Proof. Follows from Theorems 3.1, 5.1 and 5.2. □

Corollary 5.5. *A non-void MM G from X into Y is closed with respect to (X, φ_X) and regular and strongly compact (Y, φ_Y) iff it is closed-valued and USC.*

Proof. Follows from Corollaries 5.1 and 5.4 and Theorem 3.1. □

5.2 Upper closedness and upper semi-continuity^{u}

Proposition 5.1. *If $\mathcal{I}$ satisfies the conditions, for all $(x, y) \in [0,1]^2$:*

(i) $\mathcal{I}(\cdot, y)$ is left-continuous

(ii) $x \leq y \Leftrightarrow \mathcal{I}(x, y) = 1$

and is connected with $\mathcal{T}$ by the relationship:

$$(\forall (x,y,z) \in [0,1]^3)\ (\mathcal{I}(\mathcal{T}(x,y),z) = \mathcal{I}(x,\mathcal{I}(y,z)))$$

then for a closed-valued FMM F from X into Y the following properties hold, with respect to (X,φ_X) and regular (Y,φ_Y):

(i) *If F^- is surjective and F is $USC^{\mathrm{u}}_{\mathcal{I},\mathrm{e}}$ (resp. $USC^{\mathrm{u}}_{\mathcal{I},}$) then F is $UC_{\mathcal{I},\mathrm{e}}$ (resp. $UC_{\mathcal{I},}$).*

(ii) *If F is normalized and $USC^{\mathrm{u}}_{\mathcal{I}}$ then F is $UC_{\mathcal{I}}$.*

Proof. We will, for instance, prove the second property. Since the conditions of Proposition 4.5 (iii) are fulfilled, we have that F is $USC^{\mathrm{d}}_{\mathcal{T}}$. From the latter and Theorem 5.1, it follows that F is $DC_{\mathcal{T}}$. Finally, applying Proposition 3.6 (iii), we obtain that F is $UC_{\mathcal{I}}$. □

Theorem 5.4. *If $\mathcal{I}$ satisfies the condition:*

$$(\forall x \in [0,1])(\mathcal{I}(x,\cdot) \text{ is right-continuous})$$

then for an FMM F from X into Y the following properties hold, with respect to (X,φ_X) and strongly compact (Y,φ_Y):

(i) *If F is $UC_{\mathcal{I},\mathrm{e}}$ (resp. $UC_{\mathcal{I},}$) then F is $USC^{\mathrm{u}}_{\mathcal{I},\mathrm{e}}$ (resp. $USC^{\mathrm{u}}_{\mathcal{I},}$).*

(ii) *If F is normalized and $UC_{\mathcal{I}}$ then F is $USC^{\mathrm{u}}_{\mathcal{I}}$.*

Proof. Let us prove the first property. Suppose that F is $UC_{\mathcal{I},\mathrm{e}}$. Consider $x_\delta \subseteq \mathrm{dom}(F)$ and $O \in \varphi_Y$ such that $x_\delta \subseteq F^+_{\mathcal{I},\mathrm{e}}(O)$. For any y_ϵ in Y such that $y_\epsilon \subseteq \mathrm{co}\, O$, or equivalently $O \subseteq \mathrm{co}\, y_\epsilon$, we have that

$$x_\delta \subseteq F_{\mathcal{I},\mathrm{e}}(O) \subseteq F^+_{\mathcal{I},\mathrm{e}}(\mathrm{co}\, y_\epsilon).$$

Since F is $UC_{\mathcal{I},\mathrm{e}}$, it follows that there exist two open fuzzy neighbourhoods $\mathcal{U}^{y_\epsilon}(x_\delta)$ and $\mathcal{V}(y_\epsilon)$ such that

$$\mathcal{U}^{y_\epsilon}(x_\delta) \cap \mathrm{dom}(F) \subseteq F^+_{\mathcal{I},\mathrm{e}}\left(\mathrm{co}\, \mathcal{V}(y_\epsilon)\right).$$

As in the proof of Theorem 5.2 $\mathrm{co}\, O$ can be expressed in terms of all the fuzzy singletons contained in it, i.e.

$$\mathrm{co}\, O = \bigcup_{y_\epsilon \subseteq \mathrm{co}\, O} y_\epsilon \subseteq \bigcup_{y_\epsilon \subseteq \mathrm{co}\, O} \mathcal{V}(y_\epsilon).$$

Using the compactness of $\mathrm{co}\, O$, we obtain that there exists $\{y_{1\epsilon_1}, y_{2\epsilon_2}, \ldots, y_{n\epsilon_n}\}$ such that

$$\mathrm{co}\, O \subseteq \bigcup_{i=1}^{n} \mathcal{V}(y_{i\epsilon_i}).$$

Let us consider the following neighbourhoods

$$\mathcal{V}(\mathrm{co}\, O) = \bigcup_{i=1}^{n} \mathcal{V}(y_{i\epsilon_i}) \text{ and } \mathcal{U}(x_\delta) = \bigcap_{i=1}^{n} \mathcal{U}^{y_{i\epsilon_i}}(x_\delta).$$

Taking into account the right continuity of the second partial mappings of $\mathcal{I}$, we obtain that

$$\begin{aligned}
\mathcal{U}(x_\delta) \cap \mathrm{dom}(F) &= \left(\bigcap_{i=1}^{n} \mathcal{U}^{y_{i\epsilon_i}}(x_\delta)\right) \cap \mathrm{dom}(F) \\
&= \bigcap_{i=1}^{n} (\mathcal{U}^{y_{i\epsilon_i}}(x_\delta) \cap \mathrm{dom}(F)) \\
&\subseteq \bigcap_{i=1}^{n} F^{+}_{\mathcal{I},\mathrm{e}}\left(\mathrm{co}\, \mathcal{V}(y_{i\epsilon_i})\right) = F^{+}_{\mathcal{I},\mathrm{e}}\left(\bigcap_{i=1}^{n} \mathrm{co}\, \mathcal{V}(y_{i\epsilon_i})\right) \\
&= F^{+}_{\mathcal{I},\mathrm{e}}\left(\mathrm{co} \bigcup_{i=1}^{n} \mathcal{V}(y_{i\epsilon_i})\right) = F^{+}_{\mathcal{I},\mathrm{e}}(\mathrm{co}\, \mathcal{V}(\mathrm{co}\, O)) \subseteq F^{+}_{\mathcal{I},\mathrm{e}}(O).
\end{aligned}$$

Hence F is $USC^{\mathrm{u}}_{\mathcal{I},\mathrm{e}}$ at x_δ. □

Corollary 5.6. *If $\mathcal{I}$ satisfies the condition:*

$$(\forall x \in [0,1])(\mathcal{I}(x,\cdot) \text{ is right-continuous})$$

then for an FMM F from X into Y the following properties hold, with respect to (X, φ_X) and compact (Y, π_Y):

(i) If F is $UC_{\mathcal{I},\mathrm{e}}$ (resp. $UC_{\mathcal{I},}$) then F is $USC^{\mathrm{u}}_{\mathcal{I},\mathrm{e}}$ (resp. $USC^{\mathrm{u}}_{\mathcal{I},}$).

(ii) If F is normalized and $UC_{\mathcal{I}}$ then F is $USC^{\mathrm{u}}_{\mathcal{I}}$.

Proof. Can be done similarly to the proof of Theorem 5.4 since in a compact ordinary topological space every closed set is compact. □

Proposition 5.2. *If $\mathcal{I}$ satisfies the conditions, for all $(x, y) \in [0,1]^2$:*

(i) $\mathcal{I}(x,\cdot)$ *is right-continuous*
(ii) $\mathcal{I}(\cdot,y)$ *is left-continuous*
(iii) $x \le y \Leftrightarrow \mathcal{I}(x,y) = 1$

and is connected with $\mathcal{T}$ *by the relationship:*

$$(\forall(x,y,z) \in [0,1]^3)\ (\mathcal{I}(\mathcal{T}(x,y),z) = \mathcal{I}(x,\mathcal{I}(y,z)))$$

then for an FMM F from X into Y the following properties hold, with respect to (X,φ_X) *and regular and strongly compact* (Y,φ_Y)*:*

(i) *If* F^- *is surjective then F is* $UC_{\mathcal{I},\mathrm{e}}$ *(resp.* $UC_{\mathcal{I},}$*) iff it is closed-valued and* $USC^{\mathrm{u}}_{\mathcal{I},\mathrm{e}}$ *(resp.* $USC^{\mathrm{u}}_{\mathcal{I},}$*).*
(ii) *If F is normalized then F is* $UC_{\mathcal{I}}$ *iff it is closed-valued and* $USC^{\mathrm{u}}_{\mathcal{I}}$*.*

Proof. Straightforward. □

6 PROPERTIES DERIVED FROM THE INTERACTION OF CLOSEDNESS AND UPPER SEMI-CONTINUITY

The study of the properties presented in this section is not possible without being preceded by the study of the interaction of closedness and upper semi-continuity of fuzzy multivalued mappings.

6.1 Mixed intersection of closed and upper semi-continuous fuzzy multivalued mappings

Consider three FMMs F, F_1, and F_2 from X into Y and three MMs G, G_1, and G_2 from X into Y. The different types of closedness and upper semi-continuity considered in this subsection are with respect to (X,φ_X) and regular (Y,φ_Y).

Proposition 6.1. *If* $\mathcal{T}$ *is continuous then the intersection* $F_1 \cap F_2$ *of* $DC_{\mathcal{T}}$ F_1 *and closed-valued and* $USC^{\mathrm{d}}_{\mathcal{T}}$ F_2 *is* $DC_{\mathcal{T}}$*.*

Proof. According to Theorem 5.1, F_2 is $DC_{\mathcal{T}}$ and since $\mathcal{T}$ is continuous, it follows from Theorem 3.2 (i) that $F_1 \cap F_2$ is $DC_{\mathcal{T}}$. □

Proposition 6.2. *The non-void intersection $G_1 \cap G_2$ of non-void and closed G_1 and non-void, closed-valued, and USC G_2 is closed.*

Proof. Follows immediately from Corollary 5.1 and Theorem 3.4. □

Proposition 6.3. *If $\mathcal{I}$ satisfies the condition:*

$$(\forall (x, y) \in [0, 1]^2)(\mathcal{I}(x, \cdot) \text{ and } \mathcal{I}(\cdot, y) \text{ are right-continuous})$$

then for the intersection $F \cap G$ of F and non-void, closed-valued, and USC G the following properties hold:

(i) If F and $F \cap G$ are non-void and F is $UC_{\mathcal{I},e}$ then $F \cap G$ is $UC_{\mathcal{I},e}$.

(ii) If F^- and $(F \cap G)^-$ are surjective and F is $UC_{\mathcal{I},}$ then $F \cap G$ is $UC_{\mathcal{I},}$.

(iii) If F and $F \cap G$ are normalized and F is $UC_{\mathcal{I}}$ then $F \cap G$ is $UC_{\mathcal{I}}$.

Proof. We will, for instance, prove the second property. From Corollary 5.1, it follows that G is closed and since the partial mappings of $\mathcal{I}$ are right-continuous, we obtain from Theorem 3.3 (ii) that $F \cap G$ is $UC_{\mathcal{I},}$. □

Proposition 6.4. *If $\mathcal{I}$ satisfies the conditions, for all $(x, y) \in [0, 1]^2$:*

(i) $\mathcal{I}(x, \cdot)$ and $\mathcal{I}(\cdot, y)$ are right-continuous

(ii) $\mathcal{I}(\cdot, y)$ is left-continuous

(iii) $x \leq y \Leftrightarrow \mathcal{I}(x, y) = 1$

and is connected with $\mathcal{T}$ by the relationship:

$$(\forall (x, y, z) \in [0, 1]^3)\ (\mathcal{I}(\mathcal{T}(x, y), z) = \mathcal{I}(x, \mathcal{I}(y, z)))$$

then for the intersection $F_1 \cap F_2$ of F_1 and closed-valued F_2 the following properties hold:

(i) If F_1^-, F_2^- and $(F_1 \cap F_2)^-$ are surjective, F_1 is $UC_{\mathcal{I},e}$ (resp. $UC_{\mathcal{I},}$) and F_2 is $USC^{u}_{\mathcal{I},e}$ (resp. $USC^{u}_{\mathcal{I},}$) then $F_1 \cap F_2$ is $UC_{\mathcal{I},e}$ (resp. $UC_{\mathcal{I},}$).

(ii) If F_1, F_2 and $F_1 \cap F_2$ are normalized, F_1 is $UC_{\mathcal{I}}$ and F_2 is $USC^{u}_{\mathcal{I}}$ then $F_1 \cap F_2$ is $UC_{\mathcal{I},\mathcal{T}}$.

Proof. The proof is similar to the one of Proposition 6.1 and is based on Proposition 5.1 and Theorem 3.3. □

6.2 Intersection of upper semi-continuous fuzzy multivalued mappings

Proposition 6.5. *If $\mathcal{T}$ is continuous then the intersection of a finite family of closed-valued and $USC^{\mathrm{d}}_{\mathcal{T}}$ FMMs from X into Y with respect to (X, φ_X) and regular (Y, φ_Y) is $DC_{\mathcal{T}}$.*

Moreover, if (Y, φ_Y) is strongly compact then the intersection is $USC^{\mathrm{d}}_{\mathcal{T}}$.

Proof. Let $(F_i)_{i=1}^{n}$ be a finite family of closed-valued and $USC^{\mathrm{d}}_{\mathcal{T}}$ FMMs from X into Y. For all mappings of that family we can apply Theorem 5.1, thus $(\forall i \in [1, n])(F_i \text{ is } DC_{\mathcal{T}})$. Since $\mathcal{T}$ is continuous, it follows from Theorem 3.2 (i) that $\bigcap_{i=1}^{n} F_i$ is also $DC_{\mathcal{T}}$. Finally, if (Y, φ_Y) is strongly compact, we obtain from Theorem 5.2, that $\bigcap_{i=1}^{n} F_i$ is $USC^{\mathrm{d}}_{\mathcal{T}}$. □

Proposition 6.6. *The non-void intersection of a finite family of non-void, closed-valued, and USC MMs from X into Y with respect to (X, φ_X) and regular (Y, φ_Y) is closed.*

Moreover, if (Y, φ_Y) is strongly compact then the intersection is USC.

Proof. Follows in the same way as Proposition 6.5 using Corollaries 5.1 and 5.4 and Theorem 3.4. □

Proposition 6.7. *If $\mathcal{I}$ satisfies the conditions, for all $(x, y) \in [0,1]^2$:*

(i) $\mathcal{I}(x, \cdot)$ and $\mathcal{I}(\cdot, y)$ are right-continuous

(ii) $\mathcal{I}(\cdot, y)$ is left-continuous

(iii) $x \leq y \Leftrightarrow \mathcal{I}(x, y) = 1$

and is connected with $\mathcal{T}$ by the relationship:

$$(\forall (x, y, z) \in [0,1]^3)\ (\mathcal{I}(\mathcal{T}(x, y), z) = \mathcal{I}(x, \mathcal{I}(y, z)))$$

then the following properties hold,
with respect to (X, φ_X) and regular (Y, φ_Y):

(i) *The intersection of a finite family of closed-valued and* $USC^{u}_{\mathcal{I},e}$ *(resp.* $USC^{u}_{\mathcal{I},}$*) FMMs from X into Y with surjective inverse mappings, which has also a surjective inverse mapping is* $UC_{\mathcal{I},e}$ *(resp.* $UC_{\mathcal{I},}$*).*

(ii) *The normalized intersection of a finite family of normalized, closed-valued and* $USC^{u}_{\mathcal{I}}$ *FMMs from X into Y is* $UC_{\mathcal{I}}$*.*

Moreover, if (Y, φ_Y) *is strongly compact then the intersection is, respectively:*

(i) $USC^{u}_{\mathcal{I},e}$ *(resp.* $USC^{u}_{\mathcal{I},}$*).*

(ii) $USC^{u}_{\mathcal{I}}$*.*

Proof. We will present the proof of the second property. Let $(F_i)_{i=1}^{n}$ be a finite family of normalized, closed-valued and $USC^{u}_{\mathcal{I}}$ FMMs from X into Y. From Proposition 5.1 (ii), we have that $(\forall i \in [1,n])(F_i \text{ is } UC_{\mathcal{I}})$. Taking into account the right-continuity of the partial mappings of $\mathcal{I}$ we obtain from Theorem 3.3 (iii) that $\bigcap_{i=1}^{n} F_i$ is also $UC_{\mathcal{I}}$. Finally, if (Y, φ_Y) is strongly compact then it follows from Theorem 5.4 (ii), that $\bigcap_{i=1}^{n} F_i$ is $USC^{u}_{\mathcal{I}}$. □

6.3 Mixed union of closed and upper semi-continuous fuzzy multivalued mappings

Consider three FMMs F, F_1, and F_2 from X into Y and three MMs G, G_1, and G_2 from X into Y.

Proposition 6.8. *If* $\mathcal{T}$ *satisfies the condition:*

$$(\forall x \in [0,1])(\mathcal{T}(x,\cdot) \text{ is left-continuous})$$

then for the union $F_1 \cup F_2$ *of* $DC_{\mathcal{T}}$ F_1 *with surjective inverse mapping and* $USC^{d}_{\mathcal{T}}$ F_2 *with surjective inverse mapping the following properties hold, with respect to* (X, φ_X) *and* (Y, φ_Y)*:*

(i) *If* (Y, φ_Y) *is regular and* F_2 *is closed-valued then* $F_1 \cup F_2$ *is* $DC_{\mathcal{T}}$*.*

(ii) *If* (Y, φ_Y) *is strongly compact then* $F_1 \cup F_2$ *is* $USC^{d}_{\mathcal{T}}$*.*

Proof. We will, for instance prove, the first property. According to Theorem 5.1, F_2 is $DC_{\mathcal{T}}$. Taking into account the left-continuity of the partial mappings of $\mathcal{T}$, it holds from Theorem 3.5 (i) that $F_1 \cup F_2$ is $DC_{\mathcal{T}}$. □

Proposition 6.9. *For the union $G_1 \cup G_2$ of non-void and closed G_1 and non-void and USC G_2 the following properties hold, with respect to (X, φ_X) and (Y, φ_Y):*

(i) If (Y, φ_Y) is regular and G_2 is closed-valued then $G_1 \cup G_2$ is closed.

(ii) If (Y, φ_Y) is strongly compact then $G_1 \cup G_2$ is USC.

Proof. The proof is similar to the one of Proposition 6.8 and is based on Corollaries 5.1 and 5.4 and Theorems 3.7 and 4.5. □

Proposition 6.10. *If $\mathcal{I}$ satisfies the conditions, for all $(x, y) \in [0,1]^2$:*

(i) $\mathcal{I}(\cdot, y)$ is left-continuous

(ii) $x \leq y \Leftrightarrow \mathcal{I}(x, y) = 1$

and is connected with $\mathcal{T}$ by the relationship:

$$(\forall (x, y, z) \in [0,1]^3)\,(\mathcal{I}(\mathcal{T}(x, y), z) = \mathcal{I}(x, \mathcal{I}(y, z)))\,,$$

then for the union $F_1 \cup F_2$ of F_1 and closed-valued F_2 the following properties hold, w.r.t (X, φ_X) and regular (Y, φ_Y):

(i) If F_1^- and F_2^- are surjective, F_1 is $UC_{\mathcal{I},e}$ (resp. $UC_{\mathcal{I},}$) and F_2 is $USC^{u}_{\mathcal{I},e}$ (resp. $USC^{u}_{\mathcal{I},}$) then $F_1 \cup F_2$ is $UC_{\mathcal{I},e}$ (resp. $UC_{\mathcal{I},}$).

(ii) If F_1 and F_2 are normalized, F_1 is $UC_{\mathcal{I}}$ and F_2 is $USC^{u}_{\mathcal{I}}$ then $F_1 \cup F_2$ is $UC_{\mathcal{I}}$.

Proof. Let us prove the second property. According to Proposition 5.1 (ii), F_2 is $UC_{\mathcal{I}}$. Since the first partial mappings of $\mathcal{I}$ are left-continuous, it follows from Theorem 3.6 (ii) that $F_1 \cup F_2$ is $UC_{\mathcal{I}}$. □

Proposition 6.11. *If $\mathcal{I}$ satisfies the condition:*

$$(\forall y \in [0,1])(\mathcal{I}(\cdot, y) \text{ is left-continuous})$$

then for the union $F \cup G$ of F and and non-void, closed-valued, and USC G the following properties hold, with respect to (X, φ_X) and regular (Y, φ_Y):

(i) If F^- is surjective and F is $UC_{\mathcal{I},e}$ (resp. $UC_{\mathcal{I},}$) then $F \cup G$ is $UC_{\mathcal{I},e}$ (resp. $UC_{\mathcal{I},}$).

(ii) If F is normalized and $UC_{\mathcal{I}}$ then $F \cup G$ is $UC_{\mathcal{I}}$.

Proof. The proof can be verified in the same way as the one of Proposition 6.10 using Corollary 5.1 and Theorem 3.6. □

Proposition 6.12. *If $\mathcal{I}$ satisfies the conditions, for all $(x,y) \in [0,1]^2$:*

(i) $\mathcal{I}(x,\cdot)$ is right-continuous

(ii) $\mathcal{I}(\cdot,y)$ is left-continuous

then for the union $F_1 \cup F_2$ of F_1 and F_2 the following properties hold, w.r.t (X,φ_X) and strongly compact (Y,φ_Y):

(i) If F_1^- and F_2^- are surjective, F_1 is $UC_{\mathcal{I},e}$ (resp. $UC_{\mathcal{I},}$) and F_2 is $USC^{u}_{\mathcal{I},e}$ (resp. $USC^{u}_{\mathcal{I},}$) then $F_1 \cup F_2$ is $USC^{u}_{\mathcal{I},e}$ (resp. $USC^{u}_{\mathcal{I},}$).

(ii) If F_1 and F_2 are normalized, F_1 is $UC_{\mathcal{I}}$ and F_2 is $USC^{u}_{\mathcal{I}}$ then $F_1 \cup F_2$ is $USC^{u}_{\mathcal{I}}$.

Proof. We will, for instance, prove the first property. Suppose that F_1 is $UC_{\mathcal{I},e}$ and F_2 is $USC^{u}_{\mathcal{I},e}$. Taking into account the right-continuity of the second partial mappings of $\mathcal{I}$, we obtain from Theorem 5.4 (i) that F_1 is $USC^{u}_{\mathcal{I},e}$. Since the first partial mappings of $\mathcal{I}$ are left-continuous, it holds from Theorem 4.4 (i) that $F_1 \cup F_2$ is $USC^{u}_{\mathcal{I},e}$. □

6.4 Mixed composition of closed and upper semi-continuous fuzzy multivalued mappings

Consider an FMM F_1 from X into Y, an FMM F_2 from Y into Z, an MM G_1 from X into Y, and an MM G_2 from Y into Z. The different types of closedness and upper semi-continuity considered in this subsection are with respect to (X,φ_X) and (Y,φ_Y) when F_1 and G_1 are involved, with respect to (Y,φ_Y) and (Z,φ_Z) when F_2 and G_2 are involved, and with respect to (X,φ_X) and (Z,φ_Z) when $F_2 \circ F_1$ and $G_2 \circ G_1$ are involved. Suppose that (Y,φ_Y) and (Z,φ_Z) are strongly compact.

Proposition 6.13. *If $\mathcal{I}$ satisfies the conditions, for all $(x,y) \in [0,1]^2$:*

(i) $\mathcal{I}(x,\cdot)$ is right-continuous

(ii) $\mathcal{I}(\cdot,y)$ is left-continuous

and is connected with $\mathcal{T}$ by the relationship:

$$(\forall (x,y,z) \in [0,1]^3)\,(\mathcal{I}(\mathcal{T}(x,y),z) = \mathcal{I}(x,\mathcal{I}(y,z)))$$

then for the composition $F_2 \circ F_1$ of F_1 and F_2 the following properties hold:

(i) If $\mathcal{T}$ satisfies the condition:

$$(\forall x \in [0,1])(\mathcal{T}(x,\cdot) \text{ is left-continuous})$$

and F_1^- and F_2^- are surjective then
(a) If F_1 is $UC_{\mathcal{I},e}$ (resp. $UC_{\mathcal{I},}$) and F_2 is $USC^{u}_{\mathcal{I},e}$ (resp. $USC^{u}_{\mathcal{I},}$) then $F_2 \circ F_1$ is $USC^{u}_{\mathcal{I},e}$ (resp. $USC^{u}_{\mathcal{I},}$).
(b) Moreover, if $\mathcal{I}$ satisfies the condition:

$$(\forall (x,y) \in [0,1]^2)(x \le y \Leftrightarrow \mathcal{I}(x,y) = 1),$$

F_1 is $DC_{\mathcal{T}}$ and F_2 is $USC^{d}_{\mathcal{T}}$ then $F_2 \circ F_1$ is $USC^{d}_{\mathcal{T}}$.
(ii) If F_1 and F_2 are normalized then
(a) If F_1 is $UC_{\mathcal{I}}$ and F_2 is $USC^{u}_{\mathcal{I}}$ then $F_2 \circ F_1$ is $USC^{u}_{\mathcal{I}}$.
(b) Moreover, if $\mathcal{I}$ satisfies the condition:

$$(\forall (x,y) \in [0,1]^2)(x \le y \Leftrightarrow \mathcal{I}(x,y) = 1),$$

F_1 is $DC_{\mathcal{T}}$ and F_2 is $USC^{d}_{\mathcal{T}}$ then $F_2 \circ F_1$ is $USC^{d}_{\mathcal{T}}$.

Proof. Let us prove the first property. Suppose that F_1 is $UC_{\mathcal{I},}$ and F_2 is $USC^{u}_{\mathcal{I},}$. According to Theorem 5.4 (i), F_1 is $USC^{u}_{\mathcal{I},}$. Since the conditions of Theorem 4.1 (i)(a) are fulfilled, we obtain that $F_2 \circ F_1$ is $USC^{u}_{\mathcal{I},}$. □

Proposition 6.14. *The composition $G_2 \circ G_1$ of non-void and closed G_1 and non-void and USC G_2 is USC.*

Proof. The proof is similar to the one of Proposition 6.13 and is based on Corollary 5.4 and Theorem 4.2. □

6.5 Composition of closed fuzzy multivalued mappings

Consider an FMM F_1 from X into Y, an FMM F_2 from Y into Z, an MM G_1 from X into Y, and an MM G_2 from Y into Z. The different types of closedness and upper semi-continuity considered in this subsection are with respect to (X, φ_X) and (Y, φ_Y) when F_1 and G_1 are involved, with respect to (Y, φ_Y) and (Z, φ_Z) when F_2 and G_2 are involved, and with respect to (X, φ_X) and (Z, φ_Z) when $F_2 \circ F_1$ and $G_2 \circ G_1$ are involved. Suppose that (Y, φ_Y) and (Z, φ_Z) are strongly compact.

Proposition 6.15. *If $\mathcal{I}$ satisfies the conditions, for all $(x, y) \in [0,1]^2$:*

(i) $\mathcal{I}(x, \cdot)$ is right-continuous

(ii) $\mathcal{I}(\cdot, y)$ is left-continuous

(iii) $x \le y \Leftrightarrow \mathcal{I}(x, y) = 1$

and $\mathcal{I}$ is connected with $\mathcal{T}$ by the relationship:

$$(\forall (x, y, z) \in [0,1]^3)\,(\mathcal{I}(\mathcal{T}(x, y), z) = \mathcal{I}(x, \mathcal{I}(y, z)))\,,$$

then for the composition $F_2 \circ F_1$ of $DC_{\mathcal{T}}$ F_1 and $DC_{\mathcal{T}}$ F_2 the following properties hold:

(i) If $\mathcal{T}$ satisfies the condition:

$$(\forall x \in [0,1])(\mathcal{T}(x, \cdot) \text{ is left-continuous}),$$

and F_1^- and F_2^- are surjective then $F_2 \circ F_1$ is $USC_{\mathcal{T}}^{\mathrm{d}}$.

(ii) If F_1 and F_2 are normalized then $F_2 \circ F_1$ is $USC_{\mathcal{T}}^{\mathrm{d}}$.

Moreover, if $F_2 \circ F_1$ is closed-valued and (Z, φ_Z) is regular then $F_2 \circ F_1$ is $DC_{\mathcal{T}}$, respectively.

Proof. According to Theorem 5.2, F_1 and F_2 are $USC_{\mathcal{T}}^{\mathrm{d}}$ and further applying Theorem 4.1 (i)(b), we obtain that $F_2 \circ F_1$ is $USC_{\mathcal{T}}^{\mathrm{d}}$. Finally, if (Z, φ_Z) is regular, it follows from Theorem 5.1 that $F_2 \circ F_1$ is $DC_{\mathcal{T}}$. □

Proposition 6.16. *The composition $G_2 \circ G_1$ of non-void and closed G_1 and non-void and closed G_2 is USC.*

Moreover, if $G_2 \circ G_1$ is closed-valued and (Z, φ_Z) is regular then $G_2 \circ G_1$ is closed.

Proof. The proof can be easily verified in the same way as the one of Proposition 6.15 applying Corollaries 5.1 and 5.4 and Theorem 4.2. □

Proposition 6.17. *If $\mathcal{I}$ satisfies the conditions, for all $(x, y) \in [0,1]^2$:*

(i) $\mathcal{I}(x, \cdot)$ is right-continuous

(ii) $\mathcal{I}(\cdot, y)$ is left-continuous

and is connected with $\mathcal{T}$ *by the relationship:*

$$(\forall (x,y,z) \in [0,1]^3)\,(\mathcal{I}(\mathcal{T}(x,y),z) = \mathcal{I}(x,\mathcal{I}(y,z)))\,,$$

then for the composition $F_2 \circ F_1$ *of* F_1 *and* F_2 *the following properties hold:*

(i) *If* $\mathcal{T}$ *satisfies the condition:*

$$(\forall x \in [0,1])(\mathcal{T}(x,\cdot) \text{ is left-continuous}),$$

F_1^- *and* F_2^- *are surjective and* F_1 *and* F_2 *are* $UC_{\mathcal{I},\mathrm{e}}$ *(resp.* $UC_{\mathcal{I},}$*) then* $F_2 \circ F_1$ *is* $USC^{\mathrm{u}}_{\mathcal{I},\mathrm{e}}$ *(resp.* $USC^{\mathrm{u}}_{\mathcal{I},}$*).*

(ii) *If* F_1 *and* F_2 *are normalized and* $UC_{\mathcal{I}}$ *then* $F_2 \circ F_1$ *is* $USC^{\mathrm{u}}_{\mathcal{I}}$*.*

Moreover, if $\mathcal{I}$ *satisfies the condition:*

$$(\forall (x,y) \in [0,1]^2)(x \le y \Leftrightarrow \mathcal{I}(x,y) = 1),$$

and also $F_2 \circ F_1$ *is closed-valued, and* (Z, φ_Z) *is regular then the composition* $F_2 \circ F_1$ *is, respectively:*

(i) $UC_{\mathcal{I},\mathrm{e}}$ *(resp.* $UC_{\mathcal{I},}$*).*

(ii) $UC_{\mathcal{I}}$.

Proof. We will, for instance, prove the second property. Taking into account the right-continuity of the second partial mappings of $\mathcal{I}$, we obtain from Theorem 5.4 (ii) that F_1 and F_2 are $USC^{\mathrm{u}}_{\mathcal{I}}$. Since the conditions of Theorem 4.1 (ii)(a) are fulfilled, we obtain that $F_2 \circ F_1$ is $USC^{\mathrm{u}}_{\mathcal{I}}$. Finally, if (Z, φ_Z) is regular, it follows from Proposition 5.1 (ii) that $F_2 \circ F_1$ is $UC_{\mathcal{I}}$. □

7 EXAMPLES OF FAMILIES OF IMPLICATION OPERATORS

In this section, we present two families of implication operators defined from the t-norm. We show that these two families satisfy some of the conditions required in the propositions and theorems in sections 3,4,5 and 6, which proves that they are not that restrictive.

Consider the following two families of implication operators, defined from $\mathcal{T}$ by, for all $(x,y) \in [0,1]^2$:

$$\begin{aligned}\mathcal{I}_1^{\mathcal{T}}(x,y) &= 1 - \mathcal{T}(x, 1-y)\\ \mathcal{I}_2^{\mathcal{T}}(x,y) &= \sup\{z \in [0,1] \mid \mathcal{T}(x,z) \le y\}.\end{aligned}$$

$\mathcal{I}_1^{\mathcal{T}}$ and $\mathcal{I}_2^{\mathcal{T}}$ are hybrid monotonous and satisfy the neutrality principle [3].

Proposition 7.1. *$\mathcal{I}_1^{\mathcal{T}}$ is connected with $\mathcal{T}$ by the relationship:*

$$(\forall (x,y,z) \in [0,1]^3)\left(\mathcal{I}_1^{\mathcal{T}}(\mathcal{T}(x,y),z) = \mathcal{I}_1^{\mathcal{T}}(x,\mathcal{I}_1^{\mathcal{T}}(y,z))\right).$$

Moreover, the following properties hold, for all $(x,y) \in [0,1]^2$:

(i) If $\mathcal{T}(x,\cdot)$ is left-continuous then
(a) $\mathcal{I}_1^{\mathcal{T}}(x,\cdot)$ is right-continuous
(b) $\mathcal{I}_1^{\mathcal{T}}(\cdot,y)$ is left-continuous.

(ii) If $\mathcal{T}(x,\cdot)$ is right-continuous then
(a) $\mathcal{I}_1^{\mathcal{T}}(x,\cdot)$ is left-continuous
(b) $\mathcal{I}_1^{\mathcal{T}}(\cdot,y)$ is right-continuous.

Proof. The proof of continuity properties can be found in [5]. We will now prove the relationship between $\mathcal{I}_1^{\mathcal{T}}$ and $\mathcal{T}$. Consider $(x,y,z) \in [0,1]^3$. Taking into account the associativity of $\mathcal{T}$, we obtain

$$\begin{aligned}\mathcal{I}_1^{\mathcal{T}}(\mathcal{T}(x,y),z) &= 1-\mathcal{T}(\mathcal{T}(x,y),1-z) = 1-\mathcal{T}(x,\mathcal{T}(y,1-z))\\ &= 1-\mathcal{T}(x,1-\mathcal{I}_1^{\mathcal{T}}(y,z)) = \mathcal{I}_1^{\mathcal{T}}(x,\mathcal{I}_1^{\mathcal{T}}(y,z)). \quad \square\end{aligned}$$

Proposition 7.2. *If $\mathcal{T}$ satisfies the condition:*

$$(\forall x \in [0,1])(\mathcal{T}(x,\cdot) \text{ is left-continuous})$$

then for $\mathcal{I}_2^{\mathcal{T}}$ the following properties hold, for all $(x,y,z) \in [0,1]^3$:

(i) $\mathcal{I}_2^{\mathcal{T}}(x,\cdot)$ is right-continuous

(ii) $\mathcal{I}_2^{\mathcal{T}}(\cdot,y)$ is left-continuous

(iii) $\mathcal{I}_2^{\mathcal{T}}(\mathcal{T}(x,y),z) = \mathcal{I}_2^{\mathcal{T}}(x,\mathcal{I}_2^{\mathcal{T}}(y,z))$

(iv) $x \le y \Leftrightarrow \mathcal{I}_2^{\mathcal{T}}(x,y) = 1$.

Proof. The proof of the first three properties can be found in [5]. We will now prove the fourth one. Consider $(x,y) \in [0,1]$. It can be immediately verified that from $x \le y$, it follows that $\mathcal{I}_2^{\mathcal{T}}(x,y) = 1$. Suppose now that $\mathcal{I}_2^{\mathcal{T}}(x,y) = 1$ and consider the set

$$Z = \{z \in [0,1] \mid \mathcal{T}(x,z) \le y\}.$$

Taking into account that $(\forall z \in Z)(\mathcal{T}(x,z) \leq y)$ and the left-continuity of the partial mappings of $\mathcal{T}$, we obtain

$$y \geq \sup_{z \in Z} \mathcal{T}(x,z) = \mathcal{T}(x, \sup_{z \in Z} z) = \mathcal{T}(x, \mathcal{I}_2^{\mathcal{T}}(x,y)) = \mathcal{T}(x,1) = x. \quad \square$$

Acknowledgements

The author wishes to thank the St. Cyril and Methodius Foundation (Sofia, Bulgaria) for its financial support.

REFERENCES

[1] J.-P. Aubin and H. Frankowska, *Set-Valued Analysis*, Birkhäuser, Boston-Basel-Berlin, 1990.

[2] C. Berge, *Espaces Topologiques et Fonctions Multivoques*, Dunod, Paris, 1966.

[3] B. De Baets and E. Kerre, "The generalized modus ponens and the triangular fuzzy data model," *Fuzzy Sets and Systems* **59**, 305–317, 1993.

[4] B. De Baets and E. Kerre, "Fuzzy inclusions and the inverse problems," in *Proceedings of EUFIT '94*, H.-J. Zimmermann, ed., 940–945, Second European Congress on Intelligent Techniques and Soft Computing, Aachen, Germany, 1994.

[5] B. De Baets, *Solving fuzzy relational equations: an order theoretic approach* (PhD dissertation), University of Gent, 1995, 389 p. (in Dutch).

[6] G. Chang, "Fuzzy topological spaces," *Journal of Mathematical Analysis and Applications* **24**, 182–190, 1968.

[7] E. Kerre, "Characterization of normality in fuzzy topological spaces," *Simon Stevin, A Quarterly Journal of Pure and Applied Mathematics* **53**, 239–248, 1979.

[8] E. Kerre and P. Ottoy, "On the different notions of neighbourhood in Chang-Goguen fuzzy topological spaces," *Simon Stevin, A Quarterly Journal of Pure and Applied Mathematics* **61**, 131–146, 1987.

[9] B. Schweizer and A. Sklar, *Probabilistic metric spaces*, Elsevier Science Publishing Company, New York, 1983.

[10] E. Tsiporkova-Hristoskova, B. De Baets and E. Kerre, "Fuzzy lower semi-continuity of fuzzy multivalued mappings," in *Proceedings of EUFIT '94*, H.-J. Zimmermann, ed., 1341–1346, Second European Congress on Intelligent Techniques and Soft Computing, Aachen, Germany, 1994.

[11] E. Tsiporkova-Hristoskova, B. De Baets and E. Kerre, "A detailed study of direct and inverse images under fuzzy multivalued mappings ," *The Journal of Fuzzy Mathematics* (to appear).

[12] E. Tsiporkova-Hristoskova, B. De Baets and E. Kerre, "A fuzzy inclusion based approach to upper inverse images under fuzzy multivalued mappings," *Fuzzy Sets and Systems* (submitted August 1994).

[13] E. Tsiporkova-Hristoskova, B. De Baets and E. Kerre, "A comparative study of three upper inverse images under fuzzy multivalued mappings," in *Proceedings of IFSA '95*, Sixth IFSA World Congress, Sao Paulo, Brazil, 1995 (4 p., submitted).

[14] E. Tsiporkova-Hristoskova, B. De Baets and E. Kerre, "Closedness of fuzzy multivalued mappings," in *Proceedings of ISFL '95*, International Symposium on Fuzzy Logic, Zurich, Switzerland, 1995 (to appear).

[15] E. Tsiporkova-Hristoskova and E. Kerre, "Upper semi-continuity of fuzzy multivalued mappings," in *Proceedings of EUFIT '95*, Third European Congress on Intelligent Techniques and Soft Computing, Aachen, Germany, 1995 (5 p., submitted).

[16] E. Tsiporkova-Hristoskova and E. Kerre, "On separation axioms and compactness in fuzzy topological spaces," *Mathematica Pannonica* (submitted January 1995).

3

AN ORDER-THEORETHIC APPROACH TO SOLVING SUP-T EQUATIONS

Bernard De Baets

Research Assistant of the National Fund for Scientific Research (Belgium)
Department of Applied Mathematics and Computer Science
Universiteit Gent
Krijgslaan 281-S9
B-9000 Gent, Belgium

ABSTRACT

The study of fuzzy relational equations is one of the most appealing subjects in fuzzy set theory, both from a mathematical and a systems modelling point of view [8]. The most fundamental fuzzy relational equations are the sup-$\mathcal{T}$ equations, with $\mathcal{T}$ a triangular norm. It is well-known how to solve these equations on the real unit interval [4]. Zhao has shown how to solve sup-$\frown$ equations on complete Brouwerian lattices of which all elements have an irredundant finite decomposition in join-irreducible elements [14]. In this paper, we generalize various aspects of these results. We consider triangular norms on bounded ordered sets, and in particular on complete lattices. We then solve sup-$\mathcal{T}$ inequalities and sup-$\mathcal{T}$ equations on distributive, complete lattices of which all elements are either join-irreducible or join-decomposable. The morphism behaviour of the partial mappings of the triangular norm $\mathcal{T}$ plays an important role in this study. We also introduce the notion of a maximally surjective triangular norm and show how the solution procedures can be simplified for such triangular norms. The results in these papers are obtained in a systematic way using the new order-theoretic concept root system. We conclude the paper with a discussion of the product lattice $([0,1]^n, \leq)$: we show how to construct triangular norms on it, and how to calculate the corresponding residual operators. By means of extensive examples we illustrate and comment on the solution procedures.

1 SOME ORDER-THEORETIC NOTIONS

1.1 Join-irreducible and join-decomposable elements

We assume that the reader is familiar with the basic order-theoretic notions such as (bounded) ordered sets, chains, antichains, (complete) lattices, ... [2, 3]. The smallest and greatest element of a bounded ordered set $(P, \leq)$ will be denoted by 0_P and 1_P. For an element α of an ordered set $(P, \leq)$ we define $\downarrow \alpha = \{\beta \mid \beta \in P \wedge \beta \leq \alpha\}$. In this paper we use the notation I_n for $\{1, \ldots, n\}$.

We will only recall some results concerning distributive lattices. Notice that the meet (resp. join) operation of a lattice $(L, \leq)$ is denoted by $\frown$ (resp. $\smile$).

Definition 1. (Distributive lattices). [2] *A lattice $(L, \leq)$ is called distributive if and only if the following property holds :*

$$(\forall(\alpha, \beta, \gamma) \in L^3)(\alpha \frown (\beta \smile \gamma) = (\alpha \frown \beta) \smile (\alpha \frown \gamma)).$$

Proposition 1. [2] *Any chain is distributive.*

Proposition 2. [2] *A Cartesian product of lattices is distributive if and only if all of these lattices are distributive.*

Definition 2. (Join-irreducible elements). [2] *An element α of a lattice $(L, \leq)$ is called join-irreducible if and only if*

$$(\forall(\beta, \gamma) \in L^2)(\beta \smile \gamma = \alpha \Rightarrow (\beta = \alpha \vee \gamma = \alpha)).$$

Notice that all elements of a chain are join-irreducible. The following proposition will be applied repeatedly in the following sections.

Proposition 3. [2] *Let α be a join-irreducible element of a distributive lattice $(L, \leq)$ and $(\alpha_i \mid i \in I_n)$ be a finite family in L, then the following property holds :*

$$\alpha \leq \sup_{i \in I_n} \alpha_i \Leftrightarrow (\exists i \in I_n)(\alpha \leq \alpha_i).$$

Definition 3. (Join-decomposable elements). *An element α of a lattice $(L, \leq)$ is called join-decomposable if and only if there exists a set A of join-irreducible elements of L, with $\#A \geq 2$, such that $\alpha = \sup A$.*

In the foregoing definition, the set A is called a join-decomposition of α. This join-decomposition is called irredundant if and only if $(\forall B \in \mathcal{P}(L))(B \subset A \Rightarrow \sup B \neq \alpha)$.

1.2 Root systems

Of capital importance in this paper is the new order-theoretic concept *root system* [5, 6]. A root system is a particular union of closed intervals with the same ending point. This is formalized in the following definition.

Definition 4. (Root systems). [6] *A subset R of an ordered set $(P, \leq)$ is called a root system if and only if there exists an element σ in P and an antichain O in $\downarrow \sigma$ such that*

$$R = \bigcup_{\omega \in O} [\omega, \sigma].$$

One easily verifies that for a root system R the corresponding element σ and the corresponding antichain O are unique. The element σ is called the *stem* of the root system. The elements of the antichain O are called the *offshoots* of the root system. A root system is called *finitely generated (f.g.)* if the set of offshoots is finite.

Proposition 4. [6] *If a subset R of an ordered set $(P, \leq)$ is a root system with stem σ and set of offshoots O, then σ is the greatest element of R and the elements of O are the minimal elements of R.*

1.3 Intersection of root systems

Further on we will show that under some conditions the solution set of a sup-$\mathcal{T}$ equation can be seen as the intersection of a family of root systems. In view of this, it is indispensable to know the behaviour of root systems under intersection.

Proposition 5. (Intersection of root systems). [6] *Let* $(R_i \mid i \in I)$ *be a family of root systems of a complete lattice* $(L, \leq)$ *with stem* σ_i *and set of offshoots* O_i. *Put* $\sigma = \inf_{i \in I} \sigma_i$ *and define the sets*

$$O'_i = \{\omega \mid \omega \in O_i \ \wedge \ \omega \leq \sigma\}.$$

Also put $O = \prod_{i \in I} O_i$ *and* $O' = \prod_{i \in I} O'_i$. *The intersection of this family of root systems is given by*

$$\bigcap_{i \in I} R_i = \bigcup_{\gamma \in O'} [\sup_{i \in I} \gamma(i), \sigma].$$

Notice that in this proposition for each $\gamma \in O'$ the corresponding closed interval $[\sup_{i \in I} \gamma(i), \sigma]$ is not empty. This proposition also holds when the intersection of the family of root systems is empty; in that case O' also is empty. A necessary and sufficient condition for the intersection not be empty clearly is that σ belongs to it.

Proposition 5 immediately implies that the greatest element of a non-empty intersection of an arbitrary family of root systems is given by σ and that the minimal elements are given by the minimal elements of the set

$$\{\sup_{i \in I} \gamma(i) \mid \gamma \in O'\}.$$

Unfortunately, this set not necessarily contains minimal elements and even if it does so, then we still cannot conclude that for each non-minimal element a smaller minimal element can be found. The following theorem shows that this is the case when the root systems are finitely generated.

Theorem 1. (Intersection of f.g. root systems). [6]
Let $(R_i \mid i \in I)$ *be a family of finitely generated root systems of a complete lattice* $(L, \leq)$ *with stem* σ_i *and set of offshoots* O_i. *If the intersection* $\bigcap_{i \in I} R_i$ *is not empty, then it is a root system with stem* $\sigma = \inf_{i \in I} \sigma_i$ *and as offshoots the minimal elements of the set*

$$\{\sup_{i \in I} \omega_i \mid (\forall i \in I)(\omega_i \in O_i \ \wedge \ \omega_i \leq \sigma)\}.$$

Proposition 6. (Finite intersection of f.g. root systems). [6]
If the intersection of a finite family of finitely generated root systems of a complete lattice is not empty, then it is a finitely generated root system.

2 THE EQUATION $\mathcal{T}(\alpha, X) = \beta$

2.1 Triangular norms and implicators

Triangular norms were introduced in 1963 by Schweizer and Sklar [11, 12], in the framework of their study of probabilistic metric spaces. They were first considered in fuzzy set theory for defining the intersection of fuzzy sets by Alsina *et al.* [1] and Prade [10]. Triangular norms are increasing, associative and commutative binary operations on the real unit interval, satisfying some additional boundary conditions. Given their probabilistic origin, it is not at all surprising that they were only defined on [0,1]. De Cooman [7] has noticed that none of the defining properties is typical for operations on [0,1]. They can therefore easily be generalized to describe certain classes of operations on bounded ordered sets.

Definition 5. (Triangular norms). [7] *A triangular norm (t-norm) $\mathcal{T}$ on a bounded ordered set $(P, \leq)$ is a $P^2 \to P$ mapping that satisfies :*

(i) Boundary conditions: $(\forall \alpha \in P)(\mathcal{T}(1_P, \alpha) = \mathcal{T}(\alpha, 1_P) = \alpha)$;

(ii) $\mathcal{T}$ is increasing w.r.t. $(P^2, \leq)$ and $(P, \leq)$;

(iii) Associativity: $(\forall(\alpha, \beta, \gamma) \in P^3)(\mathcal{T}(\alpha, \mathcal{T}(\beta, \gamma)) = \mathcal{T}(\mathcal{T}(\alpha, \beta), \gamma))$;

(iv) Commutativity: $(\forall(\alpha, \beta) \in P^2)(\mathcal{T}(\alpha, \beta) = \mathcal{T}(\beta, \alpha))$.

Example 1. The meet operation $\frown$ on a bounded lattice $(L, \leq)$ (defined by $\alpha \frown \beta = \inf\{\alpha, \beta\}$), is a t-norm on $(L, \leq)$.

Definition 6. (Implicators). *An implicator $\mathcal{I}$ on a bounded ordered set $(P, \leq)$ is a $P^2 \to P$ mapping that satisfies:*

(i) The restriction of $\mathcal{I}$ to $\{0_P, 1_P\}^2$ coincides with the Boolean implication : $\mathcal{I}(0_P, 0_P) = \mathcal{I}(0_P, 1_P) = \mathcal{I}(1_P, 1_P) = 1_P$ *and* $\mathcal{I}(1_P, 0_P) = 0_P$;

(ii) $\mathcal{I}$ is hybrid monotonous : the first partial mappings of $\mathcal{I}$ are decreasing and the second partial mappings of $\mathcal{I}$ are increasing.

2.2 The operators $\mathcal{I}_{\mathcal{T}}$ and $\mathcal{L}_{\mathcal{T}}$

With a t-norm $\mathcal{T}$ on a complete lattice $(L, \leq)$ we associate two binary operators $\mathcal{I}_{\mathcal{T}}$ and $\mathcal{L}_{\mathcal{T}}$ on L defined by

$$\mathcal{I}_{\mathcal{T}}(\alpha, \beta) = \sup\{\gamma \mid \gamma \in L \wedge \mathcal{T}(\alpha, \gamma) \leq \beta\}$$

$$\mathcal{L}_{\mathcal{T}}(\alpha,\beta) \quad = \inf\{\gamma \mid \gamma \in L \wedge \mathcal{T}(\alpha,\gamma) \geq \beta\}.$$

Proposition 7. *Let $\mathcal{T}$ be a t-norm on a complete lattice $(L, \leq)$, then the operator $\mathcal{I}_{\mathcal{T}}$ satisfies :*

$$(\forall(\alpha,\beta) \in L^2)(\alpha \leq \beta \Rightarrow \mathcal{I}_{\mathcal{T}}(\alpha,\beta) = 1_L).$$

Proposition 8. *Let $\mathcal{T}$ be a t-norm on a complete lattice $(L, \leq)$, then the operator $\mathcal{I}_{\mathcal{T}}$ is an implicator on $(L, \leq)$.*

This result is well-known when $(L, \leq)$ is the real unit interval; the operator $\mathcal{I}_{\mathcal{T}}$ is then usually called the residual implication of the t-norm $\mathcal{T}$ [13].

Proposition 9. *Let $\mathcal{T}$ be a t-norm on a complete lattice $(L, \leq)$, then the operator $\mathcal{L}_{\mathcal{T}}$ satisfies :*

$$(\forall(\alpha,\beta) \in L^2)(\neg(\beta \leq \alpha) \Rightarrow \mathcal{L}_{\mathcal{T}}(\alpha,\beta) = 1_L).$$

Proposition 10. *Let $\mathcal{T}$ be a t-norm on a complete lattice $(L, \leq)$, then the operator $\mathcal{L}_{\mathcal{T}}$ satisfies :*

$$(\forall\beta \in L)(\mathcal{L}_{\mathcal{T}}(1_L,\beta) = \beta)$$
$$(\forall\alpha \in L)(\mathcal{L}_{\mathcal{T}}(\alpha,0_L) = 0_L).$$

Proposition 11. *Let $\mathcal{T}$ be a t-norm on a complete lattice $(L, \leq)$, then the operator $\mathcal{L}_{\mathcal{T}}$ is hybrid monotonous : the first partial mappings of $\mathcal{L}_{\mathcal{T}}$ are decreasing and the second partial mappings of $\mathcal{L}_{\mathcal{T}}$ are increasing.*

Corollary 1. *Let $\mathcal{T}$ be a t-norm on a complete lattice $(L, \leq)$, then the operator $\mathcal{L}_{\mathcal{T}}$ satisfies :*

$$(\forall\alpha \in L)(\forall\beta \in L \setminus \{0_L\})(\mathcal{L}_{\mathcal{T}}(\alpha,\beta) > 0_L).$$

Proof. For $\beta > 0_L$ it follows from Propositions 10 and 11 that

$$\mathcal{L}_{\mathcal{T}}(\alpha,\beta) \geq \mathcal{L}_{\mathcal{T}}(1_L,\beta) = \beta > 0_L. \quad \square$$

2.3 The inequality $\mathcal{T}(\alpha, x) \leq \beta$

We first recall the definition of a supremum-morphism.

Definition 7. (Supremum-morphisms). *Let $(L, \leq)$ be a complete lattice. An $L \to L$ mapping f is called a supremum-morphism w.r.t. $(L, \leq)$ and $(L, \leq)$ if and only if $(\forall A \in \mathcal{P}(L) \setminus \{\emptyset\})(f(\sup A) = \sup f(A))$.*

Proposition 12. *Let $\mathcal{T}$ be a t-norm on a complete lattice $(L, \leq)$ and $(\alpha, \beta) \in L^2$. If the partial mapping $\mathcal{T}(\alpha, \cdot)$ is a supremum-morphism, then the solution set of the inequality $\mathcal{T}(\alpha, x) \leq \beta$ in the unknown x is given by*

$$[0_L, \mathcal{I}_\mathcal{T}(\alpha, \beta)].$$

Proof. The monotonicity of the partial mapping $\mathcal{T}(\alpha, \cdot)$ and the definition of $\mathcal{I}_\mathcal{T}$ imply that it is sufficient to show that $\mathcal{I}_\mathcal{T}(\alpha, \beta)$ is a solution of the inequality $\mathcal{T}(\alpha, x) \leq \beta$. Since the set $\{y \mid y \in L \wedge \mathcal{T}(\alpha, y) \leq \beta\}$ is not empty (0_L belongs to it), it follows that

$$\begin{aligned} \mathcal{T}(\alpha, \mathcal{I}_\mathcal{T}(\alpha, \beta)) &= \mathcal{T}(\alpha, \sup\{y \mid y \in L \wedge \mathcal{T}(\alpha, y) \leq \beta\}) \\ &= \sup\{\mathcal{T}(\alpha, y) \mid y \in L \wedge \mathcal{T}(\alpha, y) \leq \beta\} \\ &\leq \beta. \quad \square \end{aligned}$$

2.4 The inequality $\mathcal{T}(\alpha, x) \geq \beta$

We first recall the definition of an infimum-morphism.

Definition 8. (Infimum-morphisms). *Let $(L, \leq)$ be a complete lattice. An $L \to L$ mapping f is called an infimum-morphism w.r.t. $(L, \leq)$ and $(L, \leq)$ if and only if $(\forall A \in \mathcal{P}(L) \setminus \{\emptyset\})(f(\inf A) = \inf f(A))$.*

Proposition 13. *Let $\mathcal{T}$ be a t-norm on a complete lattice $(L, \leq)$ and $(\alpha, \beta) \in L^2$. If the partial mapping $\mathcal{T}(\alpha, \cdot)$ is an infimum-morphism, then the solution set of the inequality $\mathcal{T}(\alpha, x) \geq \beta$ in the unknown x is given by*

$$\begin{cases} [\mathcal{L}_\mathcal{T}(\alpha, \beta), 1_L], & \text{if } \beta \leq \alpha \\ \emptyset, & \text{else} \end{cases}.$$

Proof. One easily verifies that the inequality has a solution if and only if 1_L is a solution, i.e. if and only if $\mathcal{T}(\alpha, 1_L) = \alpha \geq \beta$. Now assume that $\alpha \geq \beta$, then the monotonicity of the partial mapping $\mathcal{T}(\alpha, \cdot)$ and the definition of $\mathcal{L}_\mathcal{T}$ imply that it is sufficient to show that $\mathcal{L}_\mathcal{T}(\alpha, \beta)$ is a solution of the inequality $\mathcal{T}(\alpha, x) \geq \beta$. Since the set $\{\gamma \mid \gamma \in L \wedge \mathcal{T}(\alpha, \gamma) \geq \beta\}$ is not empty (1_L belongs to it), it follows that

$$\begin{aligned}\mathcal{T}(\alpha, \mathcal{L}_\mathcal{T}(\alpha, \beta)) &= \mathcal{T}(\alpha, \inf\{\gamma \mid \gamma \in L \wedge \mathcal{T}(\alpha, \gamma) \geq \beta\}) \\ &= \inf\{\mathcal{T}(\alpha, \gamma) \mid \gamma \in L \wedge \mathcal{T}(\alpha, \gamma) \geq \beta\} \\ &\geq \beta. \quad \square\end{aligned}$$

2.5 The equation $\mathcal{T}(\alpha, x) = \beta$

Definition 9. (Homomorphisms). *Let $(L, \leq)$ be a complete lattice. An $L \to L$ mapping f is called a homomorphism w.r.t. $(L, \leq)$ and $(L, \leq)$ if and only if it is both a supremum-morphism and an infimum-morphism w.r.t. $(L, \leq)$ and $(L, \leq)$.*

Proposition 14. *Let $\mathcal{T}$ be a t-norm on a complete lattice $(L, \leq)$ and $(\alpha, \beta) \in L^2$. If the partial mapping $\mathcal{T}(\alpha, \cdot)$ is a homomorphism, then the solution set of the equation $\mathcal{T}(\alpha, x) = \beta$ is given by*

$$\begin{cases} [\mathcal{L}_\mathcal{T}(\alpha, \beta), \mathcal{I}_\mathcal{T}(\alpha, \beta)], & \text{if } \beta \in \operatorname{rng}(\mathcal{T}(\alpha, \cdot)) \\ \emptyset, & \text{else} \end{cases}.$$

Proof. Follows immediately from Propositions 12 and 13. $\square$

The boundary conditions of a t-norm on a bounded ordered set $(P, \leq)$ imply that the range of the partial mapping $\mathcal{T}(\alpha, \cdot)$ always is a subset of $[0_P, \alpha]$. The purpose of the following definition then is clear.

Definition 10. (Maximally surjective triangular norms). *Let $\mathcal{T}$ be a t-norm on a bounded ordered set $(P, \leq)$ and $\alpha \in P$. The partial mapping $\mathcal{T}(\alpha, \cdot)$ is called maximally surjective if and only if $\operatorname{rng}(\mathcal{T}(\alpha, \cdot)) = [0_P, \alpha]$. The t-norm $\mathcal{T}$ is called maximally surjective if and only if all of its partial mappings are maximally surjective.*

Proposition 15. *Let $(L, \leq)$ be a bounded lattice, then the meet operation is maximally surjective.*

Proof. Consider $(\alpha, \beta) \in L^2$ such that $\beta \leq \alpha$, i.e. $\beta \in [0_P, \alpha]$. It then follows that $\alpha \frown \beta = \beta$, i.e. $\beta \in \text{rng}(\alpha \frown \cdot)$. □

Corollary 2. *Let $\mathcal{T}$ be a t-norm on a complete lattice $(L, \leq)$ and $(\alpha, \beta) \in L^2$. If the partial mapping $\mathcal{T}(\alpha, \cdot)$ is a maximally surjective homomorphism, then the solution set of the equation $\mathcal{T}(\alpha, x) = \beta$ is given by*

$$\begin{cases} [\mathcal{L}_{\mathcal{T}}(\alpha, \beta), \mathcal{I}_{\mathcal{T}}(\alpha, \beta)], & \text{if } \beta \leq \alpha \\ \emptyset, & \text{else} \end{cases}.$$

Proof. Trivial. □

3 SUP-$\mathcal{T}$ EQUATIONS

3.1 The inequality $\sup\limits_{i \in I} \mathcal{T}(A(i), X(i)) \leq \beta$

We consider a t-norm $\mathcal{T}$ on a complete lattice $(L, \leq)$. Let $(\alpha_i \mid i \in I)$ be an arbitrary family in L and $\beta \in L$, then we want to know the solution set of the inequality

$$\sup_{i \in I} \mathcal{T}(\alpha_i, x_i) \leq \beta$$

in the family of unknowns $(x_i \mid i \in I)$ in L. The families $(\alpha_i \mid i \in I)$ and $(x_i \mid i \in I)$ can be seen as $(L, \leq)$-fuzzy sets in the index set I. The problem can therefore be reformulated as follows. Given $A \in \mathcal{F}_{(L,\leq)}(I)$ (with $\mathcal{F}_{(L,\leq)}(I)$ the set of $(L, \leq)$-fuzzy sets in I [9]) and $\beta \in L$, determine the solution set of the inequality

$$\sup_{i \in I} \mathcal{T}(A(i), X(i)) \leq \beta$$

in the unknown $(L, \leq)$-fuzzy set X in I. Recall that $(\mathcal{F}_{(L,\leq)}(I), \subseteq)$ is a complete lattice, with order relation $\subseteq$ (inclusion) defined by

$$A \subseteq B \Leftrightarrow (\forall i \in I)(A(i) \leq B(i)),$$

and as infimum and supremum the intersection and union defined by

$$\left(\bigcap_{j\in J} A_j\right)(i) = \inf_{j\in J} A_j(i) \quad \text{and} \quad \left(\bigcup_{j\in J} A_j\right)(i) = \sup_{j\in J} A_j(i).$$

The solution set of the above inequality then is a subset of the complete lattice $(\mathcal{F}_{(L,\leq)}(I), \subseteq)$. Notice that a crisp subset A of I will be identified with its characteristic mapping χ_A, defined by

$$\chi_A(i) = \begin{cases} 1_L, & \text{if } i \in A \\ 0_L, & \text{elsewhere} \end{cases}.$$

Proposition 16. *Let $\mathcal{T}$ be a t-norm on a complete lattice $(L, \leq)$, $A \in \mathcal{F}_{(L,\leq)}(I)$ and $\beta \in L$. If the partial mappings of $\mathcal{T}$ are supremum-morphisms, then the solution set of the inequality*

$$\sup_{i\in I} \mathcal{T}(A(i), X(i)) \leq \beta$$

is given by $[\chi_\emptyset, G]$ where G is defined by $G(i) = \mathcal{I}_\mathcal{T}(A(i), \beta)$.

Proof. Follows immediately from Proposition 12 and the equivalences

$$\begin{aligned} \sup_{i\in I} \mathcal{T}(A(i), X(i)) \leq \beta \quad & \Leftrightarrow (\forall i \in I)(\mathcal{T}(A(i), X(i)) \leq \beta) \\ & \Leftrightarrow (\forall i \in I)(0_L \leq X(i) \leq \mathcal{I}_\mathcal{T}(A(i), \beta)). \quad \square \end{aligned}$$

3.2 The inequality $\sup_{i\in I_n} \mathcal{T}(A(i), X(i)) \geq \beta$

Compared to the previous subsection, there are three main differences. Firstly, the index set I is no longer arbitrary, but finite. Secondly, the complete lattice $(L, \leq)$ is no longer arbitrary but is assumed to be distributive, and the elements of it should either be join-irreducible or join-decomposable. Thirdly, this inequality not necessarily has solutions; we will determine the solution set under the assumption that it is not empty.

Proposition 17. (Join-irreducible right-hand side). *Let $\mathcal{T}$ be a t-norm on a distributive, complete lattice $(L, \leq)$, $A \in \mathcal{F}_{(L,\leq)}(I_n)$ and β be a join-*

irreducible element of L. If the partial mappings of $\mathcal{T}$ are infimum-morphisms then the solution set of the inequality

$$\sup_{i \in I_n} \mathcal{T}(A(i), X(i)) \geq \beta$$

is not empty if and only if $\beta \leq \sup_{i \in I_n} A(i)$. *If it is not empty, then it is a finitely generated root system with stem* χ_{I_n} *and as offshoots the elements of the set*

$$O = \{M_k \mid M_k \in \mathcal{F}_{(L,\leq)}(I_n) \ \wedge \ \beta \leq A(k)\}$$

with M_k *defined by* $M_k(i) = \begin{cases} \mathcal{L}_{\mathcal{T}}(A(k), \beta), & \textit{if } i = k \\ 0_L, & \textit{elsewhere} \end{cases}.$

Proof. Since β is join-irreducible, Proposition 3 implies that

$$\sup_{i \in I_n} \mathcal{T}(A(i), X(i)) \geq \beta \Leftrightarrow (\exists k \in I_n)(\mathcal{T}(A(k), X(k)) \geq \beta).$$

With Proposition 13 it then follows that

$$\sup_{i \in I_n} \mathcal{T}(A(i), X(i)) \geq \beta \Leftrightarrow (\exists k \in I_n)(\beta \leq A(k) \ \wedge \ \mathcal{L}_{\mathcal{T}}(A(k), \beta) \leq X(k)).$$

Again with Proposition 3 it follows that a necessary and sufficient solvability condition is given by $\beta \leq \sup_{i \in I_n} A(i)$. It now follows immediately that the solution set is given by $\bigcup_{M_k \in O} [M_k, \chi_{I_n}]$. For the solution set to be a root system, it then suffices to show that O is an antichain. When $\beta = 0_L$, the solution set is not empty, and it follows with Proposition 10 that $O = \{\chi_\emptyset\}$, an obvious result. When $\beta > 0_L$, it follows with Corollary 1 that $M_k(k) > 0_L$ and hence the elements of O form an antichain. □

Proposition 18. (Join-decomposable right-hand side). *Let $\mathcal{T}$ be a t-norm on a distributive, complete lattice* $(L, \leq)$, $A \in \mathcal{F}_{(L,\leq)}(I_n)$ *and* $\beta \in L$ *with join-decomposition* $\beta = \sup_{j \in J} \beta_j$. *If the partial mappings of $\mathcal{T}$ are infimum-morphisms and the solution set of the inequality*

$$\sup_{i \in I_n} \mathcal{T}(A(i), X(i)) \geq \beta$$

is not empty, then it is a root system with stem χ_{I_n} *and as offshoots the minimal elements of the set*

$$\{\bigcup_{j \in J} M_k^j \mid M_k^j \in \mathcal{F}_{(L,\leq)}(I_n) \ \wedge \ \beta_j \leq A(k)\}$$

with M_k^j *defined by* $M_k^j(i) = \begin{cases} \mathcal{L}_{\mathcal{T}}(A(k), \beta_j), & \text{if } i = k \\ 0_L, & \text{elsewhere} \end{cases}$;
if the join-decomposition is finite, then this root system is also finitely generated.

Proof. Notice the following equivalence

$$\sup_{i \in I_n} \mathcal{T}(A(i), X(i)) \geq \beta = \sup_{j \in J} \beta_j \Leftrightarrow (\forall j \in J)(\sup_{i \in I_n} \mathcal{T}(A(i), X(i)) \geq \beta_j).$$

With Proposition 17, Theorem 1 and Proposition 6 this proposition then follows immediately. □

3.3 The equation $\sup_{i \in I_n} \mathcal{T}(A(i), X(i)) = \beta$

The solution set of this equation clearly is the intersection of the solution sets of the two corresponding inequalities.

Theorem 2. (Join-irreducible right-hand side). *Let* $\mathcal{T}$ *be a t-norm on a distributive, complete lattice* $(L, \leq)$, $A \in \mathcal{F}_{(L,\leq)}(I_n)$ *and* β *be a join-irreducible element of L. If the partial mappings of* $\mathcal{T}$ *are homomorphisms and the solution set of the equation*

$$\sup_{i \in I_n} \mathcal{T}(A(i), X(i)) = \beta$$

is not empty, then it is a finitely generated root system with stem G defined by $G(i) = \mathcal{I}_{\mathcal{T}}(A(i), \beta)$ *and as offshoots the elements of the set*

$$\{M_k \mid M_k \in \mathcal{F}_{(L,\leq)}(I_n) \ \wedge \ \beta \leq A(k) \ \wedge \ M_k \subseteq G\}$$

with M_k *defined by* $M_k(i) = \begin{cases} \mathcal{L}_{\mathcal{T}}(A(k), \beta), & \text{if } i = k \\ 0_L, & \text{elsewhere} \end{cases}$;
if $\mathcal{T}$ *is also maximally surjective, then the set of offshoots can be written as*

$$\{M_k \mid M_k \in \mathcal{F}_{(L,\leq)}(I_n) \ \wedge \ \beta \leq A(k)\}.$$

A necessary and sufficient solvability condition therefore is that G is a solution.

Proof. Consider the set $O = \{M_k \mid M_k \in \mathcal{F}_{(L,\leq)}(I_n) \ \wedge \ \beta \leq A(k)\}$. If the solution set is not empty, then it follows from Propositions 16 and 17 that it is given by

$$[\chi_\emptyset, G] \cap \bigcup_{M_k \in O} [M_k, \chi_{I_n}]$$

which can be written as $\bigcup_{M_k \in O} [M_k, G]$ and also as

$$\bigcup_{M_k \in O'} [M_k, G],$$

with $O' = \{M_k \mid M_k \in O \ \wedge \ M_k \subseteq G\}$. If $\mathcal{T}$ is also maximally surjective, then $O = O'$, since always $\mathcal{L}_\mathcal{T}(A(k), \beta) \leq \mathcal{I}_\mathcal{T}(A(k), \beta)$ for $\beta \leq A(k)$, due to Corollary 2. □

Theorem 3. (Join-decomposable right-hand side). *Let $\mathcal{T}$ be a t-norm on a distributive, complete lattice $(L, \leq)$, $A \in \mathcal{F}_{(L,\leq)}(I_n)$ and $\beta \in L$ with join-decomposition $\beta = \sup_{j \in J} \beta_j$. If the partial mappings of $\mathcal{T}$ are homomorphisms and the solution set of the equation*

$$\sup_{i \in I_n} \mathcal{T}(A(i), X(i)) = \beta$$

is not empty, then it is a root system with stem G defined by $G(i) = \mathcal{I}_\mathcal{T}(A(i), \beta)$ and as offshoots the minimal elements of the set

$$\{\bigcup_{j \in J} M_k^j \mid M_k^j \in \mathcal{F}_{(L,\leq)}(I_n) \ \wedge \ \beta_j \leq A(k) \ \wedge \ M_k^j \subseteq G\}$$

with M_k^j defined by $M_k^j(i) = \begin{cases} \mathcal{L}_\mathcal{T}(A(k), \beta_j), & \text{if } i = k \\ 0_L, & \text{elsewhere} \end{cases}$;
if the join-decomposition is finite, then this root system is also finitely generated and if $\mathcal{L}_\mathcal{T} \leq \mathcal{I}_\mathcal{T}$, then the offshoots are the minimal elements of the set

$$\{\bigcup_{j \in J} M_k^j \mid M_k^j \in \mathcal{F}_{(L,\leq)}(I_n) \ \wedge \ \beta_j \leq A(k)\}.$$

A necessary and sufficient solvability condition therefore is that G is a solution.

Proof. Denote the set of minimal elements of the set

$$\{\bigcup_{j \in J} M_k^j \mid M_k^j \in \mathcal{F}_{(L,\leq)}(I_n) \ \wedge \ \beta_j \leq A(k)\}$$

by O, then it follows from Propositions 16 and 18 that the solution set, if not empty, is given by

$$[\chi_\emptyset, G] \cap \bigcup_{M \in O} [M, \chi_{I_n}] = \bigcup_{M \in O'} [M, G],$$

with $O' = \{M \mid M \in O \wedge M \subseteq G\}$. Let $M = \bigcup_{j \in J} M_k^j \in O$, then it easily follows that $M \subseteq G$ if and only if $(\forall j \in J)(M_k^j \subseteq G)$. It is then easy to verify that the elements of O' are the minimal elements of the set

$$\{\bigcup_{j \in J} M_k^j \mid M_k^j \in \mathcal{F}_{(L,\leq)}(I_n) \wedge \beta_j \leq A(k) \wedge M_k^j \subseteq G\}.$$

If $\mathcal{L}_\mathcal{T} \leq \mathcal{I}_\mathcal{T}$, then we always have that $M_k^j \subseteq G$. Indeed, from $\beta_j \leq \beta$ it follows with Proposition 11 that $\mathcal{L}_\mathcal{T}(A(k), \beta_j) \leq \mathcal{L}_\mathcal{T}(A(k), \beta) \leq \mathcal{I}_\mathcal{T}(A(k), \beta)$. □

Notice that in the last part of the foregoing proof it is not sufficient that $\mathcal{T}$ is maximally surjective. Indeed, from $\beta_j \leq A(k)$ it does not follow that $\beta \leq A(k)$, which implies that Corollary 2 cannot be invoked for concluding that $\mathcal{L}_\mathcal{T}(A(k), \beta_j) \leq \mathcal{I}_\mathcal{T}(A(k), \beta)$. However, when $(L, \leq)$ is a complete chain and $\mathcal{T}$ is also maximally surjective, then $\mathcal{L}_\mathcal{T} \leq \mathcal{I}_\mathcal{T}$ always holds. Indeed, from Corollary 2 it follows that for $\beta \leq \alpha$ we have that $\mathcal{L}_\mathcal{T}(\alpha, \beta) \leq \mathcal{I}_\mathcal{T}(\alpha, \beta)$, while for $\alpha < \beta$ it follows from Propositions 7 and 9 that $\mathcal{L}_\mathcal{T}(\alpha, \beta) = \mathcal{I}_\mathcal{T}(\alpha, \beta) = 1_L$.

4 THE REAL UNIT INTERVAL

4.1 Example 1: $(L, \leq) = ([0, 1], \leq)$

It is well-known that the real unit interval is a complete chain. Hence it is also distributive and all elements of it are join-irreducible. On $([0, 1], \leq)$ the following are well-known continuous t-norms :

(i) the minimum operator (meet) M: $M(x, y) = \min(x, y)$;

(ii) the product P: $P(x, y) = xy$;

(iii) the bold intersection W: $W(x, y) = \max(x + y - 1, 0)$.

The continuity of these t-norms implies that they are maximally-surjective and that their partial mappings are homomorphisms. Recalling the discussion following Theorem 3, we have for these t-norms that $\mathcal{L}_\mathcal{T} \leq \mathcal{I}_\mathcal{T}$.

The corresponding implicators $\mathcal{I}_\mathcal{T}$ are given by :

(i) the Gödel-Brouwer implicator : $\mathcal{I}_M(x,y) = \begin{cases} 1, & \text{if } x \le y \\ y, & \text{elsewhere} \end{cases}$;

(ii) the Goguen implicator : $\mathcal{I}_P(x,y) = \begin{cases} 1, & \text{if } x \le y \\ y/x, & \text{elsewhere} \end{cases}$;

(iii) the Luckasiewicz implicator : $\mathcal{I}_W(x,y) = \min(1 - x + y, 1)$.

The corresponding operators $\mathcal{L}_\mathcal{T}$ are given by :

(i) the operator $\mathcal{L}_M$: $\mathcal{L}_M(x,y) = \begin{cases} 1, & \text{if } x < y \\ y, & \text{elsewhere} \end{cases}$;

(ii) the operator $\mathcal{L}_P$: $\mathcal{L}_P(x,y) = \begin{cases} 1, & \text{if } x < y \\ y/x, & \text{if } 0 < y \le x \\ 0, & \text{elsewhere} \end{cases}$;

(iii) the operator $\mathcal{L}_W$: $\mathcal{L}_W(x,y) = \begin{cases} 1, & \text{if } x < y \\ 1 - x + y, & \text{if } 0 < y \le x \\ 0, & \text{elsewhere} \end{cases}$.

Example 2. Consider the t-norm P on the complete chain $([0,1], \le)$ and the sup-P equation

$$\sup_{i \in I_5} P(A(i), X(i)) = \beta$$

with $A = \begin{pmatrix} 0.6 & 0.4 & 0.1 & 1 & 0 \end{pmatrix}$ and $\beta = 0.4$. The potential greatest solution G is given by

$$G = \begin{pmatrix} 2/3 & 1 & 1 & 0.4 & 1 \end{pmatrix}.$$

One easily verifies that G is a solution. According to Theorem 2, there are three minimal solutions M_1, M_2 and M_4 given by :

$$\begin{aligned} M_1 &= \begin{pmatrix} 2/3 & 0 & 0 & 0 & 0 \end{pmatrix} \\ M_2 &= \begin{pmatrix} 0 & 1 & 0 & 0 & 0 \end{pmatrix} \\ M_4 &= \begin{pmatrix} 0 & 0 & 0 & 0.4 & 0 \end{pmatrix}. \end{aligned}$$

4.2 Example 2: $(L, \le) = ([0,1]^n, \le)$

In this final subsection, we consider the product lattice $([0,1]^n, \le)$. This is a distributive, complete lattice and all elements of it are either join-irreducible or have an irredundant finite join-decomposition. We will next show how t-norms on $([0,1]^n, \le)$ can be constructed.

Proposition 19. *Let $\mathcal{T}$ be a t-norm on a bounded ordered set $(P, \leq)$ and $n \in \mathbb{N}_0$, then the $P^n \times P^n \to P^n$ mapping $\mathcal{T}_n$ defined by*

$$\mathcal{T}_n((\alpha_1, \ldots, \alpha_n), (\beta_1, \ldots, \beta_n)) = (\mathcal{T}(\alpha_1, \beta_1), \ldots, \mathcal{T}(\alpha_n, \beta_n))$$

is a t-norm on $(P^n, \leq)$.

Proof. Straightforward. □

Proposition 20. *Let $(L, \leq)$ be a complete lattice and $n \in \mathbb{N}_0$, and consider the product lattice $(L^n, \leq)$. Let A be a subset of L^n and define the subsets A_i of L, $i \in I_n$, as follows: $A_i = \{\alpha \mid \alpha \in L \ \wedge \ (\exists \gamma \in A)(\gamma(i) = \alpha)\}$. Then the following properties hold:*

$$\begin{aligned} \inf A &= (\inf A_1, \ldots, \inf A_n) \\ \sup A &= (\sup A_1, \ldots, \sup A_n). \end{aligned}$$

Proof. Straightforward. □

Proposition 21. *Let $\mathcal{T}$ be a t-norm on a complete lattice $(L, \leq)$ and $n \in \mathbb{N}_0$, then the operators $\mathcal{I}_{\mathcal{T}_n}$ and $\mathcal{L}_{\mathcal{T}_n}$ corresponding to the t-norm $\mathcal{T}_n$ on $(L^n, \leq)$ are defined by*

$$\mathcal{I}_{\mathcal{T}_n}((\alpha_1, \ldots, \alpha_n), (\beta_1, \ldots, \beta_n)) = (\mathcal{I}_{\mathcal{T}}(\alpha_1, \beta_1), \ldots, \mathcal{I}_{\mathcal{T}}(\alpha_n, \beta_n))$$

$$\mathcal{L}_{\mathcal{T}_n}((\alpha_1, \ldots, \alpha_n), (\beta_1, \ldots, \beta_n)) = \begin{cases} (\mathcal{L}_{\mathcal{T}}(\alpha_1, \beta_1), \ldots, \mathcal{L}_{\mathcal{T}}(\alpha_n, \beta_n)), & \text{if } (\forall i \in I_n)(\beta_i \leq \alpha_i) \\ (1_L, \ldots, 1_L), & \text{else} \end{cases}.$$

Proof. We will for instance give the proof for $\mathcal{L}_{\mathcal{T}_n}$. Consider $(\alpha_1, \ldots, \alpha_n) \in L^n$ and $(\beta_1, \ldots, \beta_n) \in L^n$. If $\neg((\beta_1, \ldots, \beta_n) \leq (\alpha_1, \ldots, \alpha_n))$, then it follows from Proposition 9 that $\mathcal{L}_{\mathcal{T}_n}((\alpha_1, \ldots, \alpha_n), (\beta_1, \ldots, \beta_n)) = 1_{L^n} = (1_L, \ldots, 1_L)$. If $(\beta_1, \ldots, \beta_n) \leq (\alpha_1, \ldots, \alpha_n)$, then the sets

$$A_i = \{\gamma_i \mid \gamma_i \in L \ \wedge \ \mathcal{T}(\alpha_i, \gamma_i) \geq \beta_i\},$$

$i \in I_n$, are not-empty (1_L belongs to all of them). It then follows with Proposition 20 that

$$\mathcal{L}_{\mathcal{T}_n} \qquad ((\alpha_1, \ldots, \alpha_n), (\beta_1, \ldots, \beta_n)) =$$

$$
\begin{aligned}
&= \inf\{(\gamma_1,\ldots,\gamma_n) \mid \mathcal{T}_n((\alpha_1,\ldots,\alpha_n),(\gamma_1,\ldots,\gamma_n)) \geq (\beta_1,\ldots,\beta_n)\}\\
&= \inf\{(\gamma_1,\ldots,\gamma_n) \mid (\mathcal{T}(\alpha_1,\gamma_1),\ldots,\mathcal{T}(\alpha_n,\gamma_n)) \geq (\beta_1,\ldots,\beta_n)\}\\
&= \inf\{(\gamma_1,\ldots,\gamma_n) \mid (\forall i \in I_n)(\mathcal{T}(\alpha_i,\gamma_i) \geq \beta_i)\}\\
&= \inf A_1 \times \ldots \times A_n\\
&= (\inf A_1,\ldots,\inf A_n)\\
&= (\mathcal{L}_{\mathcal{T}}(\alpha_1,\beta_1),\ldots,\mathcal{L}_{\mathcal{T}}(\alpha_n,\beta_n)). \quad \square
\end{aligned}
$$

Notice that even for a t-norm $\mathcal{T}$ on $([0,1],\leq)$ the property $\mathcal{L}_{\mathcal{T}} \leq \mathcal{I}_{\mathcal{T}}$ does not imply that also $\mathcal{L}_{\mathcal{T}_n} \leq \mathcal{I}_{\mathcal{T}_n}$. Indeed, consider for instance $\mathcal{T} = M$ and $n = 2$, then it follows from Proposition 22 that:

$$
\begin{aligned}
\mathcal{L}_{M_2}((0.5,0.3),(0.4,0.6)) &= (1,1)\\
\mathcal{I}_{M_2}((0.5,0.3),(0.4,0.6)) &= (0.4,1).
\end{aligned}
$$

From the foregoing sections it should be clear that the morphism behaviour of the partial mappings of the t-norm $\mathcal{T}$ plays an important role in solving sup-$\mathcal{T}$ equations. In the following proposition, we show that the partial mappings of $\mathcal{T}_n$ and the partial mappings of $\mathcal{T}$ share the same morphism behaviour.

Proposition 22. *Let $\mathcal{T}$ be a t-norm on a complete lattice $(L,\leq)$ and $n \in \mathbb{N}_0$, then the following properties hold:*

(i) if the partial mappings of $\mathcal{T}$ are infimum-morphisms, then also the partial mappings of $\mathcal{T}_n$ are infimum-morphisms;

(i) if the partial mappings of $\mathcal{T}$ are supremum-morphisms, then also the partial mappings of $\mathcal{T}_n$ are supremum-morphisms.

Proof. For the sake of simplicity, we will give the proof of (i) for $n = 2$. Assume that the partial mappings of $\mathcal{T}$ are infimum-morphisms, then we have to show that for an arbitrary $(\alpha_1,\alpha_2) \in L^2$ and an arbitrary non-empty subset B of L^2 we have that

$$
\mathcal{T}_2((\alpha_1,\alpha_2),\inf B) = \inf_{(\beta_1,\beta_2)\in B} \mathcal{T}_2((\alpha_1,\alpha_2),(\beta_1,\beta_2)).
$$

Let $B_1 = \{\beta_1 \mid (\exists \beta \in L)((\beta_1,\beta) \in B)\}$ and $B_2 = \{\beta_2 \mid (\exists \beta \in L)((\beta,\beta_2) \in B)\}$, then it follows with Proposition 20 that $\inf B = (\inf B_1, \inf B_2)$. It then follows:

$$\begin{aligned}\mathcal{T}_2((\alpha_1,\alpha_2),\inf B) &= \mathcal{T}_2((\alpha_1,\alpha_2),(\inf B_1,\inf B_2))\\ &= (\mathcal{T}(\alpha_1,\inf B_1),\mathcal{T}(\alpha_2,\inf B_2))\\ &= (\inf_{\beta_1\in B_1}\mathcal{T}(\alpha_1,\beta_1),\inf_{\beta_2\in B_2}\mathcal{T}(\alpha_2,\beta_2)).\end{aligned}$$

Again with Proposition 20 it follows that

$$\begin{aligned}\mathcal{T}_2((\alpha_1,\alpha_2),\inf B) &= \inf_{(\beta_1,\beta_2)\in B}(\mathcal{T}(\alpha_1,\beta_1),\mathcal{T}(\alpha_2,\beta_2))\\ &= \inf_{(\beta_1,\beta_2)\in B}\mathcal{T}_2((\alpha_1,\alpha_2),(\beta_1,\beta_2)).\end{aligned}$$ □

Proposition 23. *Let $\mathcal{T}$ be a t-norm on a bounded ordered set $(P,\le)$ and $n\in\mathbb{N}_0$. If $\mathcal{T}$ is maximally surjective, then also $\mathcal{T}_n$ is maximally surjective.*

Proof. Straightforward. □

Example 3. Consider the t-norm W_2 on the distributive, complete lattice $([0,1]^2,\le)$. According to Propositions 22 and 23, the partial mappings of W_2 are homomorphisms and W_2 is maximally surjective. Consider the sup-W_2 equation

$$\sup_{i\in I_3} W_2(A(i),X(i)) = \beta$$

with $A = \big((0.4,0.6)\quad(0.7,0.4)\quad(0.6,0.5)\big)$ and $\beta=(0.3,0)$. The right-hand side β clearly is join-irreducible. We construct the potential greatest solution G:

$$\begin{aligned}G(1) &= \mathcal{I}_{W_2}((0.4,0.6),(0.3,0)) = (\mathcal{I}_W(0.4,0.3),\mathcal{I}_W(0.6,0)) = (0.9,0.4)\\ G(2) &= \mathcal{I}_{W_2}((0.7,0.4),(0.3,0)) = (\mathcal{I}_W(0.7,0.3),\mathcal{I}_W(0.4,0)) = (0.6,0.6)\\ G(3) &= \mathcal{I}_{W_2}((0.6,0.5),(0.3,0)) = (\mathcal{I}_W(0.6,0.3),\mathcal{I}_W(0.5,0)) = (0.7,0.5),\end{aligned}$$

and hence $G = \big((0.9,0.4)\quad(0.6,0.6)\quad(0.7,0.5)\big)$. One easily verifies that G is a solution. Since $(\forall i\in I_3)(\beta\le A(i))$, it follows from Theorem 2 that the solution set contains three minimal elements M_1, M_2 and M_3, given by:

$$\begin{aligned}M_1 &= \big((0.9,0)\quad(0,0)\quad(0,0)\big)\\ M_2 &= \big((0,0)\quad(0.6,0)\quad(0,0)\big)\\ M_3 &= \big((0,0)\quad(0,0)\quad(0.7,0)\big).\end{aligned}$$

Indeed, we have for instance that

$$M_1(1) = \mathcal{L}_{W_2}(A(1),\beta) = (\mathcal{L}_W(0.4,0.3),\mathcal{L}_W(0.6,0)) = (0.9,0).$$

Example 4. Consider the t-norm $\frown$ on the distributive, complete lattice $([0,1]^3, \leq)$. Notice that $\frown = M_3$. According to Propositions 22 and 23, the partial mappings of M_3 are homomorphisms and M_3 is maximally surjective. Consider the sup-$\frown$ equation

$$\sup_{i \in I_3} A(i) \frown X(i) = \beta$$

with $A = \big((0.2, 0.6, 0.5) \quad (0.7, 0.5, 0.9) \quad (1, 0.8, 0.7)\big)$ and $\beta = (0.7, 0.5, 0.7)$. The right-hand side β has as irredundant join-decomposition $\beta = \beta_1 \smile \beta_2 \smile \beta_3$ with $\beta_1 = (0.7, 0, 0)$, $\beta_2 = (0, 0.5, 0)$ and $\beta_3 = (0, 0, 0.7)$. We construct the potential greatest solution G:

$$G = \big((1, 0.5, 1) \quad (1, 1, 0.7) \quad (0.7, 0.5, 1)\big).$$

One easily verifies that G is a solution. According to Theorem 3, and taking into account that $\mathcal{L}_{M_3} \not\leq \mathcal{I}_{M_3}$, the minimal solutions are the minimal elements of the set

$$\{\bigcup_{j \in I_3} M_k^j \mid \beta_j \leq A(k) \wedge M_k^j \subseteq G\} = \{\bigcup_{j \in I_3} M_k^j \mid \beta(j) \leq A(k)(j) \wedge M_k^j \subseteq G\}$$

with M_k^j defined by $M_k^j(i) = \begin{cases} \mathcal{L}_{M_3}(A(k), \beta_j), & \text{if } i = k \\ (0,0,0), & \text{elsewhere} \end{cases}$. For instance, we have

$$M_2^1(i) = \begin{cases} \mathcal{L}_{M_3}((0.7, 0.5, 0.9), (0.7, 0, 0)), & \text{if } i = 2 \\ (0,0,0), & \text{elsewhere} \end{cases}$$

and hence $M_2^1 = ((0,0,0), (0.7,0,0), (0,0,0))$.

Notice that although $\mathcal{L}_{M_3} \not\leq \mathcal{I}_{M_3}$ we always have that $M_k^j \subseteq G$ when $\beta_j \leq A(k)$. Indeed, we have

$$\begin{aligned} M_k^j \subseteq G \quad & \Leftrightarrow \mathcal{L}_{M_3}(A(k), \beta_j) \leq \mathcal{I}_{M_3}(A(k), \beta) \\ & \Leftrightarrow \mathcal{L}_M(A(k)(j), \beta(j)) \leq \mathcal{I}_M(A(k)(j), \beta(j)). \end{aligned}$$

Since $\mathcal{L}_M \leq \mathcal{I}_M$, the last inequality obviously holds.

The minimal solutions then have to found among the following solutions:

$$\begin{aligned} M_2^1 \cup M_1^2 \cup M_2^3 &= \big((0, 0.5, 0) \quad (0.7, 0, 0.7) \quad (0,0,0)\big) \\ M_2^1 \cup M_1^2 \cup M_3^3 &= \big((0, 0.5, 0) \quad (0.7, 0, 0) \quad (0, 0, 0.7)\big) \\ M_2^1 \cup M_2^2 \cup M_2^3 &= \big((0,0,0) \quad (0.7, 0.5, 0.7) \quad (0,0,0)\big) \end{aligned}$$

$$
\begin{aligned}
M_2^1 \cup M_2^2 \cup M_3^3 &= \big((0,0,0) \quad (0.7,0.5,0) \quad (0,0,0.7) \big) \\
M_2^1 \cup M_3^2 \cup M_2^3 &= \big((0,0,0) \quad (0.7,0,0.7) \quad (0,0.5,0) \big) \\
M_2^1 \cup M_3^2 \cup M_3^3 &= \big((0,0,0) \quad (0.7,0,0) \quad (0,0.5,0.7) \big) \\
M_3^1 \cup M_1^2 \cup M_2^3 &= \big((0,0.5,0) \quad (0,0,0.7) \quad (0.7,0,0) \big) \\
M_3^1 \cup M_1^2 \cup M_3^3 &= \big((0,0.5,0) \quad (0,0,0) \quad (0.7,0,0.7) \big) \\
M_3^1 \cup M_2^2 \cup M_2^3 &= \big((0,0,0) \quad (0,0.5,0.7) \quad (0.7,0,0) \big) \\
M_3^1 \cup M_2^2 \cup M_3^3 &= \big((0,0,0) \quad (0,0.5,0) \quad (0.7,0,0.7) \big) \\
M_3^1 \cup M_3^2 \cup M_2^3 &= \big((0,0,0) \quad (0,0,0.7) \quad (0.7,0.5,0) \big) \\
M_3^1 \cup M_3^2 \cup M_3^3 &= \big((0,0,0) \quad (0,0,0) \quad (0.7,0.5,0.7) \big).
\end{aligned}
$$

One easily verifies that all of these solutions are minimal. We will next show that this result is no coincidence, and that it can be generalized. This means that in the case of $([0,1]^n, \leq)$, Theorem 3 can be improved in the sense that all minimal solutions are immediately known, provided of course the above irredundant join-decomposition is applied.

Consider two potential minimal solutions $M_{j_1}^1 \cup M_{j_2}^2 \cup M_{j_3}^3$ and $M_{k_1}^1 \cup M_{k_2}^2 \cup M_{k_3}^3$ with $(j_1, j_2, j_3) \neq (k_1, k_2, k_3)$. Suppose that $j_1 \neq k_1$. Let for instance $j_1 = 2$ and $k_1 = 3$, then

$$
\begin{aligned}
M_{j_1}^1 \cup M_{j_2}^2 \cup M_{j_3}^3 &= \big((0,?,?) \quad (\mathcal{L}_M(A(3)(1), \beta(1)),?,?) \quad (0,?,?) \big) \\
&= \big((0,?,?) \quad (\beta(1),?,?) \quad (0,?,?) \big) \\
M_{k_1}^1 \cup M_{k_2}^2 \cup M_{k_3}^3 &= \big((0,?,?) \quad (0,?,?) \quad (\mathcal{L}_M(A(3)(1), \beta(1)),?,?) \big) \\
&= \big((0,?,?) \quad (0,?,?) \quad (\beta(1),?,?) \big).
\end{aligned}
$$

Since $\beta(1) > 0$ it is immediately clear that these two solutions are incomparable. As this holds for any two of the potential minimal solutions, they are all minimal.

REFERENCES

[1] C. Alsina, E. Trillas and L. Valverde, "On non-distributive logical connectives for fuzzy sets theory," *BUSEFAL* **3**, 1980, 18–29.

[2] G. Birkhoff, *Lattice Theory*, AMS Colloquium Publications Volume XXV, American Mathematical Society, Providence, RI, 1967.

[3] B. Davey and H. Priestley, *Introduction to Lattices and Order*, Cambridge University Press, 1990.

[4] B. De Baets and E. Kerre, "A primer on solving fuzzy relational equations on the unit interval," *International Journal of Uncertainty, Fuzziness and Knowledge-Based Systems* **2**, 1994, 205–225.

[5] B. De Baets and E. Kerre, "A representation of solution sets of fuzzy relational equations," in *Cybernetics and Systems '94, Proceedings of the Twelfth European Meeting on Cybernetics and Systems Research*, R. Trappl, ed., World Scientific Publishing, Singapore, 1994, 287–294.

[6] B. De Baets, "Crowns and root systems," *Algebra Universalis*, in preparation.

[7] G. De Cooman and E. Kerre, "Order norms on bounded partially ordered sets," *The Journal of Fuzzy Mathematics* **2**, 1994, 281–310.

[8] A. Di Nola, S. Sessa, W. Pedrycz and E. Sanchez, *Fuzzy Relation Equations and their Applications to Knowledge Engineering, Theory and Decision Library. Series D. System Theory, Knowledge Engineering and Problem Solving*, Kluwer, Dordrecht, Boston en London, 1989.

[9] J. Goguen, "*L*-Fuzzy sets," *Journal of Mathematical Analysis and Applications* **18**, 1967, 145–174.

[10] H. Prade, "Unions et intersections d'ensembles flous," *BUSEFAL* **3**, 1980, 58–62.

[11] B. Schweizer and A. Sklar, "Associative functions and abstract semigroups," *Publ. Math. Debrecen* **10**, 1963, 69–84.

[12] B. Schweizer and A. Sklar, *Probabilistic Metric Spaces*, Elsevier, New York, 1983.

[13] E. Trillas and L. Valverde, "On implication and indistinguishability in the setting of fuzzy logic," in *Management decision support systems using fuzzy sets and possibility theory*, J. Kacprzyk and R. Yager, eds., Verlag TÜV Rheinland, Köln, 1983, 198–212.

[14] C. Zhao, "On matrix equations in a class of complete and completely distributive lattices," *Fuzzy Sets and Systems* **22**, 1987, 303–320.

4

TOWARDS A POSSIBILISTIC LOGIC

Gert de Cooman

Vakgroep Elektrische Energietechniek
Universiteit Gent
Technologiepark 9
B-9052 Zwijnaarde, Belgium

ABSTRACT

In this paper, we investigate how linguistic information can be incorporated into classical propositional logic. First, we show that Zadeh's extension principle can be justified and at the same time generalized by considerations about transformation of possibility measures. Using these results, we show how linguistic uncertainty about the truth value of a proposition leads to the introduction of the notion of a possibilistic truth value. Since propositions can be combined into new ones using logical operators, linguistic uncertainty about the truth values of the original propositions leads to linguistic uncertainty about the truth value of the resulting proposition. Furthermore, we show that in a number of special cases there is truth-functionality, i.e., the possibilistic truth value of the resulting proposition is a function of the possibilistic truth values of the original propositions. We are thus led to the introduction of possibilistic-logical functions, combining possibilistic truth values. Important classes of such functions, the possibilistic extension logics, directly result from the above-mentioned investigation, and are studied extensively. Finally, the relation between these logics, and Kleene's strong multi-valued systems is established.

1 INTRODUCTION

Classical propositional logic deals with *propositions*, or in other words, affirmative statements that are either true or false. Because the *truth value* of a proposition can only be either *true* or *false*, classical propositional logic is often called *two-valued*. Propositions can be combined into new ones, using so-called *logical operators*. Classical logic is *truth-functional*, because the behaviour of any logical operator, acting on propositions, can be characterized by a *logical function*, acting on truth values.

In the beginning of this century, Kleene and others (see, for instance, [25]) addressed an interesting problem. They considered propositions, which are *a priori* either true or false, but for which there is insufficient knowledge to determine precisely which value in {*true*, *false*} the truth value assumes. They asked themselves how a mathematical, logical description of this situation could be constructed. Their solution consisted in the introduction of a number of *multi-valued logics*, i.e., logical systems with more than two truth values, called *strong* multi-valued Kleene logics. At least two aspects of Kleene's approach deserve extra attention in the context of this paper. First of all, the truth values used by Kleene are *epistemological*, because they are not the actual truth values of the propositions involved (*a priori true* or *false*), but rather reflect *our knowledge* about the actual truth values. Ideally therefore, Kleene truth values are mathematical representations of the uncertainty that exists about the actual truth values of the propositions involved. Secondly, Kleene logics are *truth-functional*. This means that according to these systems, the behaviour of a logical operator is mirrored in a logical function, combining Kleene truth values. To give an example, in these systems it is possible, from the Kleene truth values of two propositions, unequivocally to determine the Kleene truth value of their conjunction.

In the present work, we investigate the problem studied by Kleene in a special case, namely when the uncertainty about the actual truth values of propositions is *linguistic*. Linguistic uncertainty is the uncertainty contained in (or conveyed by) affirmative statements such as 'Wietse is less than one year old' or 'Linde's temperature is high'. Such statements give us information about, say, Wietse's age or Linde's temperature, but not enough to determine their actual value completely. Due to the imprecision and/or vagueness of the predicates involved ('less than one year old', 'high'), such statements convey information, but also contain uncertainty. In 1978, Zadeh proposed the use of *possibility measures* to mathematically represent linguistic information and, dually, uncertainty [30]. In our doctoral dissertation [9], we have generalized Zadeh's possibility measures, developed a general measure- and integral-theoretic account of possibility theory, and shown that possibility measures are indeed able to represent linguistic information (see also [4, 5, 6, 11, 12, 13, 14, 15]). On the basis of this work, we show in this paper how a possibilistic logic, representing linguistic uncertainty in classical propositional logic, comes about, and how under a number of additional assumptions, this possibilistic logic leads to Kleene's strong multi-valued systems. In doing so, we at the same time generalize and provide the possibilistic basis for previous work in this field, reported on in [16].

In section 2 we have collected the preliminary definitions and notational conventions, necessary for the proper understanding of the material in this paper. In section 3 we study the transmission of possibilistic (or linguistic) information by mappings, and at the same time give a possibilistic justification for Zadeh's extension principle. Using the material of this section, Kleene's problem is addressed in section 4. We consider propositions of the type '(subject) is (predicate)', and show how possibilistic uncertainty about the value the subject assumes in a universe of possible values leads to uncertainty about the truth value of the proposition. This uncertainty can be represented by a possibility distribution on the set {*true*, *false*}, called a possibilistic truth value. Note that possibilistic truth values are descriptions of the uncertainty about the actual truth value (in {*true*, *false*}), and are therefore of an epistemological nature. They are the possibilistic counterparts of the Kleene truth values. The next step consists in investigating how the possibilistic truth value of a combination of propositions, using a logical operator, can be calculated. It turns out that in a number of cases, we can associate with a logical operator a *possibilistic-logical function*, in such a way that the possibilistic truth value of the combination of propositions is the image under this function of the possibilistic truth values of the participating propositions. In other words, our possibilistic logic is indeed in those special cases truth-functional. It also turns out that the possibilistic-logical function involved is a generalized Zadeh extension of the (classical-)logical function associated with the logical operator. We are thus led in section 5 to the introduction of special systems of possibilistic-logical functions, the so-called possibilistic extension logics. We also study the properties of these logics. An interesting special case is considered in section 6, and leads to a possibilistic justification for the introduction of strong multi-valued Kleene logics. A number of conclusions and open research problems are formulated in section 7.

2 PRELIMINARY DEFINITIONS

Let us start this discussion with a few preliminary definitions and notational conventions, valid in the rest of this paper, unless explicitly stated to the contrary. By $(L, \leq)$ we shall mean a complete lattice that is arbitrary but *fixed throughout the whole text.* The bottom element of $(L, \leq)$ will be denoted by 0 and the top element by 1. We also assume that $0 \neq 1$. The meet of $(L, \leq)$ will be denoted by $\frown$, the join of $(L, \leq)$ by $\smile$.

2.1 Triangular Norms

A *triangular norm* (or, shortly, *t-norm*) T on the complete lattice $(L, \leq)$ is a binary operator on L that is isotonic, associative and commutative, and that furthermore satisfies the boundary condition $(\forall \lambda \in L)(T(1, \lambda) = \lambda)$. As a corollary, we have for such a t-norm T that $(\forall \lambda \in L)(T(0, \lambda) = 0)$. Of course, $\frown$ is a triangular norm on $(L, \leq)$, and more specifically, the only one that is idempotent. For a more involved discussion of triangular norms defined on complete lattices, and more in general, on bounded partially ordered sets, we refer to [14]. A t-norm T on $(L, \leq)$ is called *completely distributive* w.r.t. supremum iff for arbitrary λ in L and an arbitrary family $(\mu_j \mid j \in J)$ of elements of L:

$$T(\lambda, \sup_{j \in J} \mu_j) = \sup_{j \in J} T(\lambda, \mu_j).$$

In this case, the structure $(L, \leq, P)$ is called a *complete lattice with t-norm* [14]. In what follows, we shall always denote by T a t-norm on $(L, \leq)$ that is completely distributive w.r.t. supremum.

2.2 Ample Fields

An *ample field* $\mathcal{R}$ on the universe X is a set of subsets of X that is closed under *arbitrary* unions and intersections, and under complementation in X. A special ample field on X, and at the same time the largest, is the *power class* $\mathcal{P}(X)$ of X, i.e. the set of all the subsets of X. In this sense, ample fields on X can be considered as immediate generalizations of this power class. For a more thorough discussion of this subject, we refer to [13, 24]. The *atom* of $\mathcal{R}$ containing the element x of X will be denoted by $[\,x\,]_{\mathcal{R}}$ and is defined by:

$$[\,x\,]_{\mathcal{R}} \stackrel{\text{def}}{=} \bigcap \{ A \mid x \in A \text{ and } A \in \mathcal{R} \}.$$

Remark that the atom of the ample field $\mathcal{P}(X)$ containing x is precisely the singleton $\{x\}$. Therefore, atoms of ample fields can be interpreted as generalizations of singletons. In this light, we also have that

$$(\forall x \in X)(\forall A \in \mathcal{R})(x \in A \Leftrightarrow [\,x\,]_{\mathcal{R}} \subseteq A).$$

Furthermore, for arbitrary x in X:

$$x \in [\,x\,]_{\mathcal{R}} \text{ and } [\,x\,]_{\mathcal{R}} \in \mathcal{R}. \tag{4.1}$$

A subset E of X will be called *$\mathcal{R}$-measurable* iff $E \in \mathcal{R}$. Interestingly,

$$E \in \mathcal{R} \Leftrightarrow E = \bigcup_{x \in E} [\,x\,]_E. \tag{4.2}$$

If we look at (4.2), we are led to the introduction of a $\mathcal{P}(X) - \mathcal{R}$-mapping $C_{\mathcal{R}}$, such that for arbitrary A in $\mathcal{P}(X)$:

$$C_{\mathcal{R}}(A) = \bigcup_{x \in A} [\,x\,]_{\mathcal{R}}.$$

It turns out that $C_{\mathcal{R}}$ is the closure operator on $\mathcal{P}(X)$, associated with the closure system $\mathcal{R}$ [3, 9, 13].

Consider an arbitrary subset $\mathcal{A}$ of $\mathcal{P}(X)$. Since the intersection of an arbitrary family of ample fields is again an ample field, we know that

$$\tau(\mathcal{A}) \stackrel{\text{def}}{=} \bigcap \{\, \mathcal{R} \mid \mathcal{R} \text{ is an ample field on } X \text{ and } \mathcal{A} \subseteq \mathcal{R} \,\}$$

is an ample field on X, called the *ample field generated by* $\mathcal{A}$. This notion can be used to introduce product ample fields. If we consider the universes X_1 and X_2 provided with the respective ample fields $\mathcal{R}_1$ and $\mathcal{R}_2$, then the *product ample field* on $X_1 \times X_2$ of $\mathcal{R}_1$ and $\mathcal{R}_1$ is defined as

$$\mathcal{R}_1 \times \mathcal{R}_2 \stackrel{\text{def}}{=} \tau(\{\, A_1 \times A_2 \mid A_1 \in \mathcal{R}_1 \text{ and } A_2 \in \mathcal{R}_2 \,\}).$$

Interestingly, for the atoms of $\mathcal{R}_1 \times \mathcal{R}_2$, it can be proven that

$$(\forall (x_1, x_2) \in X_1 \times X_2)([\,(x_1, x_2)\,]_{\mathcal{R}_1 \times \mathcal{R}_2} = [\,x_1\,]_{\mathcal{R}_1} \times [\,x_2\,]_{\mathcal{R}_2}),$$

which confirms our interpretation of an atom as a generalization of a singleton. Of course, these results are easily extended towards products of more than two ample fields.

If we consider two universe X and Y with respective ample fields $\mathcal{R}_X$ and $\mathcal{R}_Y$, then a $X - Y$-mapping f is called *$\mathcal{R}_X - \mathcal{R}_Y$-measurable* iff

$$(\forall B \in \mathcal{R}_Y)(f^{-1}(B) \in \mathcal{R}_X),$$

where $f^{-1}(B) = \{\, x \mid x \in X \text{ and } f(x) \in B \,\}$ is the *inverse image* of B under the mapping f.

2.3 Fuzzy Sets and Fuzzy Variables

With an arbitrary subset A of a universe X, we can associate its *characteristic $X - L$ mapping* χ_A, defined by

$$\chi_A(x) \stackrel{\text{def}}{=} \begin{cases} 1 & ; \quad x \in A \\ 0 & ; \quad x \in \text{co}A. \end{cases}$$

In accordance with the terminology introduced by Goguen [20], an arbitrary $X - L$ mapping will be called a *$(L, \leq)$-fuzzy set* (or simply fuzzy set) in X. It is an obvious generalization of a characteristic $X - L$-mapping. The set of the $(L, \leq)$-fuzzy sets in X will be denoted by $\mathcal{F}_{(L,\leq)}(X)$. A $([0,1], \leq)$-fuzzy set in X will also be called a *Zadeh fuzzy set*, and the set of all Zadeh fuzzy sets in X will be denoted by $\mathcal{F}(X)$. A $(L, \leq)$-fuzzy set h will be called *sup-normal* iff $\sup_{x \in X} h(x) = 1$.

A $X - L$ mapping h will be called *$\mathcal{R}$-measurable* iff it is constant on the atoms of $\mathcal{R}$. A $\mathcal{R}$-measurable $X - L$ mapping—or $(L, \leq)$-fuzzy set in X—is also called a *$(L, \leq)$-fuzzy variable* in $(X, \mathcal{R})$. Whenever we want to omit reference to the structures $(L, \leq)$ and $(X, \mathcal{R})$, we shall simply speak of fuzzy variables. A fuzzy variable can therefore be considered as a 'fuzzification' of a measurable set. Indeed, a subset E of X is $\mathcal{R}$-measurable if and only if its characteristic $X - L$-mapping is. The set of the $(L, \leq)$-fuzzy variables in $(X, \mathcal{R})$ is denoted by $\mathcal{G}^{\mathcal{R}}_{(L,\leq)}(X)$. A more detailed treatment of fuzzy variables can be found in [4, 5, 6, 9].

2.4 Possibility Measures

A *$(L, \leq)$-possibility measure* Π on $(X, \mathcal{R})$ is a complete join-morphism between the complete lattices $(\mathcal{R}, \subseteq)$ and $(L, \leq)$. By definition, this means that Π satisfies the following requirement: for an arbitrary family $(A_j \mid j \in J)$ of elements of $\mathcal{R}$:

$$\Pi(\bigcup_{j \in J} A_j) = \sup_{j \in J} \Pi(A_j).$$

This definition immediately implies that $\Pi(\emptyset) = 0$. For arbitrary A in $\mathcal{R}$, $\Pi(A)$ will be called the *$(L, \leq)$-possibility of A*. Π will be called *normalized*[1]

[1] In very much the same was as for probability measures, normalization is a very natural property for possibility measures, and expresses in some sense that the universe X is adequate, or large enough. We shall therefore in this paper always work with normalized possibility measures.

iff $\Pi(X) = 1$. Again, whenever we do not want to mention the complete lattice $(L, \leq)$ explicitly, we shall simply speak of possibility and possibility measures. A $\mathcal{R}$-measurable $X - L$-mapping π such that for arbitrary A in $\mathcal{R}$

$$\Pi(A) = \sup_{x \in A} \pi(x), \tag{4.3}$$

is called a *distribution* of Π. Such a distribution is *unique*, and satisfies

$$(\forall x \in X)(\pi(x) \stackrel{\text{def}}{=} \Pi([\, x \,]_{\mathcal{R}}))$$

The distribution of a normalized possibility measure is a sup-normal fuzzy variable. $(L, \leq)$-possibility measures are generalizations towards more general domains and codomains of Zadeh's possibility measures [30], Wang's fuzzy contactabilities [24], and the possibility measures we introduced in [17]. For a more detailed discussion of these generalizations, we refer to [9, 13, 17].

2.5 Possibilistic Variables

First, let us introduce transformation of possibility by a mapping. Let X and Y be universes, and f a $X - Y$-mapping. Let $\mathcal{R}$ be an ample field on X and Π a $(L, \leq)$-possibility measure on $(X, \mathcal{R})$. Then

$$\mathcal{R}^{(f)} \stackrel{\text{def}}{=} \{\, B \mid B \in \mathcal{P}(Y) \text{ and } f^{-1}(B) \in \mathcal{R} \,\}$$

is an ample field on Y, and the $\mathcal{R}^{(f)} - L$ mapping

$$\Pi^{(f)} \colon \mathcal{R}^{(f)} \to L \colon B \mapsto \Pi(f^{-1}(B)) \tag{4.4}$$

is a $(L, \leq)$-possibility measure on $(Y, \mathcal{R}^{(f)})$, called the *transformed* $(L, \leq)$-*possibility measure on* $(Y, \mathcal{R}^{(f)})$ *by the mapping* f. If $\mathcal{R}_Y$ is an ample field on Y, such that f is $\mathcal{R} - \mathcal{R}_Y$-measurable, then

$$\mathcal{R}_Y \subseteq \mathcal{R}^{(f)}, \tag{4.5}$$

and the restriction $\Pi^{(f)} | \mathcal{R}_Y$ of $\Pi^{(f)}$ to $\mathcal{R}_Y$ is called the *transformed* $(L, \leq)$-*possibility measure on* $(Y, \mathcal{R}_Y)$ *by the mapping* f.

It is also possible to define *possibilistic variables*, which are possibilistic equivalents of the stochastic variables in probability theory (see, for instance, [2]). We consider a universe Ω and an ample field $\mathcal{R}_\Omega$ on Ω. This universe is called a *basic space*. X is called a *sample space*. A $\Omega - X$-mapping that is $\mathcal{R}_\Omega - \mathcal{R}$-measurable, is called a *possibilistic variable in* $(X, \mathcal{R})$. If we also consider a normalized $(L, \leq)$-possibility measure Π_Ω on $(\Omega, \mathcal{R}_\Omega)$, we can use the possibilistic variable f to *transform* Π_Ω to a $(L, \leq)$-possibility measure Π_f on $(X, \mathcal{R})$, defined by $\Pi_f \stackrel{\text{def}}{=} \Pi_\Omega{}^{(f)} | \mathcal{R}$, or equivalently,

$$(\forall B \in \mathcal{R})(\Pi_f(B) \stackrel{\text{def}}{=} \Pi_\Omega(f^{-1}(B))).$$

Π_f is called the *possibility distribution*[2] of the possibilistic variable f. The distribution π_f of Π_f is called the *possibility distribution function* of f, and satisfies

$$\pi_f(x) = \sup_{f(\omega) \in [x]_{\mathcal{R}}} \pi_\Omega(\omega),$$

where x is an element of X and π_Ω is the distribution of Π_Ω. Of course, π_f is a sup-normal $(L, \leq)$-fuzzy variable in $(X, \mathcal{R})$. For a more detailed account of possibilistic variables, we refer to [6, 9].

2.6 Classical Truth Values and Their Combinations

We shall also be working with the set $\mathcal{T} \stackrel{\text{def}}{=} \{\textit{false}, \textit{true}\}$ of truth values in classical propositional logic. On $\mathcal{T}$, we define the total order relation

$$\leq \stackrel{\text{def}}{=} \{(\textit{false}, \textit{false}), (\textit{false}, \textit{true}), (\textit{true}, \textit{true})\},$$

i.e., $(\mathcal{T}, \leq)$ is a Boolean chain of length 2, with top element *true* and bottom element *false*. On this Boolean chain, we may define the complement $\neg$, called *negation*; the Boolean multiplication or meet $\wedge$, called *conjunction*; the Boolean addition or join $\vee$, called *disjunction*, the *implication* $\Rightarrow$ and the *equivalence* $\Leftrightarrow$. More explicitly, we have

$$\neg : \mathcal{T} \to \mathcal{T} : \nu \mapsto \begin{cases} \textit{false} & ; \quad \nu = \textit{true} \\ \textit{true} & ; \quad \nu = \textit{false} \end{cases}$$

$$\wedge : \mathcal{T}^2 \to \mathcal{T} : (\nu, \mu) \mapsto \begin{cases} \textit{true} & ; \quad \nu = \mu = \textit{true} \\ \textit{false} & ; \quad \text{elsewhere} \end{cases}$$

[2] Note that there is a clear distinction between the *distribution* of a possibility measure and the *possibility distribution* of a possibilistic variable.

$$\vee : \mathcal{T}^2 \to \mathcal{T} : (\nu, \mu) \mapsto \begin{cases} \textit{false} & ; \quad \nu = \mu = \textit{false} \\ \textit{true} & ; \quad \text{elsewhere} \end{cases}$$

$$\Rightarrow : \mathcal{T}^2 \to \mathcal{T} : (\nu, \mu) \mapsto \begin{cases} \textit{false} & ; \quad \nu = \textit{true} \text{ and } \mu = \textit{false} \\ \textit{true} & ; \quad \text{elsewhere} \end{cases}$$

$$\Leftrightarrow : \mathcal{T}^2 \to \mathcal{T} : (\nu, \mu) \mapsto \begin{cases} \textit{true} & ; \quad \nu = \mu \\ \textit{false} & ; \quad \nu \neq \mu. \end{cases}$$

3 POSSIBILITY THEORY AND THE EXTENSION PRINCIPLE

3.1 Zadeh's Extension Principle

In 1965 Zadeh proposed a method for extending a mapping from a universe X to a universe Y to a mapping from $\mathcal{F}(X)$ to $\mathcal{F}(Y)$ [28, 29]. This method later received the name '*Zadeh's extension principle*'. It can be formalized as follows.

Definition 1 (Extension principle) *With a $X - Y$-mapping φ we can associate a $\mathcal{F}(X) - \mathcal{F}(Y)$-mapping $\tilde{\varphi}$, defined as follows:*

$$(\forall h \in \mathcal{F}(X))(\forall y \in Y)(\tilde{\varphi}(h) \cdot y \overset{\text{def}}{=} \sup_{\varphi(x)=y} h(x)), \tag{4.6}$$

and called the Zadeh extension of φ.

Whereas φ maps an arbitrary element x of X into the element $\varphi(x)$ of Y, $\tilde{\varphi}$ maps an arbitrary $([0,1], \leq)$-fuzzy set h in X into the $([0,1], \leq)$-fuzzy set $\tilde{\varphi}(h)$ in Y. Let furthermore A be an arbitrary subset of X, then the characteristic $X - [0,1]$-mapping χ_A of A is a $([0,1], \leq)$-fuzzy set in X. It is easily verified that $\tilde{\varphi}(\chi_A) = \chi_{\varphi(A)}$, with $\varphi(A) = \{\, \varphi(x) \mid x \in A \,\}$ the direct image of A under φ. We conclude that a Zadeh extension can be interpreted as a generalization—or, more precisely, fuzzification—of the notion of a direct image in classical set theory. In the rest of this section, we want to show that Zadeh's extension method finds its origin in the transformation of possibility measures.

3.2 Possibilistic Extensions

Let us consider a universe X, provided with an ample field $\mathcal{R}$, and a variable ξ that assumes values in X. In this paper, we shall be working with the *formal mathematical*, rather than the intuitive *notion of a variable*. This means that we assume the existence of a *basic space* Ω, provided with an ample field $\mathcal{R}_\Omega$, and a *normalized* $(L, \leq)$-possibility measure Π_Ω on $(\Omega, \mathcal{R}_\Omega)$. The distribution of Π_Ω will be denoted by π_Ω. In the formal picture, X is considered as a *sample space*, and the variable ξ is a $\mathcal{R}_\Omega$ – *$\mathcal{R}$-measurable* $\Omega - X$-mapping, i.e., a possibilistic variable in $(X, \mathcal{R})$. Possibilistic information about the values ξ may take in X is given by the possibility distribution of ξ, or, in other words, by the transformed $(L, \leq)$-possibility measure $\Pi_\xi \overset{\text{def}}{=} \Pi_\Omega{}^{(\xi)}|\mathcal{R}$ of Π_Ω on $(X, \mathcal{R})$ by the mapping ξ. Thus, for arbitrary A in $\mathcal{R}$, taking into account (4.4),

$$\Pi_\xi(A) \overset{\text{def}}{=} \Pi_\Omega(\xi^{-1}(A)) = \text{ the possibility that } \xi \text{ takes a value in } A. \quad (4.7)$$

The possibility distribution function of ξ is the distribution of Π_ξ, and will be denoted by π_ξ. We have seen in the previous section (see (4.3)) that it completely characterizes the possibilistic information Π_ξ about the values that ξ can assume in X. Note that for arbitrary x in X

$$\pi_\xi(x) = \sup_{\xi(\omega)\in[x]_\mathcal{R}} \pi_\Omega(\omega). \quad (4.8)$$

Remark that, since $\xi^{-1}(X) = \Omega$ and since Π_Ω is normalized, Π_ξ is normalized as well. The possibility distribution function π_ξ of ξ moreover is a sup-normal $(L, \leq)$-fuzzy variable in $(X, \mathcal{R})$.

Besides the universe X we shall now also consider a universe Y and a $X - Y$-mapping φ. Since ξ is a variable in X, $\varphi(\xi) = \varphi \circ \xi$ can be considered as a variable in Y, and we wonder if we can deduce possibilistic information about the values that $\varphi(\xi)$ can assume in Y. From a formal mathematical point of view, Y can also be considered as a sample space, and we take a closer look at the $\Omega - Y$-mapping $\varphi(\xi)$. Since, by definition, for arbitrary B in $\mathcal{R}^{(\varphi)}$, $\varphi^{-1}(B) \in \mathcal{R}$, and since ξ is by assumption $\mathcal{R}_\Omega - \mathcal{R}$-measurable, we deduce that $(\varphi \circ \xi)^{-1}(B) = \xi^{-1}(\varphi^{-1}(B)) \in \mathcal{R}_\Omega$. This means that $\varphi(\xi)$ is a *possibilistic variable* in $(Y, \mathcal{R}^{(\varphi)})$. Since we want information about the values that this variable $\varphi(\xi)$ can assume in Y, let us take a look at the transformed possibility measure $\Pi_{\varphi(\xi)} \overset{\text{def}}{=} \Pi_\Omega{}^{(\varphi(\xi))}|\mathcal{R}^{(\varphi)}$ of Π_Ω on $(Y, \mathcal{R}^{(\varphi)})$ by $\varphi(\xi)$, i.e., the possibility distribution of the possibilistic variable $\varphi(\xi)$ in $(Y, \mathcal{R}^{(\varphi)})$. For arbitrary B in $\mathcal{R}^{(\varphi)}$, we have, taking into account (4.4) and (4.7) and the fact that, by definition, $\varphi^{-1}(B) \in \mathcal{R}$,

$$\begin{aligned}\Pi_{\varphi(\xi)}(B) &= \Pi_\Omega((\varphi\circ\xi)^{-1}(B)) = \Pi_\Omega(\xi^{-1}(\varphi^{-1}(B)))\\ &= \Pi_\Omega{}^{(\xi)}(\varphi^{-1}(B)) = \Pi_\xi(\varphi^{-1}(B)) = \Pi_\xi{}^{(\varphi)}(B).\end{aligned}$$

We may therefore conclude that $\Pi_{\varphi(\xi)} = \Pi_\xi{}^{(\varphi)}$. Furthermore, since $\varphi^{-1}(Y) = X$ and Π_ξ is normalized, $\Pi_{\varphi(\xi)}$ is normalized as well. For the possibility distribution function $\pi_{\varphi(\xi)}$ of the possibilistic variable $\varphi(\xi)$ in $(Y, \mathcal{R}^{(\varphi)})$ we have, for arbitrary y in Y

$$\pi_{\varphi(\xi)}(y) = \Pi_{\varphi(\xi)}([\,y\,]_{\mathcal{R}^{(\varphi)}}) = \Pi_\xi(\varphi^{-1}([\,y\,]_{\mathcal{R}^{(\varphi)}})) = \sup_{\varphi(x)\in[\,y\,]_{\mathcal{R}^{(\varphi)}}} \pi_\xi(x). \tag{4.9}$$

$\pi_{\varphi(\xi)}$ is clearly a sup-normal $(L, \leq)$-fuzzy variable in $(Y, \mathcal{R}^{(\varphi)})$. Also remark that, with respect to set inclusion, $\mathcal{R}^{(\varphi)}$ is the largest possible ample field $\mathcal{R}_Y$ on Y such that φ is still $\mathcal{R} - \mathcal{R}_Y$-measurable (see (4.5)). In this sense, the choice of the ample field $\mathcal{R}^{(\varphi)}$ on Y makes the transmission of possibilistic information from X to Y as detailed as possible.

We conclude that the mapping φ can be used to transform the possibilistic information π_ξ (or Π_ξ) about the values ξ may take in X, into possibilistic information $\pi_{\varphi(\xi)}$ (or $\Pi_{\varphi(\xi)}$) about the values $\varphi(\xi)$ can assume in Y. If we look at (4.9), we are led to the introduction of a special $\mathcal{G}^{\mathcal{R}}_{(L,\leq)}(X) - \mathcal{G}^{\mathcal{R}^{(\varphi)}}_{(L,\leq)}(Y)$-mapping, directly transforming π_ξ into $\pi_{\varphi(\xi)}$: $\pi_{\varphi(\xi)} = \tilde{\varphi}(\pi_\xi)$.

Definition 2 (Possibilistic extensions) *With a $X - Y$-mapping φ we can associate a $\mathcal{G}^{\mathcal{R}}_{(L,\leq)}(X) - \mathcal{G}^{\mathcal{R}^{(\varphi)}}_{(L,\leq)}(Y)$-mapping $\tilde{\varphi}$, defined as follows:*

$$(\forall h \in \mathcal{G}^{\mathcal{R}}_{(L,\leq)}(X))(\forall y \in Y)(\tilde{\varphi}(h)\cdot y \overset{\text{def}}{=} \sup_{\varphi(x)\in[\,y\,]_{\mathcal{R}^{(\varphi)}}} h(x)). \tag{4.10}$$

$\tilde{\varphi}$ is called the $(L, \leq)$-possibilistic extension of φ. If, for whatever reason, we do not want to mention the complete lattice $(L, \leq)$ explicitly, we shall simply call $\tilde{\varphi}$ a possibilistic extension.

Corollary 1 *If the $(L, \leq)$-fuzzy variable h in $(X, \mathcal{R})$ is sup-normal, then so is the $(L, \leq)$-fuzzy variable $\tilde{\varphi}(h)$ in $(Y, \mathcal{R}^{(\varphi)})$.*

Let in particular A be an element of $\mathcal{R}$, then the characteristic $X-L$-mapping χ_A of A is of course a $(L, \leq)$-fuzzy variable in $(X, \mathcal{R})$. Furthermore, for arbitrary y in Y

$$\tilde{\varphi}(\chi_A)\cdot y = \sup_{\varphi(x)\in[\,y\,]_{\mathcal{R}^{(\varphi)}}} \chi_A(x) = \begin{cases} 1 & ; \quad A\cap\varphi^{-1}([\,y\,]_{\mathcal{R}^{(\varphi)}}) \neq \varnothing \\ 0 & ; \quad A\cap\varphi^{-1}([\,y\,]_{\mathcal{R}^{(\varphi)}}) = \varnothing. \end{cases}$$

It is easily verified that $A \cap \varphi^{-1}([y]_{\mathcal{R}^{(\varphi)}}) \neq \emptyset \Leftrightarrow y \in \bigcup_{x\in A} [\varphi(x)]_{\mathcal{R}^{(\varphi)}}$, whence, taking into account the definition of the closure operator $C_{\mathcal{R}^{(\varphi)}}$ (see subsection 2.2), $\tilde{\varphi}(\chi_A) = \chi_{C_{\mathcal{R}^{(\varphi)}}(\varphi(A))}$. We conclude that a possibilistic extension is a generalization—or fuzzification—of the notion of a direct image, taking into account certain measurability aspects.

How does this relate to Zadeh's extension principle? Clearly, when $\mathcal{R} = \mathcal{P}(X)$, and therefore also $\mathcal{R}^{(\varphi)} = \mathcal{P}(Y)$, we find for arbitrary y in Y that $[y]_{\mathcal{R}^{(\varphi)}} = \{y\}$. Furthermore, $\mathcal{G}^{\mathcal{R}}_{(L,\leq)}(X) = \mathcal{F}_{(L,\leq)}(X)$ and $\mathcal{G}^{\mathcal{R}^{(\varphi)}}_{(L,\leq)}(Y) = \mathcal{F}_{(L,\leq)}(Y)$. This tells us that Zadeh's extension method (4.6) is a special case of the possibilistic extension method (4.10) for $(L,\leq) = ([0,1],\leq)$ and $\mathcal{R} = \mathcal{P}(X)$. We have thus deduced Zadeh's extension principle from considerations about the transformation of possibility measures by a mapping. This also provides us with an interpretation for the Zadeh extension of such a mapping.

Another question now comes to mind: what happens if two or more such transformations of possibility are concatenated? The following theorem provides an answer. It has a graphical representation in figure 1.

Theorem 1 (Chain rule for possibilistic extensions) *Let X, Y and Z be universes. Let $\mathcal{R}$ be an ample field on X. φ is a $X-Y$-mapping and ψ a $Y-Z$-mapping. We define the following $(L,\leq)$-possibilistic extensions:*

$$\tilde{\varphi}: \mathcal{G}^{\mathcal{R}}_{(L,\leq)}(X) \to \mathcal{G}^{\mathcal{R}^{(\varphi)}}_{(L,\leq)}(Y): h \mapsto \tilde{\varphi}(h),$$

$$\tilde{\psi}: \mathcal{G}^{\mathcal{R}^{(\varphi)}}_{(L,\leq)}(Y) \to \mathcal{G}^{(\mathcal{R}^{(\varphi)})^{(\psi)}}_{(L,\leq)}(Z): g \mapsto \tilde{\varphi}(g)$$

$$\widetilde{\psi\circ\varphi}: \mathcal{G}^{\mathcal{R}}_{(L,\leq)}(X) \to \mathcal{G}^{\mathcal{R}^{(\psi\circ\varphi)}}_{(L,\leq)}(Z): h \mapsto (\widetilde{\psi\circ\varphi})(h)$$

with, of course,

$$(\forall y \in Y)(\tilde{\varphi}(h)\cdot y \stackrel{\text{def}}{=} \sup_{x\in\varphi^{-1}([y]_{\mathcal{R}^{(\varphi)}})} h(x))$$

$$(\forall z \in Z)(\tilde{\psi}(g)\cdot z \stackrel{\text{def}}{=} \sup_{y\in\psi^{-1}([z]_{(\mathcal{R}^{(\varphi)})^{(\psi)}})} g(y))$$

$$(\forall z \in Z)((\widetilde{\psi\circ\varphi})(h)\cdot z \stackrel{\text{def}}{=} \sup_{x\in(\psi\circ\varphi)^{-1}([z]_{\mathcal{R}^{(\psi\circ\varphi)}})} h(x)).$$

Then (i) $\mathcal{R}^{(\psi\circ\varphi)} = (\mathcal{R}^{(\varphi)})^{(\psi)}$*; and (ii)* $\widetilde{\psi\circ\varphi} = \tilde{\psi}\circ\tilde{\varphi}$*. In particular, this means that the right-most diagram in figure 1 commutes*[3].

Proof. Let us first show that (i) holds. Consider an arbitrary $C \subseteq Z$. Then, taking into account (4.4),

$$\begin{aligned} C \in \mathcal{R}^{(\psi\circ\varphi)} &\Leftrightarrow (\psi\circ\varphi)^{-1}(C) \in \mathcal{R} \Leftrightarrow \varphi^{-1}(\psi^{-1}(C)) \in \mathcal{R} \\ &\Leftrightarrow \psi^{-1}(C) \in \mathcal{R}^{(\varphi)} \Leftrightarrow C \in (\mathcal{R}^{(\varphi)})^{(\psi)}. \end{aligned}$$

This indeed implies that $\mathcal{R}^{(\psi\circ\varphi)} = (\mathcal{R}^{(\varphi)})^{(\psi)}$. We now proceed to prove (ii). Consider an arbitrary $(L, \leq)$-fuzzy variable h in $(X, \mathcal{R})$ and an arbitrary z in Z. We know, taking into account (i) and (4.1), that $[\,z\,]_{\mathcal{R}^{(\psi\circ\varphi)}} \in (\mathcal{R}^{(\varphi)})^{(\psi)}$ which is, by definition, equivalent with $\psi^{-1}([\,z\,]_{\mathcal{R}^{(\psi\circ\varphi)}}) \in \mathcal{R}^{(\varphi)}$, and, taking into account (4.2), also equivalent with

$$\psi^{-1}([\,z\,]_{\mathcal{R}^{(\psi\circ\varphi)}}) = \bigcup_{y\in\psi^{-1}([\,z\,]_{\mathcal{R}^{(\psi\circ\varphi)}})} [\,y\,]_{\mathcal{R}^{(\varphi)}}.$$

This implies, taking into account the properties of inverse images, that

$$\begin{aligned} (\psi\circ\varphi)^{-1}([\,z\,]_{\mathcal{R}^{(\psi\circ\varphi)}}) &= \varphi^{-1}(\psi^{-1}([\,z\,]_{\mathcal{R}^{(\psi\circ\varphi)}})) \\ &= \bigcup_{y\in\psi^{-1}([\,z\,]_{\mathcal{R}^{(\psi\circ\varphi)}})} \varphi^{-1}([\,y\,]_{\mathcal{R}^{(\varphi)}}). \end{aligned}$$

If we take into account the associativity of supremum in the complete lattice $(L, \leq)$, (i) and the definitions of $\tilde{\varphi}$ and $\tilde{\psi}$, this means that

$$\begin{aligned} (\widetilde{\psi\circ\varphi})(h)\cdot z &= \sup_{x\in(\psi\circ\varphi)^{-1}([\,z\,]_{\mathcal{R}^{(\psi\circ\varphi)}})} h(x) \\ &= \sup_{x\in \bigcup_{y\in\psi^{-1}([\,z\,]_{\mathcal{R}^{(\psi\circ\varphi)}})} \varphi^{-1}([\,y\,]_{\mathcal{R}^{(\varphi)}})} h(x) \\ &= \sup_{y\in\psi^{-1}([\,z\,]_{\mathcal{R}^{(\psi\circ\varphi)}})} \; \sup_{x\in\varphi^{-1}([\,y\,]_{\mathcal{R}^{(\varphi)}})} h(x) \\ &= \sup_{y\in\psi^{-1}([\,z\,]_{\mathcal{R}^{(\psi\circ\varphi)}})} \tilde{\varphi}(h)\cdot y \\ &= \tilde{\psi}(\tilde{\varphi}(h))\cdot z \\ &= (\tilde{\psi}\circ\tilde{\varphi})(h)\cdot z, \end{aligned}$$

whence indeed $\widetilde{\psi\circ\varphi} = \tilde{\psi}\circ\tilde{\varphi}$. □

[3] To use the lingo of fuzzy set theory, this result tells us that the composition of mappings can be 'fuzzified'.

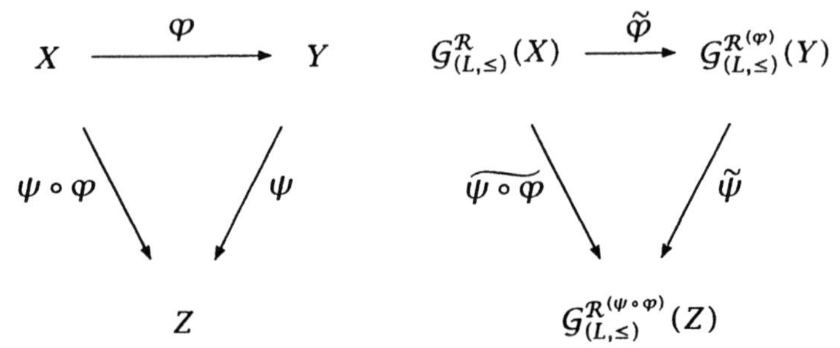

Figure 1 Commutative diagrams: the chain rule for possibilistic extensions.

3.3 Possibilistic t-norm-extensions

We shall now turn to the discussion of an interesting special case, that will play a prominent role in the discussion of possibilistic logic in the following sections. Let us consider, besides the basic space Ω, n ($n \in \mathbb{N} \setminus \{0\}$) universes X_k, each provided with an ample field $\mathcal{R}_k$ ($k = 1 \ldots, n$). Furthermore, we shall denote by ξ_k a possibilistic variable in $(X_k, \mathcal{R}_k)$, i.e. a $\mathcal{R}_\Omega - \mathcal{R}_k$-measurable $\Omega - X_k$-mapping. We may also consider the product universe $X_1 \times \cdots \times X_n$, provided with the ample field $\mathcal{R}_1 \times \cdots \times \mathcal{R}_n$, i.e., the product of the ample fields $\mathcal{R}_1, \ldots, \mathcal{R}_n$ (see subsection 2.2). It is easily proven that the $\Omega - X_1 \times \cdots \times X_n$-mapping $(\xi_1, \ldots, \xi_n)$ is a possibilistic variable in $(X_1 \times \cdots \times X_n, \mathcal{R}_1 \times \cdots \times \mathcal{R}_n)$ (see [4, 9]). On the one hand, for $k = 1, \ldots, n$ we may now consider the normalized $(L, \leq)$-possibility measure $\Pi_{\xi_k} \stackrel{\text{def}}{=} \Pi_\Omega{}^{(\xi_k)}|\mathcal{R}_k$ with distribution π_{ξ_k}. Of course, π_{ξ_k} is the possibility distribution function of the possibilistic variable ξ_k. On the other hand, we may also consider the normalized $(L, \leq)$-possibility measure $\Pi_{(\xi_1,\ldots,\xi_n)} \stackrel{\text{def}}{=} \Pi_\Omega{}^{((\xi_1,\ldots,\xi_n))}|\mathcal{R}_1 \times \cdots \times \mathcal{R}_n$ with distribution $\pi_{(\xi_1,\ldots,\xi_n)}$. $\pi_{(\xi_1,\ldots,\xi_n)}$ is the possibility distribution function of the possibilistic variable $(\xi_1, \ldots, \xi_n)$. We have shown in our treatment of *possibilistic independence* [5, 9, 15] that the possibilistic variables $\xi_1, \ldots, \xi_n$ are (Π_Ω, T)-independent (or, shortly, possibilistically independent) if and only if

$$(\forall (x_1, \ldots, x_n) \in X_1 \times \cdots \times X_n)(\pi_{(\xi_1,\ldots,\xi_n)}(x_1, \ldots, x_n) = T_{k=1}^n \pi_{\xi_k}(x_k)) \tag{4.11}$$

As a next step in this course of reasoning, we consider the $X_1 \times \cdots \times X_n - Y$-mapping φ. On the basis of the discussion in the previous subsection, with in particular $X = X_1 \times \cdots \times X_n$ and $\mathcal{R} = \mathcal{R}_1 \times \cdots \times \mathcal{R}_n$, we know that the $\Omega - Y$-mapping $\varphi(\xi_1, \ldots, \xi_n) = \varphi \circ (\xi_1, \ldots, \xi_n)$ is a possibilistic variable in $(Y, (\mathcal{R}_1 \times \cdots \times \mathcal{R}_n)^{(\varphi)})$ and furthermore $\Pi_{\varphi(\xi_1,\ldots,\xi_n)} = \Pi_{(\xi_1,\ldots,\xi_n)}{}^{(\varphi)}$. For arbitrary y in Y

$$\pi_{\varphi(\xi_1,\ldots,\xi_n)}(y) = \sup_{\varphi(x_1,\ldots,x_n)\in[y]_{(\mathcal{R}_1\times\cdots\times\mathcal{R}_n)^{(\varphi)}}} \pi_{(\xi_1,\ldots,\xi_n)}(x_1,\ldots,x_n).$$

When the possibilistic variables $\xi_1, \ldots, \xi_n$ are (Π_Ω, T)-independent, we find for arbitrary y in Y that, taking into account (4.11),

$$\pi_{\varphi(\xi_1,\ldots,\xi_n)}(y) = \sup_{\varphi(x_1,\ldots,x_n)\in[y]_{(\mathcal{R}_1\times\cdots\times\mathcal{R}_n)^{(\varphi)}}} T_{k=1}^n \pi_{\xi_k}(x_k).$$

As in the previous subsection, we are thus led to the following definition.

Definition 3 (t-norm-extensions) *With a $X_1 \times \cdots \times X_n - Y$-mapping φ we can associate a $\mathcal{G}_{(L,\leq)}^{\mathcal{R}_1}(X_1) \times \ldots \times \mathcal{G}_{(L,\leq)}^{\mathcal{R}_n}(X_n) - \mathcal{G}_{(L,\leq)}^{(\mathcal{R}_1\times\cdots\times\mathcal{R}_n)^{(\varphi)}}(Y)$-mapping $\tilde{\varphi}_T$, defined as follows:*

$$(\forall (h_1,\ldots,h_n) \in \mathcal{G}_{(L,\leq)}^{\mathcal{R}_1}(X_1) \times \ldots \times \mathcal{G}_{(L,\leq)}^{\mathcal{R}_n}(X_n))$$
$$(\forall y \in Y)(\tilde{\varphi}_T(h_1,\ldots,h_n) \cdot y \stackrel{\text{def}}{=} \sup_{\varphi(x_1,\ldots,x_n)\in[y]_{(\mathcal{R}_1\times\cdots\times\mathcal{R}_n)^{(\varphi)}}} T_{k=1}^n h_k(x_k)).$$

$\tilde{\varphi}_T$ is called the $(L,\leq)$-possibilistic T-extension of φ. If, for whatever reason, we do not want to mention the complete lattice $(L,\leq)$ and/or the triangular norm T on $(L,\leq)$ explicitly, we shall simply call $\tilde{\varphi}_T$ a possibilistic t-norm-extension.

Possibilistic t-norm-extensions are used when we have at our disposal possibilistic information in the form of the possibility distribution functions π_{ξ_k}—essentially sup-normal $(L,\leq)$-fuzzy variables in $(X_k, \mathcal{R}_k)$—of the *possibilistically independent* possibilistic variables ξ_k. This information can be transmitted from the universe $X_1 \times \cdots \times X_n$ to the universe Y by a $X_1 \times \cdots \times X_n - Y$-mapping φ. The possibility distribution function of the possibilistic variable $\varphi(\xi_1,\ldots,\xi_n)$ is then given by $\pi_{\varphi(\xi_1,\ldots,\xi_n)} = \tilde{\varphi}_T(\pi_{\xi_1},\ldots, \pi_{\xi_n})$, and constitutes information about the values that $\varphi(\xi_1,\ldots,\xi_n)$ may assume in the universe Y.

Corollary 2 *When the $(L,\leq)$-fuzzy variables $h_1, \ldots, h_n$ in $(X_1,\mathcal{R}_1), \ldots, (X_n,\mathcal{R}_n)$ respectively are sup-normal, $\tilde{\varphi}_T(h_1,\ldots,h_n)$ is a sup-normal $(L,\leq)$-fuzzy variable in $(Y, (\mathcal{R}_1 \times \cdots \times \mathcal{R}_n)^{(\varphi)})$.*

Let us now give a few examples in order better to understand the notions introduced thus far.

Example 1 Let $(L, \leq) = ([0,1], \leq)$, $X_1 = X_2 = Y = \mathbb{R}$ and $\mathcal{R}_1 = \mathcal{R}_2 = \mathcal{P}(\mathbb{R})$. A $([0,1], \leq)$-fuzzy variable in $(\mathbb{R}, \mathcal{P}(\mathbb{R}))$, or equivalently, a $([0,1], \leq)$-fuzzy set in $\mathbb{R}$ is also called a *fuzzy quantity* (see, for instance, [21] section 3.1). The $X_1 \times X_2 - Y$-mapping φ we consider here, is the additive operation on the reals, i.e., $\varphi = +$, with $([0,1], \leq)$-possibilistic T-extension $\tilde{+}_T$:

$$\tilde{+}_T \colon \mathcal{F}(\mathbb{R})^2 \to \mathcal{F}(\mathbb{R}) \colon (h, g) \mapsto h \,\tilde{+}_T\, g,$$

with, for arbitrary c in $\mathbb{R}$,

$$(h \,\tilde{+}_T\, g) \cdot c = \sup_{a+b=c} T(h(a), g(b)) = \sup_{a \in \mathbb{R}} T(h(a), g(c-a)). \tag{4.12}$$

When π_ξ and π_ζ represent possibilistic information about the values that two $([0,1], \leq)$-possibilistically T-independent possibilistic variables ξ and ζ may assume in $\mathbb{R}$, then $\pi_\xi \,\tilde{+}_T\, \pi_\zeta$ represents possibilistic information about the values that the possibilistic variable $\xi + \zeta$ may assume in $\mathbb{R}$.

Example 2 Besides possibilistic information, we may also consider probabilistic information. Let f_ξ and f_ζ be the probability density functions of two continuous real stochastic variables ξ en ζ, that are furthermore stochastically independent. It is well known that the probability density function $f_{\xi+\zeta}$ of the sum $\xi + \zeta$ can be written as (see, for instance, [2] section 12-4 and exercise 12-29)

$$f_{\xi+\zeta} \colon \mathbb{R} \to [0,1] \colon c \mapsto (f_\xi * f_\zeta) \cdot c$$

with

$$(f_\xi * f_\zeta) \cdot c = \int_{-\infty}^{+\infty} f_\xi(a) f_\zeta(c-a) da. \tag{4.13}$$

In other words, $f_{\xi+\zeta}$ is the *convolution product* $f_\xi * f_\zeta$ of f_ξ and f_ζ. Notice the striking formal analogy between (4.12) and (4.13), which is certainly not a coincidence. Both formulas are derived from considerations about transformations of measures. This again indicates that Zadeh's extension principle is by no means an isolated *ad hoc* principle, but that it rather has its natural place in a much broader measure-theoretical context.

4 POSSIBILISTIC TRUTH VALUES AND THEIR COMBINATIONS

4.1 Possibilistic Truth Values

Let us apply the results of the previous section to the problem of representing linguistic uncertainty in classical propositional logic, briefly described in the introduction. As far as we know, Van Schooten was the first to study this problem in his doctoral dissertation [27]. It must however also be mentioned that Gaines has briefly discussed it in his important article about approximate reasoning [19].

We shall first give a fairly general description of the problem. For a start, consider a *property* (or predicate) p of the elements of a universe X, that is *clear*: for every object x in X we have that x either completely satisfies or completely does not satisfy p (for more details, we refer to [7, 8, 9]). With every x in X, we may therefore associate a *proposition* $P_p(x) \overset{\text{def}}{=}$ 'x is p', i.e., an affirmative sentence that is either true or false. P_p can be considered as a mapping from the universe X to an appropriate set of propositions, and is therefore also called a *proposition function*. With P_p we can also associate the set A_{P_p} of the objects that satisfy p: $A_{P_p} \overset{\text{def}}{=} \{ x \mid x \in X \text{ and } x \text{ is } p \}$. Moreover, A_{P_p} can also be characterized by its characteristic $X - \mathcal{T}$-mapping:

$$\chi_{A_{P_p}} : X \to \mathcal{T} : x \mapsto \begin{cases} \textit{true} & ; \quad x \text{ is } p \\ \textit{false} & ; \quad x \text{ is not } p. \end{cases}$$

For arbitrary x in X, $\chi_{A_{P_p}}(x)$ is the *truth value* of the proposition 'x is p'.

Example 3 Let $X = \mathbb{R}$ and let p be the clear predicate 'greater than or equal to 10'. Then, of course, $A_{P_p} = \{ a \mid a \in \mathbb{R} \text{ and } a \geq 10 \}$ and for arbitrary a in $\mathbb{R}$

$$\chi_{A_{P_p}}(a) = \begin{cases} \textit{true} & ; \quad a \geq 10 \\ \textit{false} & ; \quad a < 10. \end{cases}$$

So far, we are able unequivocally to associate with every object x in X the truth value $\chi_{A_{P_p}}(x)$ of the proposition $P_p(x)$. In the next step of our course of reasoning, we shall introduce *linguistic uncertainty* into the picture. Let us assume that on the universe X there is defined an ample field $\mathcal{R}$ of measurable subsets of X. We consider a possibilistic variable ξ in $(X, \mathcal{R})$. As in the previous section, this means that we assume the existence of a basic space Ω, provided with an ample field $\mathcal{R}_\Omega$ and a normalized possibility measure Π_Ω. X is considered as a sample space, and ξ is a $\mathcal{R}_\Omega - \mathcal{R}$-measurable $\Omega - X$-mapping. Possibilistic information about the values that the possibilistic variable ξ may assume in X is given in the form of the possibility distribution Π_ξ (see (4.7)) or the possibility distribution function π_ξ of ξ (see (4.8)). Given this information, it is fairly natural to ask what information may be derived about the truth value $\chi_{A_{P_p}}(\xi)$ of the *proposition variable* $P_p \circ \xi = P_p(\xi) \stackrel{\text{def}}{=}$ 'ξ is p'. From the results of the previous section, we now deduce as a special case, with $Y = \mathcal{T}$ and $\varphi = \chi_{A_{P_p}}$, the following conclusions. First of all, the truth value $\chi_{A_{P_p}}(\xi)$ of the proposition variable $P_p(\xi)$ is a possibilistic variable in $(\mathcal{T}, \mathcal{R}^{(\chi_{A_{P_p}})})$. Secondly, its possibility distribution $\Pi_{\chi_{A_{P_p}}(\xi)} = \Pi_\xi^{(\chi_{A_{P_p}})}$ with distribution $\pi_{\chi_{A_{P_p}}(\xi)}$ is a normalized $(L, \leq)$-possibility measure on $(\mathcal{T}, \mathcal{R}^{(\chi_{A_{P_p}})})$, that gives possibilistic information about the values that the variable $\chi_{A_{P_p}}(\xi)$ may assume in $\mathcal{T}$, or in other words, about the truth value of the proposition variable 'ξ is p'. Finally, the $(L, \leq)$-possibilistic extension $\widetilde{\chi_{A_{P_p}}}$ of $\chi_{A_{P_p}}$ is a $\mathcal{G}^{\mathcal{R}}_{(L,\leq)}(X) - \mathcal{G}^{\mathcal{R}^{(\chi_{A_{P_p}})}}_{(L,\leq)}(\mathcal{T})$-mapping that transforms possibilistic information—in the form of $(L, \leq)$-fuzzy variables in $(X, \mathcal{R})$—about the values ξ may assume in X, into possibilistic information—in the form of $(L, \leq)$-fuzzy variables in $(\mathcal{T}, \mathcal{R}^{(\chi_{A_{P_p}})})$—about the values that $\chi_{A_{P_p}}(\xi)$ may assume in $\mathcal{T}$: $\pi_{\chi_{A_{P_p}}(\xi)} = \widetilde{\chi_{A_{P_p}}}(\pi_\xi)$. We are thus led to the following important definition.

Definition 4 (Possibilistic truth values) *If the values a variable ξ may assume in X are restricted by the possibilistic information h—in the form of a sup-normal $(L, \leq)$-fuzzy variable in $(X, \mathcal{R})$—then $\widetilde{\chi_{A_{P_p}}}(h)$—a sup-normal $(L, \leq)$-fuzzy variable in $(\mathcal{T}, \mathcal{R}^{(\chi_{A_{P_p}})})$—is called the $(L, \leq)$-possibilistic truth value of the proposition variable 'ξ is p'.*

In the next subsection, we shall take the next step in our course of reasoning, and investigate how the logical combination of propositions (and proposition variables) may lead to an appropriate *combination of possibilistic truth values*. In the rest of this subsection, we shall add a little more detail to the picture that has been sketched thus far.

First of all, it is important to note that we only work with properties p that are clear. This means that this discussion lies well within the province of *classical propositional logic*. What we are trying to do here is to add linguistic uncertainty to the classical description. Secondly, it should be clear that the information we derive about the values that $\chi_{A_{P_p}}(\xi)$ takes in $\mathcal{T}$ is really only useful if we are able to separate the information about 'ξ is p' being true on the one hand, and about its being false on the other hand. In other words, $\{true\}$ and $\{false\}$ must each be $\mathcal{R}^{(\chi_{A_{P_p}})}$-measurable sets, or equivalently, we must have that $\mathcal{R}^{(\chi_{A_{P_p}})} = \mathcal{P}(\mathcal{T})$. If not, then clearly $\mathcal{R}^{(\chi_{A_{P_p}})} = \{\emptyset, \mathcal{T}\}$ and we find for an arbitrary sup-normal $(L, \leq)$-fuzzy variable h only the following, hardly relevant, information:

$$\begin{cases} \widetilde{\chi_{A_{P_p}}}(h) \cdot true = 1 \\ \widetilde{\chi_{A_{P_p}}}(h) \cdot false = 1, \end{cases}$$

since in this case $[\, true \,]_{\mathcal{R}^{(\chi_{A_{P_p}})}} = [\, false \,]_{\mathcal{R}^{(\chi_{A_{P_p}})}} = \mathcal{T}$. The following proposition sheds more light on this observation.

Proposition 1 *Let X be a universe and let $\mathcal{R}$ be an ample field on X. Let furthermore A be an arbitrary subset of X, with characteristic $X - \mathcal{T}$-mapping χ_A. Then $\mathcal{R}^{(\chi_A)} = \mathcal{P}(\mathcal{T}) \Leftrightarrow A \in \mathcal{R}$.*

Proof. It is clear that '$\mathcal{R}^{(\chi_A)} = \mathcal{P}(\mathcal{T})$' is equivalent with '$\{true\} \in \mathcal{R}^{(\chi_A)}$', which is in turn equivalent with $\chi_A^{-1}(\{true\}) \in \mathcal{R}$. Moreover, $\chi_A^{-1}(\{true\}) = A$. □

This means that the separation of information about the proposition variable 'ξ is p' being true or its being false is only possible if A_{P_p} is $\mathcal{R}$-measurable, or, in other words, if, looking through the glasses of $\mathcal{R}$, we may distinguish between objects in X that satisfy p and objects that do not. *In the rest of this paper, we shall assume that this is indeed the case,* which in particular implies that

$$\mathcal{G}_{(L,\leq)}^{\mathcal{R}^{(\chi_{A_{P_p}})}}(\mathcal{T}) = \mathcal{G}_{(L,\leq)}^{\mathcal{P}(\mathcal{T})}(\mathcal{T}) = \mathcal{F}_{(L,\leq)}(\mathcal{T}),$$

and that $\widetilde{\chi_{A_{P_p}}}$ is a $\mathcal{G}_{(L,\leq)}^{\mathcal{R}}(X) - \mathcal{F}_{(L,\leq)}(\mathcal{T})$-mapping.

To summarize, if the information about the values that ξ may assume in X is given by a sup-normal $(L, \leq)$-fuzzy variable π_ξ in $(X, \mathcal{R})$, we now know that for arbitrary ν in $\mathcal{T}$

$$\pi_{\chi_{A_{P_p}}(\xi)}(\nu) = \widetilde{\chi_{A_{P_p}}}(\pi_\xi) \cdot \nu = \sup_{\chi_{A_{P_p}}(x)=\nu} \pi_\xi(x).$$

The following example will illustrate the course of reasoning established thus far.

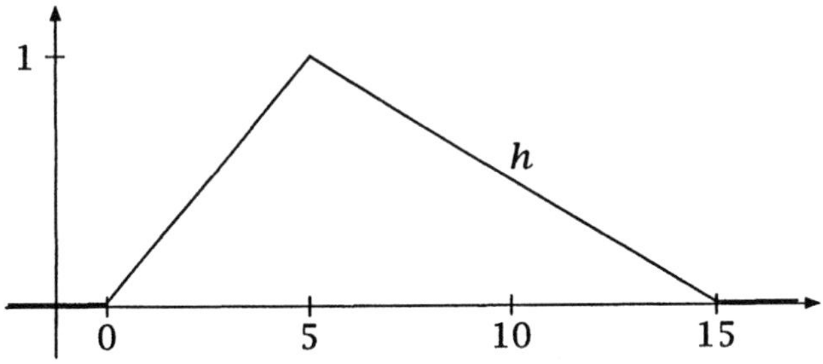

Figure 2 The fuzzy quantity h from example 4.

Example 4 Let $X = \mathbb{R}$, $\mathcal{R} = \mathcal{P}(\mathbb{R})$, $(L, \leq) = ([0,1], \leq)$ and, as in example 3, let p be the predicate 'greater than or equal to 10'. Let furthermore ξ be a real possibilistic variable, with possibility distribution function h, the triangular fuzzy quantity depicted in figure 2: for arbitrary a in $\mathbb{R}$, $h(a)$ is the $([0,1], \leq)$-possibility that ξ assumes the value a. We now ask ourselves what is the $([0,1], \leq)$-possibilistic truth value of the proposition 'ξ is greater then or equal to 10'? From the considerations above we deduce that

$$\begin{cases} \widetilde{\chi_{A_{P_p}}}(h) \cdot \mathit{true} = \sup_{a \geq 10} h(a) = 1/2 \\ \widetilde{\chi_{A_{P_p}}}(h) \cdot \mathit{false} = \sup_{a < 10} h(a) = 1. \end{cases}$$

The $([0,1], \leq)$-possibilistic truth value of the proposition '$\xi \geq 10$' is therefore $\{(\mathit{true}, 1/2), (\mathit{false}, 1)\}$, and this must be interpreted as follows: given the possibilistic information h about the values ξ can assume in $\mathbb{R}$, the $([0,1], \leq)$-possibility that the proposition '$\xi \geq 10$' is true is equal to $1/2$ and the $([0,1], \leq)$-possibility that it is false is equal to 1.

To close this subsection, let us note that, until now, we have always considered the case of just one variable ξ assuming values in a universe X. It is of course possible to extend these observations to the case of more than one, say n, variables $\xi_1, \ldots, \xi_n$ assuming values in the respective universes $X_1, \ldots, X_n$ $(n \in \mathbb{N} \setminus \{0\})$. This is not at all difficult since the treatment of n variables can be considered as a special case of the treatment of one variable by putting $\xi = (\xi_1, \ldots, \xi_n)$ and $X = X_1 \times \cdots \times X_n$. Due to limitations of space, we shall however in this paper leave such an extension implicit. For more details, we refer to our doctoral dissertation [9].

4.2 Combination of Possibilistic Truth Values

In the previous subsection, we have shown how possibilistic information about the values a variable may assume in a universe can be transformed into possibilistic information about the truth value of a proposition about this variable. To go to the next stage in our discussion, we observe that, in general, propositions can be combined to form new propositions, using so-called *logical operators*. In this way, a proposition P can be transformed by the *logical negation operator* into a new proposition NOT P. By extension, the proposition function P_p discussed in the previous subsection can be transformed into a new proposition function NOT P_p by the pointwise application of the logical negation operator:

$$(\forall x \in X)((\text{NOT } P_p)(x) \stackrel{\text{def}}{=} \text{NOT}(P_p(x)) = \text{`}x \text{ is not } p\text{'}).$$

In a completely similar way, the proposition variable $P_p(\xi) \stackrel{\text{def}}{=}$ 'ξ is p' is transformed by the logical negation operator into the new proposition variable $(\text{NOT } P_p)(\xi)$, defined as 'ξ is not p'. Analogously, the proposition variables 'ξ is p' and 'ξ is q' can be transformed into the new proposition variable $(P_p \text{ AND } P_q)(\xi)$, defined as '$\xi$ is p AND ξ is q', using the logical conjunction operator AND, and can be transformed into the new proposition variable $(P_p \text{ OR } P_q)(\xi)$, defined as '$\xi$ is p OR ξ is q', using the logical disjunction operator OR.

As is well known, classical propositional logic is *truth-functional*, and the behaviour of logical operators can be characterized by what we shall call further on classical-logical functions, i.e., in general $\mathcal{T}^n - \mathcal{T}$-mappings, with $n \in \mathbb{N} \setminus \{0\}$. To give an example, the behaviour of the *logical negation operator* is mirrored in the behaviour of the complement operator $\neg$ on the Boolean chain $(\mathcal{T}, \leq)$, in the following sense:

$$(\forall x \in X)(\chi_{A_{\text{NOT } P_p}}(x) = \neg\chi_{A_{P_p}}(x)),$$

where, of course $A_{\text{NOT } P_p} = \{x \mid x \in X \text{ and } x \text{ is not } p\} = \text{co}A_{P_p}$. In a completely analogous way, the behaviour of the *logical disjunction operator* is mirrored in the behaviour of the join $\vee$ of the Boolean chain $(\mathcal{T}, \leq)$, in the following sense:

$$(\forall x \in X)(\chi_{A_{P_p \text{ OR } P_q}}(x) = \chi_{A_{P_p}}(x) \vee \chi_{A_{P_q}}(x)),$$

where $A_{P_p \text{ OR } P_q} = \{x \mid x \in X \text{ and } (x \text{ is } p \text{ OR } x \text{ is } q)\} = A_{P_p} \cup A_{P_q}$.

Generally speaking, we can start with n ($n \in \mathbb{N} \setminus \{0\}$) clear predicates p_1, ..., p_n with associated sets $A_{P_{p_1}}, \ldots, A_{P_{p_n}}$, and with a n-ary logical operator LOP. This logical operator transforms the proposition variables $P_{p_1}(\xi)$, ..., $P_{p_n}(\xi)$ into the new proposition variable $\mathrm{LOP}(P_{p1}, \ldots, P_{p_n})(\xi)$, defined as LOP('$\xi$ is p_1',...,'ξ is p_n'), with associated set $A_{\mathrm{LOP}(P_{p1},\ldots,P_{pn})}$. The behaviour of LOP is mirrored by a $\mathcal{T}^n - \mathcal{T}$-mapping ϕ, in the following sense:

$$(\forall x \in X)(\chi_{A_{\mathrm{LOP}(P_{p1},\ldots,P_{pn})}}(x) = \phi(\chi_{A_{P_{p_1}}}(x), \ldots, \chi_{A_{P_{p_n}}}(x))),$$

or, shortly, $\chi_{A_{\mathrm{LOP}(P_{p1},\ldots,P_{pn})}} = \phi \circ (\chi_{A_{P_{p_1}}}, \ldots, \chi_{A_{P_{p_n}}})$, where, as usual, the characteristic $X - \mathcal{T}$-mapping of $A_{P_{p_k}}$ is denoted by $\chi_{A_{P_{p_k}}}$ ($k = 1, \ldots, n$), and $\chi_{A_{\mathrm{LOP}(P_{p1},\ldots,P_{pn})}}$ is the characteristic $X - \mathcal{T}$-mapping of $A_{\mathrm{LOP}(P_{p1},\ldots,P_{pn})}$.

The problem we will treat in this subsection can now be briefly formulated as follows: *how, starting with possibilistic information about the values that a possibilistic variable ξ may assume in X, can we derive the possibilistic truth value of the combined proposition variable* LOP('ξ is p_1',...,'ξ is p_n')? In order to give an answer to this question, we again apply the general results of the previous section.

For a start, the sets $A_{P_{p_1}}, \ldots, A_{P_{p_n}}$ will, as explained in the previous subsection, be assumed $\mathcal{R}$-measurable. Let us now define the $X - \mathcal{T}^n$-mapping χ as follows: $\chi \stackrel{\mathrm{def}}{=} (\chi_{A_{P_{p_1}}}, \ldots, \chi_{A_{P_{p_n}}})$. Since the sets $A_{P_{p_1}}, \ldots, A_{P_{p_n}}$ are assumed to be $\mathcal{R}$-measurable, it is easily verified that

$$\mathcal{R}^{(\chi)} = \mathcal{P}(\mathcal{T}^n). \tag{4.14}$$

Using the results of the previous section with $Y = \mathcal{T}^n$ and $\varphi = \chi$, we find that the $\Omega - \mathcal{T}^n$-mapping $\chi \circ \xi = \chi(\xi)$ is a possibilistic variable in $(\mathcal{T}^n, \mathcal{P}(\mathcal{T}^n))$. For the possibility distribution $\Pi_{\chi(\xi)}$ of this variable, we find that

$$\Pi_{\chi(\xi)} = \Pi_\xi{}^{(\chi)}, \tag{4.15}$$

and for the possibility distribution function $\pi_{\chi(\xi)}$ of $\chi(\xi)$ we have that $\pi_{\chi(\xi)} = \tilde{\chi}(\pi_\xi)$, or, for arbitrary $(v_1, \ldots, v_n)$ in $\mathcal{T}^n$:

$$\begin{aligned}\pi_{\chi(\xi)}(v_1, \ldots, v_n) &= \sup_{\chi(x)=(v_1,\ldots,v_n)} \pi_\xi(x) \\ &= \sup_{(\chi_{A_{P_{p_1}}}(x),\ldots,\chi_{A_{P_{p_n}}}(x))=(v_1,\ldots,v_n)} \pi_\xi(x) \\ &= \sup_{x \in \bigcap_{k=1}^{n} \chi_{A_{P_{p_k}}}{}^{-1}(\{v_k\})} \pi_\xi(x).\end{aligned} \tag{4.16}$$

In a second stage, we also bring the $\mathcal{T}^n - \mathcal{T}$-mapping ϕ into the picture, and apply the results of the previous section in the following special case: $\varphi = \chi_{A_{\text{LOP}(p_{p_1},\ldots,p_{p_n})}} = \phi \circ \chi$ and $Y = \mathcal{T}$. $\phi \circ \chi$ is a $X - \mathcal{T}$-mapping, for which, taking into account theorem 1 and (4.14), $\mathcal{R}^{(\phi\circ\chi)} = (\mathcal{R}^{(\chi)})^{(\phi)} = (\mathcal{P}(\mathcal{T}^n))^{(\phi)} = \mathcal{P}(\mathcal{T})$. It now follows from the course of reasoning in the previous section that the $\Omega - \mathcal{T}$-mapping $\chi_{A_{\text{LOP}(p_{p_1},\ldots,p_{p_n})}}(\xi) = (\phi \circ \chi) \circ \xi = (\phi \circ \chi)(\xi)$ is a possibilistic variable in $(\mathcal{T}, \mathcal{P}(\mathcal{T}))$. For the possibility distribution $\Pi_{\chi_{A_{\text{LOP}(p_{p_1},\ldots,p_{p_n})}}(\xi)}$ of $\chi_{A_{\text{LOP}(p_{p_1},\ldots,p_{p_n})}}(\xi)$ we find that $\Pi_{\chi_{A_{\text{LOP}(p_{p_1},\ldots,p_{p_n})}}(\xi)} = \Pi_{(\phi\circ\chi)(\xi)} = {\Pi_\xi}^{(\phi\circ\chi)}$, or equivalently, for arbitrary B in $\mathcal{P}(\mathcal{T})$, taking into account (4.15):

$$\begin{aligned}
\Pi_{\chi_{A_{\text{LOP}(p_{p_1},\ldots,p_{p_n})}}(\xi)}(B) &= {\Pi_\xi}^{(\phi\circ\chi)}(B) \\
&= \Pi_\xi((\phi\circ\chi)^{-1}(B)) \\
&= \Pi_\xi(\chi^{-1}(\phi^{-1}(B))) \\
&= \Pi_{\chi(\xi)}(\phi^{-1}(B)) \\
&= {\Pi_{\chi(\xi)}}^{(\phi)}(B),
\end{aligned}$$

since clearly $\phi^{-1}(B) \in \mathcal{P}(\mathcal{T}^n)$. Hence, $\Pi_{\chi_{A_{\text{LOP}(p_{p_1},\ldots,p_{p_n})}}(\xi)} = {\Pi_{\chi(\xi)}}^{(\phi)}$. For the possibility distribution function $\pi_{\chi_{A_{\text{LOP}(p_{p_1},\ldots,p_{p_n})}}(\xi)}$ of the possibilistic variable $\chi_{A_{\text{LOP}(p_{p_1},\ldots,p_{p_n})}}(\xi)$ we find, for arbitrary ν in $\mathcal{T}$:

$$\begin{aligned}
\pi_{\chi_{A_{\text{LOP}(p_{p_1},\ldots,p_{p_n})}}(\xi)}(\nu) &= \widetilde{\chi_{A_{\text{LOP}(p_{p_1},\ldots,p_{p_n})}}}(\pi_\xi)\cdot\nu \\
&= {\Pi_{\chi(\xi)}}^{(\phi)}(\{\nu\}) \\
&= \Pi_{\chi(\xi)}(\phi^{-1}(\{\nu\})) \\
&= \sup_{(\nu_1,\ldots,\nu_n)\in\phi^{-1}(\{\nu\})} \pi_{\chi(\xi)}(\nu_1,\ldots,\nu_n) \\
&= \sup_{\phi(\nu_1,\ldots,\nu_n)=\nu} \pi_{\chi(\xi)}(\nu_1,\ldots,\nu_n)
\end{aligned} \tag{4.17}$$

By combining (4.16) and (4.17), we find that

$$\pi_{\chi_{A_{\text{LOP}(p_{p_1},\ldots,p_{p_n})}}(\xi)}(\nu) = \sup_{\phi(\nu_1,\ldots,\nu_n)=\nu} \;\; \sup_{x\in\bigcap_{k=1}^{n}{\chi_{A_{p_{p_k}}}}^{-1}(\{\nu_k\})} \pi_\xi(x). \tag{4.18}$$

Using the appropriate possibilistic extensions, this result may also be written as

$$\pi_{\chi_{A_{\text{LOP}(p_{p_1},\ldots,p_{p_n})}}(\xi)}(\nu) = \widetilde{\chi_{A_{\text{LOP}(p_{p_1},\ldots,p_{p_n})}}}(\pi_\xi) = \widetilde{(\phi\circ\chi)}(\pi_\xi) = \tilde{\phi}(\tilde{\chi}(\pi_\xi))$$

This line of reasoning is pictorially summarized in the commuting diagrams of figure 3.

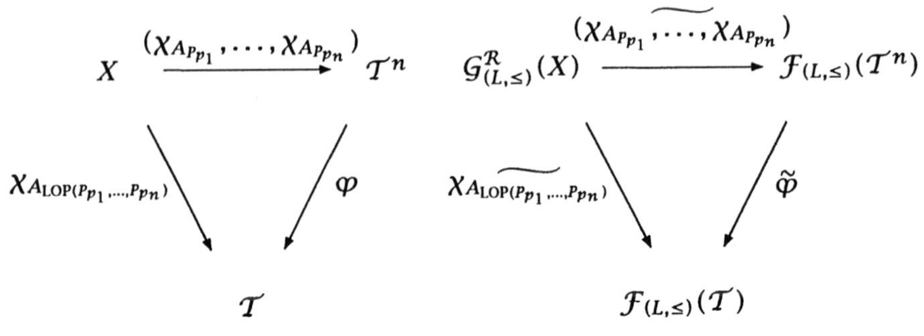

Figure 3 Commutative diagrams for the course of reasoning in subsection 4.2

4.3 A Truth-Functional Approximation

In his doctoral dissertation [27], Van Schooten uses a different way[4] to calculate the possibilistic truth value of the propositional variable $\text{LOP}(P_{p1}, \ldots, P_{p_n})(\xi)$. His approach could be called *truth-functional.* We may briefly summarize the ideas behind his method as follows. Starting with the possibilistic information π_ξ about the values ξ may assume in X, he separately calculates the $(L, \leq)$-possibilistic[5] truth value $\pi_{\chi_{A_{P_{p_k}}}(\xi)}$ of every proposition variable 'ξ is p_k' $(k = 1, \ldots, n)$. Since $\mathcal{R}^{(\chi_{A_{P_{p_k}}})} = \mathcal{P}(\mathcal{T})$ and therefore $\mathcal{G}_{(L,\leq)}^{\mathcal{R}^{(\chi_{A_{P_{p_k}}})}}(\mathcal{T}) = \mathcal{F}_{(L,\leq)}(\mathcal{T})$, we have that $\widetilde{\chi_{A_{P_{p_k}}}}$ is a $\mathcal{G}_{(L,\leq)}^{\mathcal{R}}(X) - \mathcal{F}_{(L,\leq)}(\mathcal{T})$-mapping, with, for arbitrary ν_k in $\mathcal{T}$:

$$\pi_{\chi_{A_{P_{p_k}}}(\xi)}(\nu_k) = \widetilde{\chi_{A_{P_{p_k}}}}(\pi_\xi) \cdot \nu_k = \sup_{\chi_{A_{P_{p_k}}}(x) = \nu_k} \pi_\xi(x). \tag{4.19}$$

These n $(L, \leq)$-possibilistic truth values are then combined into a new $(L, \leq)$-possibilistic truth value, using the $(L, \leq)$*-possibilistic* T*-extension* $\widetilde{\phi}_T$ *of* ϕ:

$$\widetilde{\phi}_T \colon \mathcal{F}_{(L,\leq)}(\mathcal{T})^n \to \mathcal{F}_{(L,\leq)}(\mathcal{T}) \colon (t_1, \ldots, t_n) \mapsto \widetilde{\phi}_T(t_1, \ldots, t_n)$$

[4] This statement must be seen as an *a posteriori* evaluation of Van Schooten's approach, based upon the possibility-theoretic framework we are constructing here. First of all, Van Schooten only considers the special case of unary and binary logical operators, and Zadeh's possibility measures. Secondly, his approach is more *ad hoc*, and certainly does not draw upon the rigorous mathematical account of possibility measures and possibilistic variables we have developed in the previous section, and in other papers [4, 6, 10, 15].

[5] Van Schooten only works with the special case $(L, \leq) = ([0, 1], \leq)$, and the minimum operator on $[0, 1]$ as triangular norm, but his approach can easily be extended towards the more general $(L, \leq)$ and arbitrary t-norms T on $(L, \leq)$. It is precisely this generalization that we are discussing here.

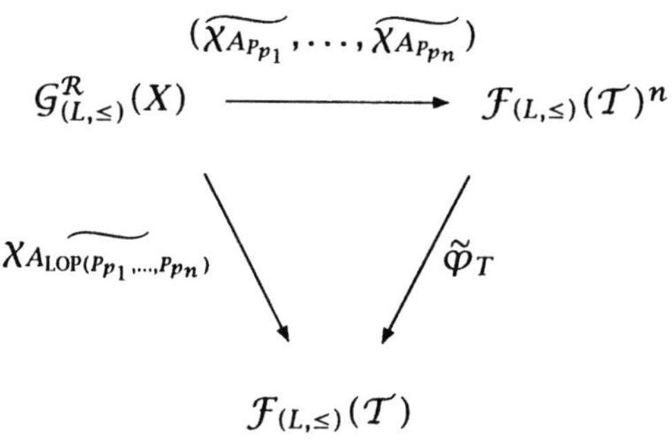

Figure 4 Commutative diagram for the course of reasoning in subsection 4.3

with, for arbitrary ν in $\mathcal{T}$

$$\widetilde{\phi}_T(t_1,\ldots,t_n)\cdot\nu = \sup_{\phi(\nu_1,\ldots,\nu_n)=\nu} T_{k=1}^n t_k(\nu_k). \tag{4.20}$$

If we denote this new $(L,\leq)$-possibilistic truth value by $\widetilde{\chi_{A_{\mathrm{LOP}(P_{p1},\ldots,P_{pn})}}}'(\pi_\xi)$, then we find, by combining (4.19) and (4.20)

$$\begin{aligned}\widetilde{\chi_{A_{\mathrm{LOP}(P_{p1},\ldots,P_{pn})}}}'(\pi_\xi) &= \widetilde{\phi}_T(\pi_{\chi_{A_{P_{p1}}}(\xi)},\ldots,\pi_{\chi_{A_{P_{pn}}}(\xi)}) \\ &= \widetilde{\phi}_T(\widetilde{\chi_{A_{P_{p1}}}}(\pi_\xi),\ldots,\widetilde{\chi_{A_{P_{pn}}}}(\pi_\xi)),\end{aligned}$$

or, equivalently, for arbitrary ν in $\mathcal{T}$

$$\widetilde{\chi_{A_{\mathrm{LOP}(P_{p1},\ldots,P_{pn})}}}'(\pi_\xi)\cdot\nu = \sup_{\phi(\nu_1,\ldots,\nu_n)=\nu} T_{k=1}^n \sup_{\chi_{A_{P_{pk}}}(x)=\nu_k} \pi_\xi(x). \tag{4.21}$$

This course of reasoning is summarized in the commuting diagram of figure 4.

In the following counter-example, we show that this truth-functional approach does not necessarily lead to the same result as the possibilistic method described in the previous subsection.

Example 5 We use the notations and conventions of examples 3 and 4. Besides the predicate p, we shall also consider the complementary predicate q, defined as 'smaller than 10'. The $(L,\leq)$-possibilistic truth value $\widetilde{\chi_{A_{P_q}}}(h)$ of the proposition variable '$\xi < 10$' is then given by

$$\begin{cases} \widetilde{\chi_{A_{P_q}}}(h)\cdot \mathit{true} = \sup_{a<10} h(a) = 1 \\ \widetilde{\chi_{A_{P_q}}}(h)\cdot \mathit{false} = \sup_{a\geq 10} h(a) = 1/2. \end{cases}$$

Let us furthermore consider the logical disjunction operator OR with associated logical function $\phi = \vee$, then clearly $A_{P_p \text{ OR } P_q} = \mathbb{R}$, since, by definition, for arbitrary a in $\mathbb{R}$, $\chi_{A_{P_p \text{ OR } P_q}}(a) = \chi_{A_{P_p}}(a) \vee \chi_{A_{P_q}}(a) = true$. Therefore $\widetilde{\chi_{A_{P_p \text{ OR } P_q}}}(h) = \{(true, 1), (false, 0)\}$, irrespective of h, which was, of course, to be expected, since the predicates p and q are complementary.

In order to apply the truth-functional method, we must calculate $\tilde{\vee}_T$. For arbitrary (t_1, t_2) in $\mathcal{F}_{(L,\leq)}(\mathcal{T})^2$:

$$\begin{aligned}
&(t_1 \,\tilde{\vee}_T\, t_2) \cdot true \\
&\quad = \sup_{v_1 \vee v_2 = true} T(t_1(v_1), t_2(v_2)) \\
&\quad = \sup(T(t_1(true), t_2(true)), T(t_1(true), t_2(false)), T(t_1(false), t_2(true)))
\end{aligned}$$

and

$$(t_1 \,\tilde{\vee}_T\, t_2) \cdot false = \sup_{v_1 \vee v_2 = false} T(t_1(v_1), t_2(v_2)) = T(t_1(false), t_2(false)).$$

If we substitute $t_1 = \widetilde{\chi_{A_{P_p}}}(h)$ and $t_2 = \widetilde{\chi_{A_{P_q}}}(h)$ in these expressions, we find

$$\begin{cases} \widetilde{\chi_{A_{P_p \text{ OR } P_q}}}'(h) \cdot true = \sup(T(1/2, 1), T(1/2, 1/2), T(1, 1)) = 1 \\ \widetilde{\chi_{A_{P_p \text{ OR } P_q}}}'(h) \cdot false = T(1, 1/2) = 1/2, \end{cases}$$

for every triangular norm T, which clearly cannot be correct.

The fact that the truth-functional method is not necessarily correct should not surprise us. Indeed, we find by comparing (4.17) and (4.21), taking into account

$$\pi_{\chi_{A_{P_{p_k}}}(\xi)}(v_k) = \widetilde{\chi_{A_{P_{p_k}}}}(\pi_\xi) \cdot v_k = \sup_{\chi_{A_{P_{p_k}}} = v_k} \pi_\xi(x),$$

that it is only correct if

$$(\forall (v_1, \ldots, v_n) \in \mathcal{T}^n)(\pi_{\chi(\xi)}(v_1, \ldots, v_n) = T_{k=1}^{n} \pi_{\chi_{A_{P_{p_k}}}(\xi)}(v_k)),$$

which is equivalent with (see [9, 15]) the (Π_Ω, T)-independence of the possibilistic variables $\chi_{A_{P_{p_1}}}(\xi), \ldots, \chi_{A_{P_{p_n}}}(\xi)$ in $(\mathcal{T}, \mathcal{P}(\mathcal{T}))$, and, taking into account $\chi_{A_{P_{p_k}}}(\xi) = \chi_{\xi^{-1}(A_{P_{p_k}})}$, also equivalent with the (Π_Ω, T)-independence of the events $\xi^{-1}(A_{P_{p_1}}), \ldots, \xi^{-1}(A_{P_{p_n}})$ in Ω. From a possibilistic point of view, this imposes certain restrictions on the truth-functional approach, as proposed by Van Schooten. Nevertheless, in the next proposition we show that the truth-functional approach is to a certain extent defendable. Its proof is obvious, taking into account the definition and properties of infimum and supremum.

Proposition 2 *Let X be a universe, $\mathcal{R}$ an ample field on X, n an element of $\mathbb{N} \setminus \{0\}$, ϕ a $\mathcal{T}^n - \mathcal{T}$-mapping, A_k an element of $\mathcal{R}$ ($k = 1, \ldots, n$), h a $(L, \leq)$-fuzzy variable in $(X, \mathcal{R})$ and v an element of $\mathcal{T}$. Then*

$$\sup_{\phi(v_1,\ldots,v_n)=v} \quad \sup_{x \in \bigcap_{k=1}^{n} X_{A_k}^{-1}(\{v_k\})} h(x) \leq \sup_{\phi(v_1,\ldots,v_n)=v} \inf_{k=1}^{n} \sup_{x \in X_{A_k}^{-1}(\{v_k\})} h(x). \tag{4.22}$$

This proposition implies that in the case $(L, \leq) = ([0,1], \leq)$ and $T = \frown$ considered by Van Schooten, the truth-functional approach results in a *conservative approximation*, because it pushes the possibilistic truth values towards $\{(true, 1), (false, 1)\}$, i.e., both truth values *true* and *false* are equally possible. To put it more concretely, assume that the possibilistic method results in a $([0,1], \leq)$-possibilistic truth value[6] $t_{pos} = \{(true, 1), (false, a)\}$, with $a \in [0,1]$. The proposition above then tells us that the truth-functional approach must yield a $([0,1], \leq)$-possibilistic truth value $t_{tf} = \{(true, 1), (false, b)\}$ with $b \in [0,1]$ and $b \geq a$. This means that the information, obtained in the truth-functional way, is only less restrictive, or in other words, there can be no contradiction, only loss of specificity. This, together with the results derived in section 6, makes the truth-functional approach for $T = \frown$ surely defendable.

5 POSSIBILISTIC EXTENSION LOGICS

5.1 Towards a Possibilistic Logic

In this section, we shall take the discussion of the previous section one step further, and look at its results from the standpoint of multi-valued logic (see, for instance, [25]). In other words, we want to investigate how the introduction of possibilistic uncertainty in classical propositional logic leads to the introduction of a special multi-valued logic, with a proper set of truth values and logical functions combining these truth values.

As in the previous sections, we consider a universe X, provided with an ample field $\mathcal{R}$ of measurable sets. Let us briefly summarize what we already know. In the previous sections, we have seen that possibilistic information

[6]Normalization implies that at least one of both numbers $t_{pos}(true)$ and $t_{pos}(false)$ must be equal to 1.

about the values that a variable ξ may assume in X, can be represented by a sup-normal $(L, \leq)$-fuzzy variable in $(X, \mathcal{R})$, interpreted as the distribution of a normalized $(L, \leq)$-possibility measure on $(X, \mathcal{R})$. When p is a clear property, and the associated subset A_{P_p} of X is $\mathcal{R}$-measurable, this possibilistic information can be transformed into possibilistic information about the truth value of the proposition variable 'ξ is p'. This information can be represented by a sup-normal $(L, \leq)$-fuzzy variable in $(\mathcal{T}, \mathcal{P}(\mathcal{T}))$. Such a $(L, \leq)$-fuzzy variable in $(\mathcal{T}, \mathcal{P}(\mathcal{T}))$, in other words a $(L, \leq)$-fuzzy set in $\mathcal{T}$, can be generally called a $(L, \leq)$-possibilistic truth value. This is formalized in the following definition.

Definition 5 (Possibilistic truth values) *We call $(L, \leq)$-possibilistic truth value any sup-normal $(L, \leq)$-fuzzy set in $\mathcal{T} = \{true, false\}$. The set of the $(L, \leq)$-possibilistic truth values will be denoted by $\widetilde{\mathcal{T}}$. If, for whatever reason, we do not want to mention the complete lattice $(L, \leq)$ explicitly, we shall simply speak of possibilistic truth values.*

Since the complete lattice $(L, \leq)$ is bounded, we immediately arrive at the following general definition.

Definition 6 *We introduce three $(L, \leq)$-possibilistic truth values with a special meaning: $\widetilde{false} \stackrel{\text{def}}{=} \{(true, 0), (false, 1)\}$, $\widetilde{true} \stackrel{\text{def}}{=} \{(true, 1), (false, 0)\}$ and $\widetilde{unknown} \stackrel{\text{def}}{=} \{(true, 1), (false, 1)\}$.*

These special possibilistic truth values can be interpreted as follows. When a proposition variable 'ξ is p' has the $(L, \leq)$-possibilistic truth value $\widetilde{true}$, this means that it cannot be false, and is therefore *necessarily true*, taking into account the information we have about the values that ξ may assume in X. An analogous (dual) interpretation can be given to $\widetilde{false}$. When, on the other hand, the proposition variable 'ξ is p' has the $(L, \leq)$-possibilistic truth value $\widetilde{unknown}$, this means that, taking into account the information we have about the values that ξ may assume, it is completely possible that the proposition variable is true, and equally possible that it is false. In other words, the truth value of this proposition variable is *completely unknown*, because of insufficient information about the values that ξ may assume in X.

An important property of classical propositional logic is what could be called its *truth-functionality*. This means that propositions can be combined to form new propositions using *logical operators*, the behaviour of which is mirrored in *logical functions* that turn the *truth values* of those propositions into the truth values of the new, combined propositions. In other words, with every logical operator, acting on propositions, there can be associated a unique logical function, acting on truth values, that completely characterizes its behaviour. The study of logical functions is of course an important part of classical propositional logic. In the rest of this section, we shall concentrate on the introduction and study of logical functions for the new type of (possibilistic) logic we are creating here, and that is used to model possibilistic uncertainty in classical logic. In the following definition we explicitly repeat the classical definition of a logical function, using our notations and terminology (see, for instance, [23] sections 1.6 and 1.7).

Definition 7 (Classical-logical functions) *Let n be an element of $\mathbb{N} \setminus \{0\}$. A $\mathcal{T}^n - \mathcal{T}$-mapping is called a classical-logical function of arity n. The set of the classical-logical functions of arity n is denoted by $\mathcal{L}_n$. The set of classical-logical functions of arbitrary arity is given the notation $\mathcal{L}$.*

Example 6 The conjunction $\wedge$, the disjunction $\vee$ and the implication $\Rightarrow$, defined on $\mathcal{T}$, are classical-logical functions of arity 2, characterizing the truth-functional behaviour of respectively the logical conjunction, disjunction and implication operator in classical propositional logic. The negation $\neg$, defined on $\mathcal{T}$, is a classical-logical function of arity 1, characterizing the truth-functional behaviour of the logical negation operator in that logic.

What we now want to do is to extend the classical, truth-functional approach: we formally consider $\widetilde{\mathcal{T}}$ as a set of truth values, and look at how such possibilistic truth values can be combined into new ones. After that, we intend to show that at least for some of these combinations, there is a clear and definite link with combinations of propositions. In this way, we intend to prove that, in some cases, our possibilistic logic is also truth-functional.

Definition 8 (Possibilistic-logical functions) *Let n be an element of $\mathbb{N} \setminus \{0\}$. A $(\widetilde{\mathcal{T}})^n - \widetilde{\mathcal{T}}$-mapping is called a $(L, \leq)$-possibilistic-logical function of arity n. The set of the $(L, \leq)$-possibilistic-logical functions of arity n is denoted by $\widetilde{\mathcal{L}}_n$. The set of the $(L, \leq)$-possibilistic-logical functions of arbitrary arity is given the notation $\widetilde{\mathcal{L}}$. If, for whatever reason, we do not want to mention the complete lattice $(L, \leq)$ explicitly, we shall simply speak of possibilistic-logical functions.*

If we look at the previous section, we at once see that we can associate a $(L, \leq)$-possibilistic-logical function with every classical-logical function, simply by looking at its $(L, \leq)$-possibilistic T-extension. Of course, this extension must be properly restricted, because we only work with elements of $\widetilde{\mathcal{T}}$—and not $\mathcal{F}_{(L,\leq)}(\mathcal{T})$—as possibilistic truth values.

Definition 9 *Let n be an element of $\mathbb{N} \setminus \{0\}$ and let ϕ be a classical-logical function of arity n. The $(L, \leq)$-possibilistic-logical T-extension $\tilde{\phi}_{\ell T}$ of ϕ is defined as the restriction of the $(L, \leq)$-possibilistic T-extension $\tilde{\phi}_T$ of ϕ to the set $(\widetilde{\mathcal{T}})^n$, i.e., $\tilde{\phi}_{\ell T} \stackrel{\text{def}}{=} \tilde{\phi}_T | (\widetilde{\mathcal{T}})^n$.*

Corollary 3 *$(L, \leq)$-possibilistic-logical T-extensions of classical-logical functions are $(L, \leq)$-possibilistic-logical functions: $(\forall \phi \in \mathcal{L})(\tilde{\phi}_{\ell T} \in \tilde{\mathcal{L}})$.*

Definition 10 *We call $(L, \leq)$-possibilistic T-extension logic the set $\tilde{\mathcal{L}}_T$ of the $(L, \leq)$-possibilistic-logical T-extensions of the classical-logical functions of any arity, i.e., $\tilde{\mathcal{L}}_T \stackrel{\text{def}}{=} \{\tilde{\phi}_{\ell T} \mid \phi \in \mathcal{L}\}$. If, for whatever reason, we do not want to mention the complete lattice $(L, \leq)$ and/or the t-norm T explicitly, we shall simply speak of possibilistic extension logics.*

The rationale for the introduction of these extension logics has been given in subsection 4.3. Borrowing the notations from that subsection, we know that if the events $\xi^{-1}(A_{p_{p_1}}), \ldots, \xi^{-1}(A_{p_{p_n}})$ in Ω are (Π_Ω, T)-independent, the $(L, \leq)$-possibilistic truth value $\pi_{\chi_{A_{\text{LOP}(p_{p_1},\ldots,p_{p_n})}}}(\xi) = \widetilde{\chi_{A_{\text{LOP}(p_{p_1},\ldots,p_{p_n})}}}(\pi_\xi)$ of the proposition variable LOP('ξ is p_1', ..., 'ξ is p_n'), is given by

$$\pi_{\chi_{A_{\text{LOP}(p_{p_1},\ldots,p_{p_n})}}}(\xi) = \tilde{\phi}_{\ell T}(\pi_{\chi_{A_{p_{p_1}}}}(\xi), \ldots, \pi_{\chi_{A_{p_{p_n}}}}(\xi)),$$

where $\pi_{\chi_{A_{p_{p_k}}}}(\xi) = \widetilde{\chi_{A_{p_{p_k}}}}(\pi_\xi)$ is the $(L, \leq)$-possibilistic truth value of the proposition variable 'ξ is p_k' $(k = 1, \ldots, n)$. Indeed, in the case of possibilistic independence, there is truth-functionality for our possibilistic logic.

5.2 Some Interesting Restrictions

We have already introduced three special possibilistic truth values $\widetilde{\mathit{true}}$, $\widetilde{\mathit{false}}$ and $\widetilde{\mathit{unknown}}$, and have briefly discussed their meaning. If we define the sets $\mathcal{W}_1 \stackrel{\text{def}}{=} \{\widetilde{\mathit{true}}, \widetilde{\mathit{false}}\}$ and $\mathcal{W}_2 \stackrel{\text{def}}{=} \{\widetilde{\mathit{true}}, \widetilde{\mathit{false}}, \widetilde{\mathit{unknown}}\}$, and properly restrict a possibilistic extension logic to these sets, a number of interesting observations can be made. The proofs of these observations are straightforward, and will be omitted here. These proofs, and more details, can be found in [9], and for the special case $(L, \leq) = ([0, 1], \leq)$ in [16]. For a start, all elements of $\widetilde{\mathcal{L}}_T$ are internal in the sets $\mathcal{W}_1$ and $\mathcal{W}_2$, or in other words,

$$\begin{cases} (\forall n \in \mathbb{N} \setminus \{0\})(\forall \phi \in \mathcal{L}_n)(\forall t \in (\mathcal{W}_1)^n)(\widetilde{\phi}_{\ell T}(t) \in \mathcal{W}_1) \\ (\forall n \in \mathbb{N} \setminus \{0\})(\forall \phi \in \mathcal{L}_n)(\forall t \in (\mathcal{W}_2)^n)(\widetilde{\phi}_{\ell T}(t) \in \mathcal{W}_2). \end{cases}$$

Furthermore, $\widetilde{\mathcal{L}}_T$ restricted to $\mathcal{W}_1$ is essentially the same as—or isomorphic to—$\mathcal{L}$, via an isomorphism that identifies $\widetilde{true}$ and *true* on the one hand, and $\widetilde{false}$ en *false* on the other hand. In particular, this also implies that the Boolean algebras $(\mathcal{T}, \wedge, \vee, \neg)$ and $(\mathcal{W}_1, \widetilde{\wedge}_{\ell T}|(\mathcal{W}_1)^2, \widetilde{\vee}_{\ell T}|(\mathcal{W}_1)^2, \widetilde{\neg}_{\ell T}|\mathcal{W}_1)$ are isomorphic.

Since all elements of $\widetilde{\mathcal{L}}_T$ are internal in $\mathcal{W}_2$, restriction of the truth domain of these possibilistic-logical functions to $\mathcal{W}_2$ yields a *three-valued logic*. Since all triangular norms on $(L, \leq)$ have the same behaviour in the subset $\{0, 1\}^2$ of L^2, we will find *the same ternary logic* for every choice of T. For different choices of $(L, \leq)$, the corresponding ternary logics are furthermore isomorphic. It is easily shown [16] that the truth tables for $\widetilde{\neg}_{\ell T}|\mathcal{W}_2$, $\widetilde{\wedge}_{\ell T}|(\mathcal{W}_2)^2$, $\widetilde{\vee}_{\ell T}|(\mathcal{W}_2)^2$ and $\widetilde{\Rightarrow}_{\ell T}|(\mathcal{W}_2)^2$ are identical to the corresponding truth tables of the so-called *strong ternary logic of Kleene* (see, for instance, [25] section 2.5). We shall return to this interesting fact in the following section.

5.3 A Few Properties

In the rest of this section, we shall study the most important properties of some special $(L, \leq)$-possibilistic-logical functions of arity 1 and 2: $\widetilde{\neg}_{\ell T}$, $\widetilde{\wedge}_{\ell T}$, $\widetilde{\vee}_{\ell T}$ and $\widetilde{\Rightarrow}_{\ell T}$. First of all, it will help us if we can find simple expressions for these operators. This is the subject of the next proposition. Its proof is straightforward, and is therefore omitted.

Proposition 3 *(i)* $\widetilde{\neg}_{\ell T}: \widetilde{\mathcal{T}} \to \widetilde{\mathcal{T}} : t \mapsto \widetilde{\neg}_{\ell T}\, t$, *with*

$$\begin{cases} (\widetilde{\neg}_{\ell T}\, t) \cdot true = t(false) \\ (\widetilde{\neg}_{\ell T}\, t) \cdot false = t(true). \end{cases}$$

(ii) $\widetilde{\wedge}_{\ell T}: (\widetilde{\mathcal{T}})^2 \to \widetilde{\mathcal{T}} : (t_1, t_2) \mapsto t_1 \,\widetilde{\wedge}_{\ell T}\, t_2$, *with*

$$\begin{cases} (t_1 \,\widetilde{\wedge}_{\ell T}\, t_2) \cdot true = T(t_1(true), t_2(true)) \\ (t_1 \,\widetilde{\wedge}_{\ell T}\, t_2) \cdot false = t_1(false) \smile t_2(false)). \end{cases}$$

(iii) $\tilde{\vee}_{\ell T}: (\widetilde{\mathcal{T}})^2 \to \widetilde{\mathcal{T}}: (t_1, t_2) \mapsto t_1 \tilde{\vee}_{\ell T} t_2$, *with*

$$\begin{cases} (t_1 \tilde{\vee}_{\ell T} t_2) \cdot true = t_1(true) \smile t_2(true) \\ (t_1 \tilde{\vee}_{\ell T} t_2) \cdot false = T(t_1(false), t_2(false)). \end{cases}$$

(iv) $\tilde{\Rightarrow}_{\ell T}: (\widetilde{\mathcal{T}})^2 \to \widetilde{\mathcal{T}}: (t_1, t_2) \mapsto t_1 \tilde{\Rightarrow}_{\ell T} t_2$, *with*

$$\begin{cases} (t_1 \tilde{\Rightarrow}_{\ell T} t_2) \cdot true = t_1(false) \smile t_2(true) \\ (t_1 \tilde{\Rightarrow}_{\ell T} t_2) \cdot false = T(t_1(true), t_2(false)). \end{cases}$$

Let us now give a brief survey of the most important properties of the above-mentioned possibilistic-logical functions. The proofs of these properties are fairly simple, and we have consequently omitted them. It should nevertheless be noted that the equalities that appear in these properties, are equalities of $(L, \leq)$-possibilistic truth values, and therefore pointwise equalities of $\mathcal{T} - L$-mappings.

Property 1 (Commutativity) *For arbitrary* t_1 *and* t_2 *in* $\widetilde{\mathcal{T}}$:

$$\begin{cases} t_1 \tilde{\wedge}_{\ell T} t_2 = t_2 \tilde{\wedge}_{\ell T} t_1 \\ t_1 \tilde{\vee}_{\ell T} t_2 = t_2 \tilde{\vee}_{\ell T} t_1. \end{cases}$$

Property 2 (Neutral elements) *For arbitrary* t *in* $\widetilde{\mathcal{T}}$:

$$\begin{cases} t \tilde{\wedge}_{\ell T} \widetilde{true} = t \\ t \tilde{\vee}_{\ell T} \widetilde{false} = t. \end{cases}$$

Property 3 (Associativity) *For arbitrary* t_1, t_2 *and* t_3 *in* $\widetilde{\mathcal{T}}$:

$$\begin{cases} t_1 \tilde{\wedge}_{\ell T} (t_2 \tilde{\wedge}_{\ell T} t_3) = (t_1 \tilde{\wedge}_{\ell T} t_2) \tilde{\wedge}_{\ell T} t_3 \\ t_1 \tilde{\vee}_{\ell T} (t_2 \tilde{\vee}_{\ell T} t_3) = (t_1 \tilde{\vee}_{\ell T} t_2) \tilde{\vee}_{\ell T} t_3. \end{cases}$$

Property 4 (De Morgan's Laws) *For arbitrary* t_1 *and* t_2 *in* $\widetilde{\mathcal{T}}$:

$$\begin{cases} \tilde{\neg}_{\ell T} (t_1 \tilde{\wedge}_{\ell T} t_2) = (\tilde{\neg}_{\ell T} t_1) \tilde{\vee}_{\ell T} (\tilde{\neg}_{\ell T} t_2) \\ \tilde{\neg}_{\ell T} (t_1 \tilde{\vee}_{\ell T} t_2) = (\tilde{\neg}_{\ell T} t_1) \tilde{\wedge}_{\ell T} (\tilde{\neg}_{\ell T} t_2). \end{cases}$$

Property 5 (Absorbing elements) *For arbitrary* t *in* $\widetilde{\mathcal{T}}$:

$$\begin{cases} t \tilde{\wedge}_{\ell T} \widetilde{false} = \widetilde{false} \\ t \tilde{\vee}_{\ell T} \widetilde{true} = \widetilde{true}. \end{cases}$$

Property 6 (Involutivity) *For arbitrary* t *in* $\widetilde{\mathcal{T}}$:

$$\tilde{\neg}_{\ell T} (\tilde{\neg}_{\ell T} t) = t.$$

Property 7 (Implication) *For arbitrary t_1 and t_2 in $\widetilde{\mathcal{T}}$:*
$t_1 \mathbin{\tilde{\Rightarrow}_{\ell T}} t_2 = (\tilde{\neg}_{\ell T}\, t_1) \mathbin{\tilde{\vee}_{\ell T}} t_2$.

Property 8 (Complementation) *For arbitrary t in $\widetilde{\mathcal{T}}$:*

$$\begin{cases} t \mathbin{\tilde{\wedge}_{\ell T}} (\tilde{\neg}_{\ell T}\, t) = \{(\mathit{true}, T(t(\mathit{true}), t(\mathit{false}))), (\mathit{false}, 1)\} \\ t \mathbin{\tilde{\vee}_{\ell T}} (\tilde{\neg}_{\ell T}\, t) = \{(\mathit{true}, 1), (\mathit{false}, T(t(\mathit{true}), t(\mathit{false})))\} \end{cases}$$

Property 9 (Contrapositive symmetry) *For arbitrary t_1 and t_2 in $\widetilde{\mathcal{T}}$:*
$t_1 \mathbin{\tilde{\Rightarrow}_{\ell T}} t_2 = (\tilde{\neg}_{\ell T}\, t_2) \mathbin{\tilde{\Rightarrow}_{\ell T}} (\tilde{\neg}_{\ell T}\, t_1)$.

Property 10 (Neutrality principle) *For arbitrary t in $\widetilde{\mathcal{T}}$:*
$(\widetilde{\mathit{true}} \mathbin{\tilde{\Rightarrow}_{\ell T}} t) = t$.

Property 11 (Exchange principle) *For arbitrary t_1, t_2 and t_3 in $\widetilde{\mathcal{T}}$:*
$t_1 \mathbin{\tilde{\Rightarrow}_{\ell T}} (t_2 \mathbin{\tilde{\Rightarrow}_{\ell T}} t_3) = t_2 \mathbin{\tilde{\Rightarrow}_{\ell T}} (t_1 \mathbin{\tilde{\Rightarrow}_{\ell T}} t_3)$.

Property 12 (Boundary conditions) *For arbitrary t_1 and t_2 in $\widetilde{\mathcal{T}}$:*

$$\begin{cases} (t_1 \mathbin{\tilde{\wedge}_{\ell T}} t_2 = \widetilde{\mathit{true}}) \Leftrightarrow (t_1 = \widetilde{\mathit{true}} \text{ and } t_2 = \widetilde{\mathit{true}}) \\ (t_1 \mathbin{\tilde{\vee}_{\ell T}} t_2 = \widetilde{\mathit{false}}) \Leftrightarrow (t_1 = \widetilde{\mathit{false}} \text{ and } t_2 = \widetilde{\mathit{false}}) \\ (\tilde{\neg}_{\ell T}\, t = \widetilde{\mathit{true}}) \Leftrightarrow (t = \widetilde{\mathit{false}}) \\ (\tilde{\neg}_{\ell T}\, t = \widetilde{\mathit{false}}) \Leftrightarrow (t = \widetilde{\mathit{true}}) \end{cases}$$

It is important to note that $\tilde{\wedge}_{\ell T}$ and $\tilde{\vee}_{\ell T}$ are idempotent if and only if T is, or, in other words, if and only if $T = \frown$ (see, for instance, [9, 14]). Furthermore, $\tilde{\wedge}_{\ell T}$ and $\tilde{\vee}_{\ell T}$ are mutually distributive if and only if T and sup are mutually distributive. This is only possible if $T = \frown$ (see, for instance, [9, 14]). Thus, it appears that the choice $T = \frown$ is a rather special[7] one. We therefore devote the next section to the study of this special case.

[7]In this respect, it should also be noted that if we consider the lattice $(\widetilde{\mathcal{T}}, \tilde{\leq})$, where $\tilde{\leq}$ is the partial order relation on $\widetilde{\mathcal{T}}$, introduced in the following section, then $\tilde{\wedge}_{\ell T}$ is a t-norm and $\tilde{\vee}_{\ell T}$ is a t-conorm [9, 14] on this structure. These operators are dual [9, 14] w.r.t. the negation $\tilde{\neg}_{\ell T}$ on $(\widetilde{\mathcal{T}}, \tilde{\leq})$. Remark that $\tilde{\wedge}_{\ell T}$ is the meet and $\tilde{\vee}_{\ell T}$ the join of the lattice $(\widetilde{\mathcal{T}}, \tilde{\leq})$ if and only if $T = \frown$.

6 AN INTERESTING SPECIAL CASE

6.1 A Few Algebraic Results

In this section, we intend to take a closer look at the notions, introduced in the previous section, in the special case $T = \frown$. This means that we shall assume that $(L, \leq)$ is a complete Brouwerian lattice (see [1] section V.10). In particular, this implies that the binary operators $\frown$ and $\smile$ are mutually distributive (see, for instance, [1] section I.6). In this subsection, we have collected a number of algebraic and order-theoretic results, the first of which is given is proposition 4.

Proposition 4 *$(\widetilde{\mathcal{T}}, \widetilde{\wedge}_{\ell\frown}, \widetilde{\vee}_{\ell\frown})$ is a bounded distributive lattice (as an algebra) with top element $\widetilde{true}$ and bottom element $\widetilde{false}$. The natural partial order relation $\stackrel{\sim}{\leq}$ on $\widetilde{\mathcal{T}}$ that corresponds with this structure, satisfies*

$$(\forall (t_1, t_2) \in (\widetilde{\mathcal{T}})^2) \left(t_1 \stackrel{\sim}{\leq} t_2 \Leftrightarrow \begin{cases} t_1(true) \leq t_2(true) \\ t_1(false) \geq t_2(false) \end{cases} \right). \tag{4.23}$$

Proof. Let us first show that $(\widetilde{\mathcal{T}}, \widetilde{\wedge}_{\ell\frown}, \widetilde{\vee}_{\ell\frown})$ is a lattice (as an algebra) (see, for instance, [1] section I.5). It must be proven that $\widetilde{\wedge}_{\ell\frown}$ and $\widetilde{\vee}_{\ell\frown}$ satisfy the fundamental properties of meet and join in lattices (see [1] theorem I.8). Indeed, making use of the simplified expressions for $\widetilde{\wedge}_{\ell\frown}$ and $\widetilde{\vee}_{\ell\frown}$, derived in proposition 3 (with the special choice $T = \frown$), it is easy to prove that $\widetilde{\wedge}_{\ell\frown}$ and $\widetilde{\vee}_{\ell\frown}$ are idempotent, commutative and associative, and that they satisfy the absorption laws. Next, it is a well-known result from lattice theory (see, for instance, [1] theorem I.8 and lemma I.1) that for any lattice as an algebra, and in particular also for the structure $(\widetilde{\mathcal{T}}, \widetilde{\wedge}_{\ell\frown}, \widetilde{\vee}_{\ell\frown})$, a natural partial order relation $\stackrel{\sim}{\leq}$ can be defined on $\widetilde{\mathcal{T}}$, in such a way that $(\widetilde{\mathcal{T}}, \stackrel{\sim}{\leq})$ is an order-theoretic lattice with meet $\widetilde{\wedge}_{\ell\frown}$ and join $\widetilde{\vee}_{\ell\frown}$. Let t_1 and t_2 be elements of $\widetilde{\mathcal{T}}$, then we must have, taking into account the consistency property of the meet $\widetilde{\wedge}_{\ell\frown}$ in the lattice $(\widetilde{\mathcal{T}}, \stackrel{\sim}{\leq})$ (see, for instance, [1] section I.5), that

$$\begin{aligned} t_1 \stackrel{\sim}{\leq} t_2 &\Leftrightarrow t_1 \, \widetilde{\wedge}_{\ell\frown} \, t_2 = t_1 \\ &\Leftrightarrow \begin{cases} t_1(true) \frown t_2(true) = t_1(true) \\ t_1(false) \smile t_2(false) = t_1(false), \end{cases} \\ &\Leftrightarrow \begin{cases} t_1(true) \leq t_2(true) \\ t_1(false) \geq t_2(false), \end{cases} \end{aligned}$$

taking into account the consistency property of $\frown$ and $\smile$ in $(L, \leq)$. This proves (4.23). From property 5 for $T = \frown$ and the consistency property of the meet $\tilde{\wedge}_{\ell_\frown}$ and the join $\tilde{\vee}_{\ell_\frown}$ in $(\tilde{\mathcal{T}}, \lesssim)$ it then follows that the lattice (as an algebra) $(\tilde{\mathcal{T}}, \tilde{\wedge}_{\ell_\frown}, \tilde{\vee}_{\ell_\frown})$ is bounded, with top element $\widetilde{true}$ and bottom element $\widetilde{false}$. We proceed to show that the lattice (as an algebra) $(\tilde{\mathcal{T}}, \tilde{\wedge}_{\ell_\frown}, \tilde{\vee}_{\ell_\frown})$ is distributive (see, for instance [1] section I.6). Let t_1, t_2 and t_3 be elements of $\tilde{\mathcal{T}}$. Then, since $(L, \leq)$ is by assumption in particular a distributive lattice,

$$\begin{aligned}
(t_1 \tilde{\wedge}_{\ell_\frown} (t_2 \tilde{\vee}_{\ell_\frown} t_3)) \cdot true &= t_1(true) \frown (t_2 \tilde{\vee}_{\ell_\frown} t_3) \cdot true \\
&= t_1(true) \frown (t_2(true) \smile t_3(true)) \\
&= (t_1(true) \frown t_2(true)) \smile (t_1(true) \frown t_3(true)) \\
&= (t_1 \tilde{\wedge}_{\ell_\frown} t_2) \cdot true \smile (t_1 \tilde{\wedge}_{\ell_\frown} t_3) \cdot true \\
&= ((t_1 \tilde{\wedge}_{\ell_\frown} t_2) \tilde{\vee}_{\ell_\frown} (t_1 \tilde{\wedge}_{\ell_\frown} t_3)) \cdot true.
\end{aligned}$$

Analogously,

$$(t_1 \tilde{\wedge}_{\ell_\frown} (t_2 \tilde{\vee}_{\ell_\frown} t_3)) \cdot false = ((t_1 \tilde{\wedge}_{\ell_\frown} t_2) \tilde{\vee}_{\ell_\frown} (t_1 \tilde{\wedge}_{\ell_\frown} t_3)) \cdot false,$$

whence $t_1 \tilde{\wedge}_{\ell_\frown} (t_2 \tilde{\vee}_{\ell_\frown} t_3) = (t_1 \tilde{\wedge}_{\ell_\frown} t_2) \tilde{\vee}_{\ell_\frown} (t_1 \tilde{\wedge}_{\ell_\frown} t_3)$. This implies that the structure $(\tilde{\mathcal{T}}, \tilde{\wedge}_{\ell_\frown}, \tilde{\vee}_{\ell_\frown})$ is a distributive lattice (as an algebra). □

Besides the binary operators meet $\tilde{\wedge}_{\ell_\frown}$ and join $\tilde{\vee}_{\ell_\frown}$ of the bounded distributive lattice $(\tilde{\mathcal{T}}, \lesssim)$, there also exists the unary operator $\tilde{\neg}_{\ell_\frown}$. The properties of this operator are studied in the next proposition. By a negation operator on a bounded poset, we shall mean a dual order-automorphism on that structure (for more details, see [9, 14]).

Proposition 5 *$\tilde{\neg}_{\ell_\frown}$ is an involutive negation operator on $(\tilde{\mathcal{T}}, \lesssim)$, but not a complement operator on $(\tilde{\mathcal{T}}, \lesssim)$.*

Proof. Let us first show that $\tilde{\neg}_{\ell_\frown}$ is an involutive negation operator on $(\tilde{\mathcal{T}}, \lesssim)$. The involutivity of $\tilde{\neg}_{\ell_\frown}$ follows from property 6 for $T = \frown$. An involutive transformation is furthermore always a permutation. Let t_1 and t_2 be elements of $\tilde{\mathcal{T}}$. Then, taking into account (4.23) and proposition 3(i),

$$\begin{aligned}
(\tilde{\neg}_{\ell_\frown} t_1) \lesssim (\tilde{\neg}_{\ell_\frown} t_2) &\Leftrightarrow \begin{cases} (\tilde{\neg}_{\ell_\frown} t_1) \cdot true \leq (\tilde{\neg}_{\ell_\frown} t_2) \cdot true \\ (\tilde{\neg}_{\ell_\frown} t_1) \cdot false \geq (\tilde{\neg}_{\ell_\frown} t_2) \cdot false \end{cases} \\
&\Leftrightarrow \begin{cases} t_1(false) \leq t_2(false) \\ t_1(true) \geq t_2(true) \end{cases} \\
&\Leftrightarrow t_2 \lesssim t_1.
\end{aligned}$$

We conclude that $\tilde{\neg}_{\ell_\frown}$ is a dual order-automorphism of $(\tilde{\mathcal{T}}, \lesssim)$, or equivalently, a negation operator on $(\tilde{\mathcal{T}}, \lesssim)$. As mentioned above, $\tilde{\neg}_{\ell_\frown}$ is furthermore involutive. In order to complete the proof, we must show that $\tilde{\neg}_{\ell_\frown}$

is not a complement operator on $(\widetilde{\mathcal{T}}, \widetilde{\leq})$. $\widetilde{\neg}_{\ell_\frown}$ is a complement operator on $(\widetilde{\mathcal{T}}, \widetilde{\leq})$ if and only if $(\forall t \in \widetilde{\mathcal{T}})(\widetilde{\neg}_{\ell_\frown} t$ is a complement of $t)$. Taking into account the definition of a complement (see, for instance, [1] section I.9), this is equivalent with $(\forall t \in \widetilde{\mathcal{T}})(t \,\widetilde{\wedge}_{\ell_\frown} (\widetilde{\neg}_{\ell_\frown} t) = \widetilde{false}$ and $t \,\widetilde{\vee}_{\ell_\frown} (\widetilde{\neg}_{\ell_\frown} t) = \widetilde{true})$, which, taking into account property 12 for $T = \frown$, is also equivalent with $(\forall t \in \widetilde{\mathcal{T}})(t(true) \frown t(false) = 0)$. Since, surely, $\widetilde{unknown} \in \widetilde{\mathcal{T}}$ and furthermore $\widetilde{unknown}(true) \frown \widetilde{unknown}(false) = 1 \frown 1 = 1 \neq 0$, we deduce that $\widetilde{\neg}_{\ell_\frown}$ cannot be a complement operator on $(\widetilde{\mathcal{T}}, \widetilde{\leq})$. □

The following proposition immediately follows from proposition 5 and property 4 for $T = \frown$.

Proposition 6 *$(\widetilde{\mathcal{T}}, \widetilde{\wedge}_{\ell_\frown}, \widetilde{\vee}_{\ell_\frown}, \widetilde{\neg}_{\ell_\frown})$ is a Morgan algebra[8], i.e., $(\widetilde{\mathcal{T}}, \widetilde{\wedge}_{\ell_\frown}, \widetilde{\vee}_{\ell_\frown})$ is a bounded distributive lattice (as an algebra), with a unary operator $\widetilde{\neg}_{\ell_\frown}$ satisfying (i) $\widetilde{\neg}_{\ell_\frown}$ is involutive; and (ii) $\widetilde{\wedge}_{\ell_\frown}$, $\widetilde{\vee}_{\ell_\frown}$ and $\widetilde{\neg}_{\ell_\frown}$ satisfy de Morgan's laws.*

In the next proposition, we establish the relationship between our possibilistic $\frown$-extension logics and a class of *multi-valued logics*, studied extensively in the literature (see, for instance, [25]).

Proposition 7 *$(\widetilde{\mathcal{T}}, \widetilde{\wedge}_{\ell_\frown}, \widetilde{\vee}_{\ell_\frown}, \widetilde{\neg}_{\ell_\frown})$ is a Kleene algebra[9], i.e., the structure $(\widetilde{\mathcal{T}}, \widetilde{\wedge}_{\ell_\frown}, \widetilde{\vee}_{\ell_\frown}, \widetilde{\neg}_{\ell_\frown})$ is a Morgan algebra with furthermore*

$$(\forall (t_1, t_2) \in (\widetilde{\mathcal{T}})^2)(t_1 \,\widetilde{\wedge}_{\ell_\frown} (\widetilde{\neg}_{\ell_\frown} t_1) \,\widetilde{\leq}\, t_2 \,\widetilde{\vee}_{\ell_\frown} (\widetilde{\neg}_{\ell_\frown} t_2)).$$

Proof. We know from proposition 6 that $(\widetilde{\mathcal{T}}, \widetilde{\wedge}_{\ell_\frown}, \widetilde{\vee}_{\ell_\frown}, \widetilde{\neg}_{\ell_\frown})$ is indeed a Morgan algebra. Let furthermore t_1 and t_2 be elements of $\widetilde{\mathcal{T}}$. Then, taking into account property 8 for $T = \frown$,

$$\begin{cases} t_1 \,\widetilde{\wedge}_{\ell_\frown} (\widetilde{\neg}_{\ell_\frown} t_1) = \{(true, t_1(true) \frown t_1(false)), (false, 1)\} \\ t_2 \,\widetilde{\vee}_{\ell_\frown} (\widetilde{\neg}_{\ell_\frown} t_2) = \{(true, 1), (false, t_2(true) \frown t_2(false))\}. \end{cases}$$

Since, for $k = 1, 2$, $t_k(true) \frown t_k(false) \leq 1$, we find that

$$\begin{cases} (t_1 \,\widetilde{\wedge}_{\ell_\frown} (\widetilde{\neg}_{\ell_\frown} t_1)) \cdot true \leq (t_2 \,\widetilde{\vee}_{\ell_\frown} (\widetilde{\neg}_{\ell_\frown} t_2)) \cdot true \\ (t_1 \,\widetilde{\wedge}_{\ell_\frown} (\widetilde{\neg}_{\ell_\frown} t_1)) \cdot false \geq (t_2 \,\widetilde{\vee}_{\ell_\frown} (\widetilde{\neg}_{\ell_\frown} t_2)) \cdot false, \end{cases}$$

[8] For the introduction a Morgan algebra, we refer to [26].

[9] For the introduction of a Kleene algebra and a discussion of its meaning, we refer to [26].

whence $t_1 \mathbin{\tilde{\wedge}_{\ell\frown}} (\tilde{\neg}_{\ell\frown} t_1) \mathrel{\tilde{\leq}} t_2 \mathbin{\tilde{\vee}_{\ell\frown}} (\tilde{\neg}_{\ell\frown} t_2)$. □

This proposition can be interpreted as follows: the operators $\tilde{\neg}_{\ell\frown}$, $\tilde{\wedge}_{\ell\frown}$ and $\tilde{\vee}_{\ell\frown}$ on $\widetilde{\mathcal{T}}$ satisfy the characteristic properties of the negation, conjunction and disjunction operator in the *multi-valued strong Kleene logics*[10] with truth domain $(\widetilde{\mathcal{T}}, \leq)$ (see, for instance, [25] section 2.5; appendix section 11). This correspondence is also apparent in the definitions of a number of other important operators, combining possibilistic truth values in possibilistic $\frown$-extension logics: for the *conjunction*, the meet $\tilde{\wedge}_{\ell\frown}$ of $(\widetilde{\mathcal{T}}, \tilde{\leq})$ is used; for the *disjunction*, the join $\tilde{\vee}_{\ell\frown}$ of $(\widetilde{\mathcal{T}}, \tilde{\leq})$ is used; for the *implication* we have, taking into account property 7 for $T = \frown$, that $t_1 \mathrel{\tilde{\Rightarrow}_{\ell\frown}} t_2 = (\tilde{\neg}_{\ell\frown} t_1) \mathbin{\tilde{\vee}_{\ell\frown}} t_2$ for arbitrary t_1 and t_2 in $\widetilde{\mathcal{T}}$, which implies that this implication operator is a typical instance of what is called a *Kleene-Dienes implication* in the literature (see, for instance, [25] appendix section 11, [21] section 5.3); for the *equivalence*, it is easily verified, using the mutual distributivity of $\frown$ and $\smile$ in $(L, \leq)$, that $t_1 \mathrel{\tilde{\Leftrightarrow}_{\ell\frown}} t_2 = (t_1 \mathrel{\tilde{\Rightarrow}_{\ell\frown}} t_2) \mathbin{\tilde{\wedge}_{\ell\frown}} (t_2 \mathrel{\tilde{\Rightarrow}_{\ell\frown}} t_1)$ for arbitrary t_1 and t_2 in $\widetilde{\mathcal{T}}$. At the same time, it should be noted that if $(L, \leq)$ is a Boolean chain (of length 2), we recover Kleene's strong ternary logic (see, in this respect, subsection 5.2). The exact relationship between our possibilistic $\frown$-extension logics and Kleene's strong ternary logic is studied in detail in the following subsection.

6.2 Classical Possibility and Kleene's Strong Ternary Logic

Let us consider a universe X and two clear properties p and q. As always, we consider an ample field $\mathcal{R}$ of measurable subsets of X. We also assume that the sets $A_{P_p} \stackrel{\text{def}}{=} \{ x \mid x \in X \text{ and } x \text{ is } p \}$ and $A_{P_q} \stackrel{\text{def}}{=} \{ x \mid x \in X \text{ and } x \text{ is } q \}$ are $\mathcal{R}$-measurable. Finally, we consider a possibilistic variable ξ in $(X, \mathcal{R})$. Let us assume that we have the following information about the values that ξ may assume in X: ξ must be an element of A, with $A \in \mathcal{R} \setminus \{\emptyset\}$. This information can be represented in the form of the normalized $(\{0,1\}, \leq)$-

[10]Kleene [22] was the first to introduce the ternary logic satisfying these properties. The extension towards general multi-valued logics is mainly due to Dienes [18]. That explains why these logics are often called 'Kleene-Dienes logics'. It must be noted that on the one hand the multi-valued logics of Kleene and Dienes, and on the other hand for instance the Lukasiewicz logics do not differ as far as the negation, conjunction and disjunction operators are concerned. They do differ, however, in their implication operator (see, for instance, [25], sections 2.6 and 2.7).

possibility measure (or classical possibility measure, see [9]) Π_A on $(X, \mathcal{R})$, with, for arbitrary B in $\mathcal{R}$:

$$\Pi_A(B) = \begin{cases} 1 & ; \quad B \cap A \neq \emptyset \\ 0 & ; \quad B \cap A = \emptyset \end{cases}$$

the $(\{0,1\}, \leq)$-possibility (or simply possibility) that ξ belongs to B. Indeed, if $B \cap A = \emptyset$, then ξ cannot belong to B, since we already know that $\xi \in A$. With Π_A we can associate the dual $(\{0,1\}, \leq)$-necessity measure (or classical necessity measure, see [9]) N_A, with, for arbitrary B in $\mathcal{R}$:

$$\mathrm{N}_A(B) = \begin{cases} 1 & ; \quad A \subseteq B \\ 0 & ; \quad A \not\subseteq B \end{cases}$$

the $(\{0,1\}, \leq)$-necessity (or simply necessity) that ξ belongs to B. Indeed, if $A \subseteq B$, then ξ must belong to B, since we already know that $\xi \in A$. Remark that the distribution of Π_A, and therefore also the possibility distribution function of ξ, is the characteristic $X - \{0,1\}$-function χ_A of A.

Starting from this possibilistic information χ_A, we now ask ourselves what can be deduced about the truth values of the proposition variables 'ξ is p', 'ξ is q' and a few of their combinations. In order to answer this question, we simply apply the theory, developed in the previous sections, in the special case $(L, \leq) = (\{0,1\}, \leq)$. The only triangular norm on $(\{0,1\}, \leq)$ is the binary infimum operator or meet $\frown$ [14], which immediately leads us to the special case, discussed in this section. It should be noted that in this particular case

$$\begin{aligned} \widetilde{\mathcal{T}} &= \{\widetilde{false}, \widetilde{unknown}, \widetilde{true}\} \\ \widetilde{\leq} &= \{(\widetilde{false}, \widetilde{false}), (\widetilde{false}, \widetilde{unknown}), (\widetilde{false}, \widetilde{true}), \\ &\quad\ (\widetilde{unknown}, \widetilde{unknown}), (\widetilde{unknown}, \widetilde{true}), (\widetilde{true}, \widetilde{true})\}, \end{aligned}$$

which implies that $(\widetilde{\mathcal{T}}, \widetilde{\leq})$ is a chain of length 3, with bottom element $\widetilde{false}$, top element $\widetilde{true}$ and in between $\widetilde{unknown}$. In this chain, $\widetilde{\wedge}_{\ell\frown}$ is the binary infimum operator or meet, $\widetilde{\vee}_{\ell\frown}$ is the binary supremum operator or join, and $\widetilde{\neg}_{\ell\frown}$ is the unique, involutive negation operator. The structure $(\widetilde{\mathcal{T}}, \widetilde{\wedge}_{\ell\frown}, \widetilde{\vee}_{\ell\frown}, \widetilde{\neg}_{\ell\frown})$ is for this choice of $(L, \leq)$ and T isomorphic to the structure $(\mathcal{W}_2, \widetilde{\wedge}_{\ell T}|(\mathcal{W}_2)^2, \widetilde{\vee}_{\ell T}|(\mathcal{W}_2)^2, \widetilde{\neg}_{\ell T}|\mathcal{W}_2)$, mentioned in subsection 5.2. The structure $(\widetilde{\mathcal{T}}, \widetilde{\wedge}_{\ell\frown}, \widetilde{\vee}_{\ell\frown}, \widetilde{\neg}_{\ell\frown})$ is a Kleene algebra and is as such isomorphic to the corresponding structure of the strong ternary logic introduced by Kleene (see, for instance, [25]).

The $(\{0,1\},\leq)$-possibilistic truth value $t_{P_p} \overset{\text{def}}{=} \widetilde{\chi_{A_{P_p}}}(\chi_A)$ of the proposition variable 'ξ is p' is determined by

$$\begin{cases} t_{P_p}(\mathit{true}) = \sup\limits_{\chi_{A_{P_p}}(x)=\mathit{true}} \chi_A(x) = \begin{cases} 1; & A_{P_p} \cap A \neq \varnothing \\ 0; & A_{P_p} \cap A = \varnothing \end{cases} = \Pi_A(A_{P_p}) \\ t_{P_p}(\mathit{false}) = \sup\limits_{\chi_{A_{P_p}}(x)=\mathit{false}} \chi_A(x) = \begin{cases} 1; & \mathrm{co}A_{P_p} \cap A \neq \varnothing \\ 0; & \mathrm{co}A_{P_p} \cap A = \varnothing \end{cases} = \Pi_A(\mathrm{co}A_{P_p}), \end{cases}$$

where χ_A is the characteristic $X - \{0,1\}$-mapping of A and $\chi_{A_{P_p}}$ the characteristic $X - \mathcal{T}$-mapping of A_{P_p}[11]. For the $(\{0,1\},\leq)$-possibilistic truth value t_{P_p} there are *three possibilities*, since $t_{P_p} \in \widetilde{\mathcal{T}}$. We have that

$$t_{P_p} = \widetilde{\mathit{true}} \Leftrightarrow \begin{cases} \Pi_A(A_{P_p}) = 1 \\ \mathrm{N}_A(A_{P_p}) = 1 \end{cases} \Leftrightarrow \mathrm{N}_A(A_{P_p}) = 1 \Leftrightarrow A \subseteq A_{P_p},$$

or equivalently, it is *necessary* that ξ is p. On the other hand,

$$t_{P_p} = \widetilde{\mathit{false}} \Leftrightarrow \begin{cases} \Pi_A(A_{P_p}) = 0 \\ \mathrm{N}_A(A_{P_p}) = 0 \end{cases} \Leftrightarrow \Pi_A(A_{P_p}) = 0 \Leftrightarrow A \cap A_{P_p} = \varnothing,$$

or equivalently, it is *impossible* that ξ is p. Finally, we have that

$$t_{P_p} = \widetilde{\mathit{unknown}} \Leftrightarrow \begin{cases} \Pi_A(A_{P_p}) = 1 \\ \mathrm{N}_A(A_{P_p}) = 0 \end{cases} \Leftrightarrow \begin{cases} A \cap A_{P_p} \neq \varnothing \\ A \cap \mathrm{co}A_{P_p} \neq \varnothing, \end{cases}$$

or equivalently, it is possible but not necessary that ξ is p, in other words, it is *uncertain* whether ξ is p. These observations completely agree with the interpretation of the possibilistic truth values $\widetilde{\mathit{true}}$, $\widetilde{\mathit{unknown}}$ and $\widetilde{\mathit{false}}$, given in section 5.

For the possibilistic truth value t_{P_q} of the proposition variable 'ξ is q', completely analogous observations can be made. Let us now turn our attention to the $(\{0,1\},\leq)$-possibilistic truth value of the proposition variable 'NOT(ξ is p)', or equivalently, (NOT P_p)(ξ), or 'ξ is not p'. It is obvious that $A_{\mathrm{NOT}\,P_p} = \mathrm{co}A_{P_p}$, whence

$$\begin{cases} t_{\mathrm{NOT}\,P_p}(\mathit{true}) = \Pi_A(A_{\mathrm{NOT}\,P_p}) = \Pi_A(\mathrm{co}A_{P_p}) = t_{P_p}(\mathit{false}) \\ t_{\mathrm{NOT}\,P_p}(\mathit{false}) = \Pi_A(\mathrm{co}A_{\mathrm{NOT}\,P_p}) = \Pi_A(A_{P_p}) = t_{P_p}(\mathit{true}). \end{cases}$$

[11]The reader will have noticed that we work with two Boolean chains: $(\{0,1\},\leq)$ for the representation of the uncertainty about the values ξ can assume in X, and $(\mathcal{T},\leq)$ for the truth values of propositions.

We may therefore write, taking into account proposition 3(i) for $T = \frown$, that $t_{\text{NOT } P_p} = \tilde{\neg}_{\ell\frown} t_{P_p}$. We conclude that for the logical negation operator of classical propositional logic, there is *always truth-functionality* as far as the $(\{0, 1\}, \leq)$-possibilistic truth values are concerned.

Let us now investigate the proposition variable 'ξ is p AND ξ is q', or equivalently, $(P_p \text{ AND } P_q)(\xi)$, where P_p AND P_q is a proposition function that is the pointwise conjunction of the proposition functions P_p and P_q. It is obvious that $A_{P_p \text{ AND } P_q} = A_{P_p} \cap A_{P_q}$, whence

$$t_{P_p \text{ AND } P_q}(true) = \Pi_A(A_{P_p \text{ AND } P_q}) = \Pi_A(A_{P_p} \cap A_{P_q})$$

and, also taking into account proposition 3(ii) for $T = \frown$,

$$\begin{aligned} t_{P_p \text{ AND } P_q}(false) &= \Pi_A(\text{co}A_{P_p \text{ AND } P_q}) \\ &= \Pi_A(\text{co}(A_{P_p} \cap A_{P_q})) \\ &= \Pi_A(\text{co}A_{P_p} \cup \text{co}A_{P_q}) \\ &= \Pi_A(\text{co}A_{P_p}) \smile \Pi_A(\text{co}A_{P_q}) \\ &= t_{P_p}(false) \smile t_{P_q}(false) \\ &= (t_{P_p} \tilde{\wedge}_{\ell\frown} t_{P_q}) \cdot false. \end{aligned}$$

Only if

$$\Pi_A(A_{P_p} \cap A_{P_q}) = \Pi_A(A_{P_p}) \frown \Pi_A(A_{P_q}) \tag{4.24}$$

we have, taking into account proposition 3(ii) for $T = \frown$,

$$\begin{aligned} t_{P_p \text{ AND } P_q}(true) &= \Pi_A(A_{P_p} \cap A_{P_q}) = \Pi_A(A_{P_p}) \frown \Pi_A(A_{P_q}) \\ &= t_{P_p}(true) \frown t_{P_q}(true) = (t_{P_p} \tilde{\wedge}_{\ell\frown} t_{P_q}) \cdot true. \end{aligned}$$

We conclude that only in this case there is *truth-functionality* for the logical conjunction operator in classical propositional logic as far as the possibilistic truth values are concerned, or equivalently,

$$t_{P_p \text{ AND } P_q} = t_{P_p} \tilde{\wedge}_{\ell\frown} t_{P_q}, \tag{4.25}$$

Let us also investigate the proposition variable 'ξ is p OR ξ is q', or equivalently, $(P_p \text{ OR } P_q)(\xi)$, where P_p OR P_q is a proposition function that is the pointwise disjunction of the proposition functions P_p and P_q. It is obvious that $A_{P_p \text{ OR } P_q} = A_{P_p} \cup A_{P_q}$, whence, taking into account proposition 3(iii) for $T = \frown$,

$$\begin{aligned} t_{P_p \text{ OR } P_q}(true) &= \Pi_A(A_{P_p \text{ OR } P_q}) \\ &= \Pi_A(A_{P_p} \cup A_{P_q}) \\ &= \Pi_A(A_{P_p}) \smile \Pi_A(A_{P_q}) \\ &= t_{P_p}(true) \smile t_{P_q}(true) \\ &= (t_{P_p} \tilde{\vee}_{\ell\frown} t_{P_q}) \cdot true. \end{aligned}$$

and

$$t_{P_p \text{ OR } P_q}(false) = \Pi_A(\text{co}A_{P_p \text{ OR } P_q}) = \Pi_A(\text{co}A_{P_p} \cap \text{co}A_{P_q}).$$

Only if

$$\Pi_A(\text{co}A_{P_p} \cap \text{co}A_{P_q}) = \Pi_A(\text{co}A_{P_p}) \frown \Pi_A(\text{co}A_{P_q}) \tag{4.26}$$

we have, also taking into account proposition 3(iii) for $t = \frown$, that

$$\begin{aligned} t_{P_p \text{ OR } P_q}(false) &= \Pi_A(\text{co}A_{P_p} \cap \text{co}A_{P_q}) = \Pi_A(\text{co}A_{P_p}) \frown \Pi_A(\text{co}A_{P_q}) \\ &= t_{P_p}(false) \frown t_{P_q}(false) = (t_{P_p} \tilde{\vee}_{\ell\frown} t_{P_q}) \cdot false \end{aligned}$$

We conclude that only in this case there is *truth-functionality* for the logical disjunction operator in classical propositional logic as far as the possibilistic truth values are concerned, or equivalently,

$$t_{P_p \text{ OR } P_q} = t_{P_p} \tilde{\vee}_{\ell\frown} t_{P_q}. \tag{4.27}$$

Let us now briefly discuss the meaning of (4.24)–(4.27). It is easily shown that (4.24), and therefore also (4.25), does not hold if and only if $A \cap A_{P_p \text{ AND } P_q} = \emptyset$ and at the same time

$$A \cap A_{P_p} \neq \emptyset \text{ and } A \cap \text{co}A_{P_p} \neq \emptyset \text{ and } A \cap A_{P_q} \neq \emptyset \text{ and } A \cap \text{co}A_{P_q} \neq \emptyset,$$

in other words, if and only if it is *uncertain* (i.e., not impossible and not necessary) whether ξ is p and whether ξ is q, and at the same time *impossible* that ξ is p AND ξ is q. Indeed, in that case, we have that $t_{P_p \text{ AND } P_q} = \widetilde{false}$, whereas $t_{P_p} \tilde{\wedge}_{\ell\frown} t_{P_q} = \widetilde{unknown} \tilde{\wedge}_{\ell\frown} \widetilde{unknown} = \widetilde{unknown}$. A similar argument can be given for the disjunction. We conclude that there is *not necessarily truth-functionality* for the logical disjunction and conjunction operators of classical propositional logic, as far as the $(\{0,1\}, \leq)$-possibilistic truth values are concerned. In other words, (4.25) and (4.27) are not necessarily valid for arbitrary clear properties p and q, with $\mathcal{R}$-measurable A_{P_p} and A_{P_q}. Our possibilistic approach therefore only results in a strong ternary Kleene logic if a number of independence properties are satisfied[12]. In some cases these conditions are not satisfied, and our possibilistic approach is therefore *not truth-functional*, and therefore does not lead to a strong ternary Kleene logic. In these cases however, the strong ternary Kleene logic does provide us with a conservative approximation, since wherever it goes wrong, it will result in the possibilistic truth value $\widetilde{unknown}$, where our possibilistic approach would yield the possibilistic truth values $\widetilde{true}$ or $\widetilde{false}$ (see also proposition 2).

[12]It is shown in [9, 15] that conditions (4.24) and (4.26) are related to the conditions for the possibilistic (or logical) independence of the events A_{P_p} and A_{P_q}.

7 CONCLUSION

In the previous sections, we have shown how a possibilistic logic can be constructed. Possibilistic logic can be described as a set of techniques that enable us to incorporate linguistic (possibilistic) uncertainty in classical propositional logic. It turns out that under a number of independence assumptions, possibilistic logic leads to the special case of a possibilistic extension logic. A special subclass of these, the possibilistic-logical $\frown$-extensions, are related with strong multi-valued Kleene logics. Thus, a possibilistic justification is given for the introduction and use of these Kleene systems.

There are a number of problems, however, which have not been dealt with in this paper. Among them, we explicitly mention the decomposability problem. In classical propositional logic, there exist basic sets of classical-logical functions, such that all other classical-logical functions can be expressed in terms of these functions. The following question can then be asked: is a similar result valid in a possibilistic extension logic, i.e., can any member of this logic be decomposed in terms of a basic set? And if so, what is the relationship between the decomposition of a classical-logical function and the decomposition of its possibilistic-logical extension? A partial answer in the special case $(L, \leq) = ([0, 1], \leq)$ and $T = \min$ has been given in [16], theorem 4.1, although it must be mentioned that the conditions imposed in this theorem are to weak. For a better and more general formulation of this theorem, we refer to [9] chapter 8. Also, in the related domain of reliability theory, we have proven a decomposability property for possibilistic structure functions, which are, for fixed $(L, \leq)$ and T, isomorphic to an isotonic subclass of the corresponding $(L, \leq)$-possibilistic T-extension logic. These results will be reported on elsewhere (see, however, also [10]).

Acknowledgements

Gert de Cooman is a Senior Research Assistant of the Belgian National Fund for Scientific Research. He would like to thank this institution for funding the research, reported on in this article.

At the same time, he would like to dedicate this work to Prof. dr. Etienne Kerre who, first as a mentor and a teacher, and later as a friend, has influenced his views on what the science of mathematics is, and what it should be.

REFERENCES

[1] G. Birkhoff. *Lattice Theory*, volume XXV. A.M.S. Colloquium Publications, Providence, Rhode Island, 1967.

[2] C. W. Burrill. *Measure, Integration and Probability*. McGraw-Hill, New York, 1972.

[3] B. A. Davey and H. A. Priestley. *Introduction to Lattices and Order*. Cambridge University Press, Cambridge, 1990.

[4] G. De Cooman. Possibility theory: Conditional possibility. In preparation.

[5] G. De Cooman. Possibility theory: Possibilistic independence. In preparation.

[6] G. De Cooman. Possibility theory: The measure- and integral-theoretic groundwork. In preparation.

[7] G. De Cooman. Evaluation sets: An order-theoretic approach to fuzzy set theory. *CC-AI*, 8:81–101, 1991.

[8] G. De Cooman. Evaluation sets and mappings — the order-theoretic aspect of the meaning of properties. In E. E. Kerre, editor, *Introduction to the Basic Principles of Fuzzy Set Theory and Some of Its Applications*, pages 159–213. Communication & Cognition, Ghent, 1991.

[9] G. De Cooman. *Evaluatieverzamelingen en -afbeeldingen - Een ordetheorische benadering van vaagheid en onzekerheid [Evaluation Sets and Mappings—An Order-Theoretic Approach to Vagueness and Uncertainty]*. PhD thesis, Universiteit Gent, 1993.

[10] G. De Cooman and B. Cappelle. A possibilistic uncertainty model in classical reliability theory. In *Fuzzy Logic and Intelligent Technologies in Nuclear Science (Proceedings of the First International FLINS Workshop)*, pages 19–25, Mol, Belgium, September 14–16, 1994.

[11] G. De Cooman and E. E. Kerre. *An Order-Theoretic Description of Vagueness, Imprecision and Uncertainty*. In preparation.

[12] G. De Cooman and E. E. Kerre. Possibility and necessity integrals. Accepted for publication in Fuzzy Sets and Systems.

[13] G. De Cooman and E. E. Kerre. Ample fields. *Simon Stevin*, 67:235–244, 1993.

[14] G. De Cooman and E. E. Kerre. Order norms on bounded partially ordered sets. *The Journal of Fuzzy Mathematics*, 2:281–310, 1994.

[15] G. De Cooman and E. E. Kerre. A new approach to possibilistic independence. In *Proceedings of Third IEEE Conference on Fuzzy Systems (FUZZ-IEEE'94, IEEE World Congress on Computational Intelligence)*, volume 2, pages 1446–1451, Orlando, Florida, June 26–29, 1994.

[16] G. De Cooman, E. E. Kerre, B. Cappelle, D. Ruan, and F. Vanmassenhove. On the extension of classical propositional logic by means of a triangular norm. *International Journal of Intelligent Systems*, 5:307–322, 1990.

[17] G. De Cooman, E. E. Kerre, and F. Vanmassenhove. Possibility theory: An integral theoretic approach. *Fuzzy Sets and Systems*, 46:287–300, 1992.

[18] Z. P. Dienes. On an implication function in many-valued systems of logic. *The Journal of Symbolic Logic*, 14:95–97, 1949.

[19] B. R. Gaines. Foundations of fuzzy reasoning. *International Journal of Man-Machine Studies*, 8:623–668, 1976.

[20] J. A. Goguen. *L*-fuzzy sets. *Journal of Mathematical Analysis and Applications*, 18:145–174, 1967.

[21] E. E. Kerre. Basic principles of fuzzy set theory for the representation and manipulation of imprecision and uncertainty. In E. E. Kerre, editor, *Introduction to the Basic Principles of Fuzzy Set Theory and Some of Its Applications*, pages 1–158. Communication & Cognition, Ghent, 1991.

[22] S. C. Kleene. *Introduction to Metamathematics.* Van Nostrand, New York, 1952.

[23] E. Mendelson. *Theory and Problems of Boolean Algebra and Switching Circuits.* McGraw-Hill, New York, 1970.

[24] Wang Pei-Zhang. Fuzzy contactability and fuzzy variables. *Fuzzy Sets and Systems*, 8:81–92, 1982.

[25] N. Rescher. *Many-Valued Logic.* McGraw-Hill, New York, 1969.

[26] H. J. Skala. On many-valued logics, fuzzy sets, fuzzy logics and their applications. *Fuzzy Sets and Systems,* 1:129–149, 1978.

[27] A. Van Schooten. *Ontwerp en implementatie van een model voor de representatie en manipulatie van onzekerheid en imprecisie in databanken en expertsystemen [Design and Implementation of a Model for the Representation and Manipulation of Uncertainty and Imprecision in Databases and Expert Systems].* PhD thesis, Universiteit Gent, 1989.

[28] L. A. Zadeh. Fuzzy sets. *Information and Control,* 8:338–353, 1965.

[29] L. A. Zadeh. Calculus of fuzzy restrictions. In L. A. e.a. Zadeh, editor, *Fuzzy Sets and Their Applications To Cognitive and Decision Processes,* pages 1–39. Academic Press, New York, 1975.

[30] L. A. Zadeh. Fuzzy sets as a basis for a theory of possibility. *Fuzzy Sets and Systems,* 1:3–28, 1978.

Part 2

DECISION MAKING AND SYSTEMS SCIENCE

5

FUZZY DYNAMIC PROGRAMMING: A NEW QUALITY THROUGH FUZZY SETS

Janusz Kacprzyk

Systems Research Institute
Polish Academy of Sciences
ul. Newelska 6
01-447 Warsaw, Poland

ABSTRACT

We briefly show how fuzzy sets theory has provided dynamic programming, one of the most powerful tools of applied mathematics, with new qualities and prospects, by making it possible to more adequately formulate inherently "soft" real world problems.

1 INTRODUCTION

Dynamic programming (DP), introduced by Bellman in the mid-1950s, is one of the most powerful tools of applied mathematics, which has found applications in a variety of fields exemplified by automatic control, operations research, data analysis, etc.

One of the crucial elements of real world problems is their inherent "softness," the relevant aspect of which is uncertainty and imprecision to be reflected in the problem's model. Initially, uncertainty was solely equated with randomness leading to stochastic DP. With the inception of fuzzy sets (Zadeh, 1965), a formal tool became available for dealing with imprecision.

And, indeed, DP was one of the earliest fundamental methodologies to which fuzzy sets theory was applied (Chang, 1969; Bellman and Zadeh, 1970), leading to fuzzy dynamic programming (FDP). Then, numerous works on FDP followed, with a recent renewal of interest. This progress in FDP is documented in the surveys by, e.g., Esogbue and Bellman (1982,

1984), Esogbue, Fedrizzi and Kacprzyk (1988) and Kacprzyk (1994), and Kacprzyk's (1983b) book.

The purpose of this paper is to show how fuzzy sets theory has provided DP with new qualities to better account for the softness of real-world problems. Our exposition will be in the form of a readable survey of more relevant FDP models covering specific soft aspects of problem formulations.

2 DECISION MAKING AND CONTROL UNDER FUZZINESS

Virtually all works related to decision making under fuzziness, including those on FDP, take Bellman and Zadeh's (1970) framework as a point of departure. Its basic elements are: a fuzzy goal G in X, a fuzzy constraint C in X, and a fuzzy decision D in X; X is a (nonfuzzy) space of options (alternatives, variants, ...).

Basically, we wish to "satisfy C and attain G," which leads to the fuzzy decision

$$\mu_D(x) = \mu_C(x) \wedge \mu_G(x), \tag{5.1}$$

which yields the "goodness" of an $x \in X$ as a solution: from 1 for definitely good (perfect) to 0 for definitely bad (unacceptable), through all intermediate values. The '$\wedge$' (min) is commonly used, and is assumed here, but may be replaced by, e.g, a t-norm.

For an optimal (nonfuzzy) solution, an $x^* \in X$ such that

$$\mu_D(x^*) = \sup_{x \in X} \mu_D(x) = \sup_{x \in X}(\mu_C(x) \wedge \mu_G(x)) \tag{5.2}$$

is a natural (but not the only possible) choice.

Suppose now that we have a deterministic system under described by a state transition equation

$$x_{t+1} = f(x_t, u_t), \qquad t = 0, 1, \ldots \tag{5.3}$$

where $x_t, x_{t+1} \in X = \{x\} = \{s_1, \ldots, s_n\}$ are the states (equated here for simplicity with outputs) at time (control stage) t and $t + 1$, respectively, and $u_t \in U = \{u\} = \{c_1, \ldots, c_m\}$ is the control (input) at t.

At each t, u_t is subjected to a fuzzy constraint $\mu_{C^t}(u_t)$ and a fuzzy goal $\mu_{G^{t+1}}(x_{t+1})$ is imposed on x_{t+1}. The fuzzy decision is therefore

$$\begin{aligned}\mu_D(u_0,\ldots,u_{N-1}|x_0) &= \mu_{C^0}(u_0) \wedge \mu_{G^1}(x_1) \wedge \ldots \\ &\qquad \wedge \mu_{C^{N-1}}(u_{N-1}) \wedge \mu_{G^N}(x_N) \\ &= \bigwedge_{t=0}^{N-1} (\mu_{C^t}(u_t) \wedge \mu_{G^{t+1}}(x_{t+1})), \end{aligned} \qquad (5.4)$$

where $x_0 \in X$ is an initial state, the x_{t+1}'s are given by (3), and N is a termination time (fixed and specified for now).

We seek an optimal sequence of controls $u_0^*,\ldots,u_{N-1}^*$ such that

$$\begin{aligned}\mu_D(u_0^*,\ldots,u_{N-1}^*|x_0) &= \max_{u_0,\ldots,u_{N-1}} \mu_D(u_0,\ldots,u_{N-1}|x_0) \\ &= \max_{u_0,\ldots,u_{N-1}} \bigwedge_{t=0}^{N-1} (\mu_{C^t}(u_t) \wedge \mu_{G^{t+1}}(x_{t+1})). \end{aligned} \qquad (5.5)$$

It may readily be seen that this general problem formulation may be a starting point for numerous extensions. A convenient classification of problem classes is here with respect to (Kacprzyk, 1983b):

- the type of the termination time: (1) fixed and specified in advance, (2) fuzzy, (3) implicitly given, and (4) infinite;
- the type of the system under control: (1) deterministic, (2) stochastic, and (3) fuzzy.

In virtually all these cases, an FDP algorithm may be devised, and they will be presented in this paper.

3 FDP FORMULATION FOR A FIXED AND SPECIFIED TERMINATION TIME

The case of a fixed and specified (in advance) termination time is basic, hence will be discussed in more detail.

3.1 The case of a deterministic system under control

A deterministic system under control is described by (3), that is, $x_{t+1} = f(x_t, u_t)$; $x_t, x_{t+1} \in X = \{s_1, \ldots, s_n\}, u_t \in U = \{c_1, \ldots, c_m\}; t = 0, 1, \ldots, N-1$. The fuzzy constraints are $\mu_{C^0}(u_0), \ldots, \mu_{C^{N-1}}(u_{N-1})$ and the fuzzy goal is $\mu_{G^N}(x_N)$. The fuzzy decision is

$$\mu_D(u_0, \ldots, u_{N-1}|x_0) = \mu_{C^0}(u_0) \wedge \ldots \wedge \mu_{C^{N-1}}(u_{N-1}) \wedge \mu_{G^N}(x_N) \tag{5.6}$$

and the problem is to find $u_0^*, \ldots, u_{N-1}^*$ such that

$$\mu_D(u_0^*, \ldots, u_{N-1}^*|x_0) = \max_{u_0, \ldots, u_{N-1}} (\mu_{C^0}(u_0) \wedge \ldots \wedge \mu_{C^{N-1}}(u_{N-1}) \wedge \mu_{G^N}(x_N)). \tag{5.7}$$

It is easy to see that only the two last terms in (7), i.e., $\mu_{C^{N-1}}(u_{N-1}) \wedge \mu_{G^N}(f(x_{N-1}, u_{N-1}))$, depend upon u_{N-1}, hence (7) can be rewritten as

$$\begin{aligned} &\max_{u_0, \ldots, u_{N-1}} (\mu_{C^0}(u_0) \wedge \ldots \wedge \mu_{C^{N-1}}(u_{N-1}) \wedge \mu_{G^N}(x_N)) = \\ &\quad \max_{u_0, \ldots, u_{N-2}} (\mu_{C^0}(u_0) \wedge \ldots \wedge \mu_{C^{N-2}}(u_{N-2}) \\ &\quad \wedge \max_{u_{N-1}}(\mu_{C^{N-1}}(u_{N-1}) \wedge \mu_{G^N}(f(x_{N-1}, u_{N-1})))). \end{aligned} \tag{5.8}$$

On repeating this backward iteration for $u_{N-2}, \ldots, u_0$, which is a simple example of DP, we obtain the set of FDP recurrence equations

$$\begin{cases} \mu_{G^{N-i}}(x_{N-i}) = \max\limits_{u_{N-i}}(\mu_{C^{N-i}}(u_{N-i}) \wedge \mu_{G^{N-i+1}}(x_{N-i+1})) \\ x_{N-i+1} = f(x_{N-1}, u_{N-1}), \quad i = 1, \ldots, N \end{cases} \tag{5.9}$$

where $\mu_{G^{N-i}}(x_{N-i})$ is the membership function of a fuzzy goal at $t = N - i$ induced by a fuzzy goal at $t = N - i + 1$.

An optimal sequence of controls sought, $u_0^*, \ldots, u_{N-1}^*$, is given by the successive maximizing values of u_{N-i} in (9) or, what is more convenient, by an optimal policy (function) $a_t^*: X \to U$ such that $u_t^* = a_t^*(x_t), t = 0, 1, \ldots, N-1$, relating an optimal decision (control) to the current state.

3.2 The case of a stochastic system under control

The system under control is assumed to be a Markov chain the temporal evolution of which is described by a conditional probability $p(x_{t+1}|x_t, u_t)$;

$x_t, x_{t+1} \in X = \{s_1, \ldots, s_n\}; u_t \in U = \{c_1, \ldots, c_m\}$. At each t, a fuzzy constraint $\mu_{C^t}(u_t)$ is imposed on $u_t, t = 0, 1, \ldots, N-1$, and a fuzzy goal $\mu_{G^N}(x_N)$ is imposed on x_N.

In the two problem formulations, we will use Zadeh's (1968) classic (non-fuzzy) probability of a fuzzy event.

We have now the following two problem formulations:

- due to Bellman and Zadeh (1970): find an optimal sequence of controls $u_0^*, \ldots, u_{N-1}^*$ to maximize the probability of attainment of the fuzzy goal subjected to the fuzzy constraints, written

$$\mu_D(u_0^*, \ldots, u_{N-1}^* | x_0) = \max_{u_0, \ldots, u_{N-1}} (\mu_{C^0}(u_0) \wedge \ldots \wedge \mu_{C^{N-1}}(u_{N-1}) \wedge E\mu_{G^N}(x_N)) \tag{5.10}$$

  where (cf. Zadeh, 1968)

$$E\mu_{G^N}(x_N) = \sum_{x_N \in X} p(x_N | x_{N-1}, u_{N-1}) \mu_{G^N}(x_N); \tag{5.11}$$

- due to Kacprzyk and Staniewski (1980): find an optimal sequence of controls $u_0^*, \ldots, u_{N-1}^*$ to maximize the expectation of the fuzzy decision's membership function; this may be written as

$$\mu_D(u_0^*, \ldots, u_{N-1}^* | x_0) = \max_{u_0, \ldots, u_{N-1}} E(\mu_{C^0}(u_0) \wedge \ldots \wedge \mu_{C^{N-1}}(u_{N-1}) \wedge \mu_{G^N}(x_N)) \tag{5.12}$$

Notice that the two formulations are clearly not equivalent.

For Bellman and Zadeh's formulation, an FDP algorithm may be devised. Namely, it is easy to see that the two right-most terms of (10), namely $\mu_{C^{N-1}}(u_{N-1}) \wedge E\mu_{G^N}(f(x_{N-1}, u_{N-1}))$, depend only upon u_{N-1}, the next two right-most terms depend only upon u_{N-2}, etc.; the structure of (10) is essentially the same as that of (7), hence the FDP set of recurrence equations is

$$\begin{cases} \mu_{G^{N-i}}(x_{N-i}) = \max\limits_{u_{N-i}} (\mu_{C^{N-i}}(u_{N-i}) \wedge E\mu_{G^{N-i+1}}(x_{N-i+1})) \\ E\mu_{G^{N-i+1}}(x_{N-i+1}) = \sum\limits_{x_{N-i+1} \in X} p(x_{N-i+1} | x_{N-i}, u_{N-i}) \mu_{G^{N-i+1}}(x_{N-i+1}) \\ i = 1, \ldots, N \end{cases} \tag{5.13}$$

By solving this set, we consecutively obtain u_{N-i}^*, or in fact optimal policies a_{N-i}^* such that $u_{N-i}^* = a_{N-i}^*(x_{N-i}), i = 1, \ldots, N$.

Kacprzyk and Staniewski's (1980) approach, which is clearly more difficult, may be solved by an iterative scheme, and we refer the reader to the source paper or Kacprzyk's (1983b) book.

3.3 The case of a fuzzy system under control

The fuzzy system under control is described by a fuzzy state transition equation

$$X_{t+1} = F(X_t, U_t), \qquad t = 0, 1, \ldots \tag{5.14}$$

where $X_t, X_{t+1} \subseteq X = \{x\}$ are fuzzy states at time (control stage) t and $t+1$, respectively, and $U_t \subseteq U = \{u\}$ is a fuzzy control at t; X_t, X_{t+1}, and U_t are characterized by their membership functions $\mu_{X_t}(x_t)$, $\mu_{X_{t+1}}(x_{t+1})$, and $\mu_{U_t}(u_t)$, respectively; (14) is equivalent to a conditioned fuzzy set $\mu_{X_{t+1}}(x_{t+1}|x_t, u_t)$.

An FDP algorithm was here proposed by Baldwin and Pilsworth (1982). As previously, at each t, a fuzzy constraint $\mu_{C^t}(u_t)$ is imposed on U_t , and X_{t+1} is subjected to a fuzzy goal $\mu_{G^{t+1}}(x_{t+1}), t = 0, 1, \ldots, N-1$.

For each $t = 0, 1, \ldots, N-1$, we construct a fuzzy relation $R^t = C^t \times G^{t+1}$, such that

$$\mu_{R^t}(u_t, x_{t+1}) = \mu_{C^t}(u_t) \wedge \mu_{G^{t+1}}(x_{t+1}) \tag{5.15}$$

and the degree to which U_t and X_{t+1} satisfy C^t and G^{t+1} is

$$\begin{aligned}\mu_T(u_t, \mu_{R^t}(u_t, x_{t+1}), x_{t+1}) &= \max_{x_{t+1}}(\max_{u_t}(\mu_{U_t}(u_t) \\ &\quad \wedge \mu_{R^t}(u_t, x_{t+1})) \wedge \mu_{X_{t+1}}(x_{t+1})) \\ &= \max_{u_t}((\mu_{U_t}(u_t) \wedge \mu_{C^t}(u_t)) \\ &\quad \wedge \max_{x_{t+1}}(\mu_{X_{t+1}}(x_{t+1}) \wedge \mu_{G^{t+1}}(x_{t+1}))) \end{aligned} \tag{5.16}$$

The fuzzy decision is

$$\begin{aligned}\mu_D(U_0, \ldots, U_{N-1}|X_0) &= \mu_T(u_0, \mu_{R^0}(u_0, x_1), x_1) \wedge \ldots \\ &\quad \wedge \mu_T(u_{N-1}, \mu_{R^{N-1}}(u_{N-1}, x_N), x_N) \\ &= \max_{u_0}((\mu_{U_0}(u_0) \wedge \mu_{C^0}(u_0)) \\ &\quad \wedge \max_{x_1}(\mu_{X_1}(x_1) \wedge \mu_{G^1}(x_1))) \wedge \ldots \\ &\quad \wedge \max_{u_{N-1}}((\mu_{U_{N-1}}(u_{N-1}) \wedge \mu_{C^{N-1}}(u_{N-1}))\end{aligned}$$

$$\wedge \max_{x_N}(\mu_{X_N}(x_N) \wedge \mu_{G^N}(x_N))) \tag{5.17}$$

and we seek $U_0^*, \ldots, U_{N-1}^*$ such that

$$\mu_D(U_0^*, \ldots, U_{N-1}^* | X_0) = \max_{U_0, \ldots, U_{N-1}} \mu_D(U_0, \ldots, U_{N-1} | X_0). \tag{5.18}$$

Assuming now for simplicity that the fuzzy constraints are imposed at $t = 0, 1, \ldots, N-1$ but the fuzzy goal is imposed only at $t = N$, we obtain the fuzzy decision

$$\begin{aligned} &\mu_D(U_0, \ldots, U_{N-1} | X_0) = \\ &\quad \max_{u_0}(\mu_{U_0}(u_0) \wedge \mu_{C^0}(u_0)) \wedge \ldots \wedge \max_{u_{N-1}}((\mu_{U_{N-1}}(u_{N-1}) \\ &\quad \wedge \mu_{C^{N-1}}(u_{N-1})) \wedge \max_{x_N}(\mu_{X_N}(x_N) \wedge \mu_{G^N}(x_N))) \end{aligned} \tag{5.19}$$

and the problem is to find $U_0^*, \ldots, U_{N-1}^*$ such that

$$\begin{aligned} \mu_D(U_0^*, \ldots, U_{N-1}^* | X_0) &= \max_{U_0, \ldots, U_{N-1}} \mu_D(U_0, \ldots, U_{N-1} | X_0) \\ &= \max_{U_0} \max_{u_0}(\mu_{U_0}(u_0) \wedge \mu_{C^0}(u_0)) \wedge \ldots \\ &\quad \wedge \max_{U_{N-1}}(\max_{u_{N-1}}(\mu_{U_{N-1}}(u_{N-1}) \wedge \mu_{C^{N-1}}(u_{N-1})) \\ &\quad \wedge \max_{X_N}(\max_{x_N}(\mu_{X_N}(x_N) \wedge \mu_{G^N}(x_N)))) \end{aligned} \tag{5.20}$$

The structure of (20) is similar to that of (11), hence the set of FDP recurrence equations is

$$\begin{cases} \mu_{G^{1N}}(X_N) = \max_{x_N}(\mu_{X_N}(x_N) \wedge \mu_{G^N}(x_N)) \\ \mu_{G^{1N-i}}(X_{N-i}) = \max_{U_{N-i}}(\max_{u_{N-i}}(\mu_{U_{N-i}}(u_{N-i}) \wedge \mu_{C^{N-i}}(u_{N-i})) \\ \qquad\qquad \wedge \mu_{G^{1N-i+1}}(X_{N-i+1})) \\ \mu_{X_{N-i+1}}(X_{N-i+1}) = \max_{x_{N-i}}(\max_{u_{N-i}}(\mu_{U_{N-i}}(u_{N-i}) \wedge \mu_{X_{N-i+1}}(x_{N-i+1} | x_{N-i}, u_{N-i})) \\ \qquad\qquad \wedge \mu_{X_{N-i}}(x_{N-i})), \\ i = 1, \ldots, N-1 \end{cases} \tag{5.21}$$

This set may eventually be solved in principle. However, $\mu_{G^{1N-i}}(x_{N-i})$ must be specified for all possible X_{N-i}'s and the maximization is to proceed over all possible U_{N-i}'s. To make the problem solvable, Baldwin and Pilsworth (1982) predefine some (sufficiently small) number of reference (standard) fuzzy states and fuzzy controls, and redefine the problem in terms of the reference fuzzy states and fuzzy controls to make it solvable. This idea is close to Kacprzyk and Staniewski's (1982), and is related to Kacprzyk's (1993a–c) recent works on interpolative reasoning for multistage fuzzy control.

In general, Baldwin and Pilsworth's (1982) approach is complicated, and the simple branch-and-bound approach by Kacprzyk (1979, 1983b) may be a better choice.

4 FDP FORMULATION WITH A FUZZY TERMINATION TIME

In practice, e.g., in socioeconomic problems, it may be more adequate to assume a fuzzy termination time as, say, "more or less 5 years," as was proposed by Fung and Fu (1977) and Kacprzyk (1977).

Let $R = \{0, 1, \ldots, K-1, K, K+1, \ldots, N\}$ be the set of control stages. At each $t \in R$, we have a fuzzy constraint $\mu_{C^t}(u_t)$; a fuzzy goal $\mu_{G^\nu}(x_\nu), \nu \in R$, is imposed on the final state. The fuzzy termination time is given by $\mu_T(\nu), \nu \in R$, viewed as a degree of how preferable ν is as the termination time: from 1 for the most preferable (perfect) to 0 for unacceptable, through all intermediate values.

The fuzzy decision is now

$$\mu_D(u_0, \ldots, u_{\nu-1} | x_0) = \mu_{C^0}(u_0) \wedge \ldots \wedge \mu_{C^{\nu-1}}(u_{\nu-1}) \wedge \mu_{G^\nu}(x_\nu) \tag{5.22}$$

and the problem is to find an optimal termination time ν^* and an optimal sequence of controls $u_0^*, \ldots, u_{\nu^*-1}^*$ such that

$$\begin{aligned} \mu_D(u_0^*, \ldots, u_{\nu^*-1}^* | x_0) &= \max_{\nu, u_0, \ldots, u_{\nu-1}} (\mu_{C^0}(u_0) \wedge \ldots \\ &\quad \wedge \mu_{C^{\nu-1}}(u_{\nu-1}) \wedge \mu_T(\nu) \mu_{G^\nu}(x_\nu)). \end{aligned} \tag{5.23}$$

4.1 The case of a deterministic system under control

In Kacprzyk's (1977, 1978c) original formulation, the set of possible termination times is $\{\nu \in R: \mu_T(\nu) > 0\} = \{K, K+1, \ldots, N\} \subseteq R$, hence an optimal sequence of controls is $u_0^*, \ldots, u_{K-2}^*, u_{K-1}^*, u_K^*, \ldots, u_{\nu^*-1}^*$. Its second part, i.e., $u_{K-1}^*, u_K^*, \ldots, u_{\nu^*-1}^*$, is determined by solving

$$\begin{cases} \mu_{G^{\nu-i}}(x_{\nu-i}, \nu) = \max\limits_{u_{\nu-i}}(\mu_{C^{\nu-i}}(u_{\nu-i}) \wedge \mu_{G^{\nu-i+1}}(x_{\nu-i+1}, \nu)) \\ x_{\nu-i+1} = f(x_{\nu-i}, u_{\nu-i}), \quad i = 1, 2, \ldots, \nu - K + 1; \nu = K, K+1, \ldots, N-1 \end{cases} \tag{5.24}$$

where $\mu_{G^\nu}(x_\nu, \nu) = \mu_T(\nu)\mu_{G^\nu}(x_\nu)$.

An optimal termination time ν^* is then determined by the maximizing ν in

$$\mu_{G^{K-1}}(x_{K-1}) = \max_{\nu} \mu_{G^{K-1}}(x_{K-1}, \nu). \tag{5.25}$$

The first part of the optimal sequence of control, i.e., $u_0^*, \ldots, u_{K-2}^*$, is then determined by solving

$$\begin{cases} \mu_{G^{K-1-i}}(x_{K-1-i}) = \max\limits_{u_{K-1-i}}(\mu_{C^{K-1-i}}(u_{K-1-i}) \wedge \mu_{G^{K-i}}(x_{K-i})) \\ x_{K-i} = f(x_{K-1-i}, u_{K-1-i}), \quad i = 1, 2, \ldots, K-1 \end{cases} \tag{5.26}$$

Later on, Stein (1980) presented a computationally more efficient FDP approach, the idea of which is as follows. At $t = N - i, i \in \{1, \ldots, N-1\}$, we can either stop and attain $\mu_{G^{N-i}}(x_{N-i}) = \mu_T(N-i)\mu_{G^{N-i}}(x_{N-i})$, or apply u_{N-i} and attain $\mu_{C^{N-i}}(u_{N-i}) \wedge \mu_{G^{N-i+1}}(x_{N-i+1})$. The better alternative should evidently be chosen, $t = N - i - 1, 2, \ldots, 0$.

The set of FDP recurrence equations is then

$$\begin{cases} \mu_{G^{N-i}}(x_{N-i}) = \mu_{G^{N-i}}(x_{N-i}) \vee \max\limits_{u_{N-i}}(\mu_{C^{N-i}}(u_{N-i}) \wedge \mu_{G^{N-i+1}}(x_{N-i+1})) \\ x_{N-i+1} = f(x_{N-i}, u_{N-i}), \quad i = 1, 2, \ldots, N \end{cases} \tag{5.27}$$

and an optimal termination time is such a $t = N - i$ at which the terminating control, $u_{\nu^*-1}^*$, occurs, i.e., when

$$\mu_{G^{N-i}}(x_{N-i}) > \max_{u_{N-i}}(\mu_{C^{N-i}}(u_{N-i}) \vee \mu_{G^{N-i+1}}(x_{N-i+1})). \tag{5.28}$$

4.2 The case of a stochastic system under control

The problem is to find an optimal termination time ν^* and an optimal sequence of controls $u_0^*, \ldots, u_{\nu^*-1}^*$ such that

$$\mu_D(u_0^*, \ldots, u_{\nu^*-1}^* \mid x_0) = \max_{\nu, u_0, \ldots, u_{\nu-1}} (\mu_{C^0}(u_0) \wedge \ldots \wedge \mu_{C^{\nu-1}}(u_{\nu-1}) \wedge E\mu_{G^\nu}(x_\nu)). \tag{5.29}$$

where $\mu_{G^\nu}(x_\nu) = \mu_T(\nu)\mu_{G^\nu}(x_\nu)$; $\{\nu: \mu_T(\nu) > 0\} = \{K, K+1, \ldots, N\}$.

As formerly, we find ν^* and $u_{K-1}^*, u_K^*, \ldots, u_{\nu^*-1}^*$ by solving

$$\begin{cases} \mu_{G^{\nu-i}}(x_{\nu-i}, \nu) = \max\limits_{u_{\nu-i}} (\mu_{C^{\nu-i}}(u_{\nu-i}) \wedge E\mu_{G^{\nu-i+1}}(x_{\nu-i+1}, \nu)) \\ E\mu_{G^{\nu-i+1}}(x_{\nu-i+1}, \nu) = \sum\limits_{x_{\nu-i+1}} p(x_{\nu-i+1} \mid x_{\nu-i}, u_{\nu-i}) \mu_{G^{\nu-i+1}}(x_{\nu-i+1}, \nu), \\ i = 1, 2, \ldots, \nu - K + 1; \nu = K, K+1, \ldots, N-1 \end{cases} \tag{5.30}$$

and ν^* is given by the maximizing ν in

$$\mu_{G^{K-1}}(x_{K-1}) = \max_{\nu} \mu_{G^{K-1}}(x_{K-1}, \nu). \tag{5.31}$$

The remaining $u_{K-2}^*, u_{K-3}^*, \ldots, u_0^*$ are obtained by solving

$$\begin{cases} \mu_{G^{K-1-i}}(x_{K-1-i}) = \max\limits_{u_{K-1-i}} (\mu_{C^{K-1-i}}(u_{K-1-i}) \wedge E\mu_{G^{K-i}}(x_{K-i})) \\ E\mu_{G^{K-i}}(x_{K-i}) = \sum\limits_{x_{K-i}} p(x_{K-i} \mid x_{K-1-i}, u_{K-1-i}) \mu_{G^{K-i}}(x_{K-i}), \\ i = 1, 2, \ldots, K-1 \end{cases} \tag{5.32}$$

In Stein's (1980) formulation, the problem is solved by

$$\begin{cases} \mu_{G^{N-i}}(x_{N-i}) = \mu_{G^{N-i}}(x_{N-i}) \vee \max\limits_{u_{N-i}} (\mu_{C^{N-i}}(u_{N-i}) \wedge E\mu_{G^{N-i+1}}(x_{N-i+1})) \\ E\mu_{G^{N-i+1}}(x_{N-i+1}) = \sum\limits_{x_{N-i+1}} p(x_{N-i+1} \mid x_{N-i}, u_{N-i}) \mu_{G^{N-i+1}}(x_{N-i+1}), \\ i = 1, 2, \ldots, N \end{cases} \tag{5.33}$$

where ν^* occurs when

$$\mu_{G^{N-i}}(x_{N-i}) > \max_{u_{N-i}} (\mu_{C^{N-i}}(u_{N-i}) \wedge E\mu_{G^{N-i+1}}(x_{N-i+1}). \tag{5.34}$$

To obtain an FDP formulation for the case of a fuzzy termination time and a fuzzy system under control, we fix some (finite and relatively small) number of reference fuzzy states (and possibly controls), and obtain an auxiliary approximate system whose state transitions are of a deterministic type (cf. Section 3.3; Kacprzyk, 1983b; or Kacprzyk and Staniewski, 1982). Then, Kacprzyk's or Stein's approach can be employed. In many cases, however, a simple branch-and-bound algorithm due to Kacprzyk (1983b) is a better choice.

5 REMARKS ON THE CASES OF AN IMPLICITLY GIVEN AND INFINITE TERMINATION TIMES

The problem with an implicit termination time was first discussed by Bellman and Zadeh (1970) with a deterministic system described by $x_{t+1} = f(x_t, u_t); x_t, x_{t+1} \in X = \{s_1, \ldots, s_n\}; u_t \in U = \{c_1, \ldots, c_m\}; t = 0, 1, \ldots$ The process is to terminate when x_{t+1} attains for the first time a terminating set of states $W = \{s_{p+1}, s_{p+2}, \ldots, s_n\} \subset X = \{s_1, \ldots, s_p, s_{p+1}, \ldots, s_n\}; x_0 \in X \backslash W$.

At each $t = 0, 1, \ldots, N-1$, u_t is subjected to a fuzzy constraint $\mu_C(u_t | x_t)$ which is not time-dependent as in the previous sections but state-dependent because the termination time is not known in advance; similarly, the fuzzy goal on x_N is $\mu_G(x_N)$.

The problem is to find $u_0^*, \ldots, u_{N-1}^*$ such that

$$\mu_D(u_0^*, \ldots, u_{N-1}^* | x_0) = \max_{u_0, \ldots, u_{N-1}} (\mu_C(u_0 | x_0) \wedge \ldots \wedge \mu_C(u_{N-1} | x_{N-1}) \wedge \mu_G(x_N) \tag{5.35}$$

where N is such that $x_{N-k} \notin W$ for each $1 < k < N$, and $x_N \in W$.

Evidently, we are interested in time-invariant optimal policies $a, a^*: X \backslash W \to U$ such that $u_t = a(x_t)$ and $u_t^* = a^*(x_t); t = 0, 1, \ldots, N-1$.

This problem can be solved by

- an iterative approach by Bellman and Zadeh (1970),
- a graph-theoretic approach by Komolov *et al.* (1979), and
- a branch-and-bound approach by Kacprzyk (1983b).

The problem with an infinite termination time was first formulated and solved by Kacprzyk, Safteruk and Staniewski (1981), and Kacprzyk and Staniewski (1982, 1983). Formulations with an infinite termination time are in general used for problems with long planning horizons, low varying goals and constraints, etc.

The problem with a deterministic system under control was first formulated and solved by Kacprzyk and Staniewski (1983). At each t, we have a state-dependent fuzzy constraint $\mu_C(u_t|x_t)$, and a fuzzy goal $\mu_G(x_t)$, both the same for all t. The deterministic system is given by $x_{t+1} = f(x_t, u_t), t = 0, 1, \ldots$

The fuzzy decision is now

$$\begin{aligned} \mu_D(a|x_0) &= \mu_C(a(x_0)|x_0) \wedge \mu_G(x_1) \wedge \mu_C(a(x_1)|x_1) \wedge \mu_G(x_2) \wedge \ldots \\ &= \bigwedge_{t=0}^{\infty} (\mu_C(a(x_t)|x_t) \wedge \mu_G(f(x_t, a(x_t)))) \end{aligned} \tag{5.36}$$

and we seek an optimal time-invariant policy $a^*: X \to U$, such that

$$\mu_D(a^*|x_0) = \max_a (\bigwedge_{t=0}^{\infty} (\mu_C(a(x_t) \wedge \mu_G(f(x_t, a(x_t)))). \tag{5.37}$$

Such an optimal time-invariant policy exists and can be found in a finite number of iterations using a policy-iteration-type algorithm.

The problem with a stochastic system under control and an infinite termination time was first formulated and solved by Kacprzyk, Staniewski and Safteruk (1981) and Kacprzyk, Safteruk and Staniewski (1981). Basically, we seek an optimal stationary policy a^* such that

$$\begin{aligned} \mu_D(a^*|x_0) &= \max_a E\mu_D(a|x_0) \\ &= \max_a E(\bigwedge_{t=0}^{\infty} (\mu_C(a(x_t)|x_t) \wedge \mu_G(f(x_t, a(x_t))))). \end{aligned} \tag{5.38}$$

Unfortunately, the solution of (38) turns out to be difficult. In principle, we need to consider stationary policies of the type $a: X \times [0, 1] \to U$, i.e., relating the control to the current state and a "summary" ($\in [0, 1]$) of the trajectory, and though a policy iteration algorithm improves the policies, it has not been proven that it will give an optimal policy sought.

The problem with a fuzzy system under control was first formulated and solved by Kacprzyk and Staniewski (1982) by using the concept of reference fuzzy states and controls (cf. Section 3.3) to obtain an auxiliary approximate deterministic system type problem which was structurally equivalent to that with a deterministic system (37).

6 REMARKS ON SOME FURTHER EXTENSIONS

In Kacprzyk (1983a) and Kacprzyk and Iwański (1987), a new problem formulation was proposed to seek an optimal sequence of controls which would best satisfy fuzzy goals and constraints at most (almost all, much more than 75%, . . .) of the earlier control stages. Notice that this is a considerable departure from "... at all the control stages ... " which is explicitly or implicitly assumed in the conventional problem formulations including those considered so far here. Moreover, in Kacprzyk and Iwański (1987), that approach was extended to include a "soft" discounting by formulating the problem as to seek an optimal sequence of controls best satisfying fuzzy constraints and goals at, say, most of the earlier control stages. Fuzzy logic with linguistic quantifiers (see, e.g., Zadeh, 1983) was used, and some simpler cases of the problem boiled down to the solution of some FDP recurrence equations.

Recently, to avoid numerical complexity in FDP with a fuzzy system under control, Kacprzyk (1994a-c) proposed an interpolative reasoning scheme in which, by using reference fuzzy controls and states (cf. Section 3.3), a simpler auxiliary problem is first formulated and solved, and its solution is then "adjusted" (by a possibilistic interpolative reasoning scheme) to yield a "real" solution.

7 APPLICATIONS OF FUZZY DYNAMIC PROGRAMMING

FDP has found successful applications in modeling a large spectrum of problems. The best accounts of these applications are Esogbue and Bellman (1984) and Esogbue (1983b), while below we will only provide a short list of them:

- resource allocation processes in general (Esogbue and Ramesh, 1970), and in the context of R&D (Esogbue, 1983a);
- health care as, e.g., budgeting in cancer research (Esogbue, 1977), anesthesia administration (Esogbue, 1985);
- data analysis, mainly clustering (Esogbue, 1986);
- environmental and water systems (Esogbue and Bellman, 1981; Esogbue, Theologidu and Guo, 1992);
- energy systems (Huang, Lin and Huang, 1992; Su and Hsu, 1991);
- chemical engineering (Kraslawski, Gorak and Vogelpohl, 1989);
- regional development (Kacprzyk and Straszak, 1980, 1984), etc.

8 CONCLUDING REMARKS

In this paper, we briefly presented some relevant classes of FDP, and indicated how fuzzy sets theory can provide new insights, prospects, etc. We hope that, in view of a recent increase of interest in FDP, it will be a topic of more and more active research in the years to come.

Acknowledgements

This paper is dedicated to my friend, Professor Etienne E. Kerre, who has always managed to effectively combine brilliant mathematics with relevant applications in diverse areas.

REFERENCES

Baldwin J. F. and B. W. Pilsworth (1982) "Dynamic programming for fuzzy systems with fuzzy environment," *J. Math. Anal. and Appls.* **85**, 1–23.

Bellman R. E. and L. A. Zadeh (1970) "Decision-making in a fuzzy environment," *Management Sci.* **17**, B141–B164.

Chang S. S. L. (1969) "Fuzzy dynamic programming and the decision making process," *Proc. 3rd Princeton Conf. on Inf. Sci. and Syst.* (Princeton, USA), 200–203.

Esogbue A. O. (1983a) "Dynamic programming, fuzzy sets, and the modeling of R&D management control systems," *IEEE Trans. on Syst., Man and Cybern.* **SMC–13**, 18–30.

Esogbue A. O. (1983b) "Some novel applications of fuzzy dynamic programming," *Proc. IEEE*, 501–505.

Esogbue A. O. (1985) "A fuzzy dynamic programming model of intra-operative anesthesia administration." In J. Kacprzyk and R. R. Yager (Eds.), *Management Decision—Support Systems Using Fuzzy Sets and Possibility Theory*, Verlag TUV Rheinland, Cologne, 155–161.

Esogbue A. O. (1986) "Optimal clustering of fuzzy data via fuzzy dynamic programming," *Fuzzy Sets and Systems* **18**, 283–298.

Esogbue A. O. and R. E. Bellman (1981) "A fuzzy dynamic programming algorithm for clustering non-quantitative data arising in water pollution control planning," *Proc. 3rd International Conference on Mathematical Modeling* (Los Angeles, USA).

Esogbue A. O. and R. E. Bellman (1984) "Fuzzy dynamic programming and its extensions." In H.-J. Zimmermann, L. A. Zadeh and B. R. Gaines (Eds.), *Fuzzy Sets and Decision Analysis*, Elsevier, Amsterdam, 147–167.

Esogbue A. O., M. Fedrizzi and J. Kacprzyk (1988) "Fuzzy dynamic programming with stochastic systems." In J. Kacprzyk and M. Fedrizzi (Eds.), *Combining Fuzzy Imprecision with Probabilistic Uncertainty in Decision Making*, Springer-Verlag, Berlin / New York / Tokyo, 266–287.

Esogbue A. O. and V. Ramesh (1970) "Dynamic programming and fuzzy allocation processes," *Memo No. 202, Dept. of Op. Res., Case Western Reserve University*, Cleveland (USA).

Esogbue A. O., M. Theologidu and K. Guo (1992) "On the application of fuzzy sets theory to the optimal flood control problem arising in water resources systems," *Fuzzy Sets and Systems* **48**, 155–172.

Fung L. W. and K. S. Fu (1977) "Characterization of a class of fuzzy optimal control problems." In M. M. Gupta, G. N. Saridis and B. R. Gaines (Eds.), *Fuzzy Automata and Decision Processes*, North-Holland, Amsterdam, 205-220.

Hojo T., T. Terano and S. Masui (1993) "Design of quasi-optimal fuzzy controller by fuzzy dynamic programming," *Proc. of IEEE International Conference on Fuzzy Systems*, vol. 2, 1253–1258.

Howard R. A. (1960) *Dynamic Programming and Markov Processes*, MIT Press, Cambridge (USA).

Huang C. J., C. E. Lin and C. L. Huang (1992) "Fuzzy approach for generator maintenance scheduling," *Electric Power Systems Research*, 31–38.

Kacprzyk J. (1977) "Control of a nonfuzzy system in a fuzzy environment with fuzzy termination time," *Systems Sci.* **3**, 325–341.

Kacprzyk J. (1878a) "A branch-and-bound algorithm for the multistage control of a nonfuzzy system in a fuzzy environment," *Control and Cybern.* **7**, 51–64.

Kacprzyk J. (1978b) "Control of a stochastic system in a fuzzy environment with fuzzy termination time," *Systems Sci.* **4**, 291–300.

Kacprzyk J. (1978c) "Decision making in a fuzzy environment with fuzzy termination time," *Fuzzy Sets and Syst.* **1**, 169–179.

Kacprzyk J. (1979) "A branch-and-bound algorithm for the multistage control of a fuzzy system in a fuzzy environment," *Kybernetes* **8**, 139–147.

Kacprzyk J. (1983a) "A generalization of fuzzy multistage decision making and control via linguistic quantifiers," *Int. J. of Control* **38**, 1249–1270.

Kacprzyk J. (1983b) *Multistage Decision-Making under Fuzziness*, Verlag TUV Rheinland, Cologne.

Kacprzyk J. (1991) "Fuzzy linguistic quantifiers in decision making and control," *Proc. of International Fuzzy Engineering Symposium—IFES '91* (Yokohama, Japan), vol. 2, 800–811.

Kacprzyk J. (1992a) "Fuzzy optimal control revisited: toward a new generation of fuzzy control?" *Proc. of Second Int. Conference on Fuzzy Logic and Neural Networks IIZUKA '92* (Iizuka, Japan), vol. 1, 429-432.

Kacprzyk J. (1992b) "Fuzzy logic with linguistic quantifiers in decision making and control," *Archives of Control Sciences* 1 (XXXVII), 127-141.

Kacprzyk J. (1993a) "Fuzzy control with an explicit performance function using dynamic programming and interpolative reasoning," *Proceedings of EUFIT'93—First European Congress on Fuzzy and Intelligent Technologies* (Aachen, Germany), vol. 3, 1459-1463.

Kacprzyk J. (1993b) "In search for a new generation of fuzzy control: can a prescriptive approach based on interpolative reasoning and neural networks help?" *Proc. of ANZIIS '93—Australian and New Zealand Conference on Intelligent Information Systems* (Perth, Australia), 402-406.

Kacprzyk J. (1993c) "Interpolative reasoning for computationally efficient optimal fuzzy control," *Proc. of Fifth Int. IFSA World Congress* (Seoul, Korea), vol. II, 1270-1273.

Kacprzyk J. (1993d) "A prescriptive approach to fuzzy control: a step toward a 'more mature' fuzzy control?," *Proc. of First Asian Fuzzy Systems Symposium* (Singapore), 360-365.

Kacprzyk J. (1994) "Fuzzy dynamic programming-basic issues." In M. Delgado, J. Kacprzyk, J.-L. Verdegay and M. A. Vila (Eds.), *Fuzzy Optimization: Recent Advances*, Physica-Verlag, Heidelberg (A Springer-Verlag Company), 321-331.

Kacprzyk J. and C. Iwański (1987) "A generalization of discounted multistage decision making and control via fuzzy linguistic quantifiers," *Int. J. of Control* 45, 1909-1930.

Kacprzyk J., K. Safteruk and P. Staniewski (1981) "On the control of stochastic systems in a fuzzy environment over infinite horizon," *Systems Sci.* 7, 121-131.

Kacprzyk J. and P. Staniewski (1980) "A new approach to the control of stochastic systems in a fuzzy environment," *Archiwum Automatyki i Telemechaniki XXV*, 443-444.

Kacprzyk J. and P. Staniewski (1982) "Long-term inventory policy making through fuzzy decision-making models," *Fuzzy Sets and Syst.* **8**, 117–132.

Kacprzyk J. and P. Staniewski (1983) "Control of a deterministic system in a fuzzy environment over infinite horizon," *Fuzzy Sets and Syst.* **10**, 291–298.

Kacprzyk J. and A. Straszak (1980) "Optimal policies for 'stable' integrated regional development through fuzzy decision making models." In P. P. Wang and S. K. Chang (Eds.), *Fuzzy Set Theory and Applications.* Plenum, New York, 321–328.

Kacprzyk J. and A. Straszak (1984) "Determination of 'stable' integrated regional development strategies via fuzzy decision-making models," *IEEE Trans. on Syst., Man and Cybern.* **SMC–14**, 310–313.

Komolov S. V., S. P. Makeev, G. P. Serov and J. F. Shakhnov (1979) "On the problem of optimal control of a finite automaton with fuzzy constraints and goals (in Russian)," *Kybernetika (Kiev)* **6**, 30–34.

Kraslawski A., A. Gorak and A. Vogelpohl (1989) "Fuzzy dynamic programming in the synthesis of distillation column systems," *Computers and Chemical Engineering* **13**, 611–618.

Stein W. E. (1980) "Optimal stopping in a fuzzy environment," *Fuzzy Sets and Syst.* **3**, 252–259.

Su C. C. and Y. Y. Hsu (1991) "Fuzzy dynamic programming: An application to unit commitment," *IEEE Transactions on Power Systems* **PS–6**, 1231–1237.

Yuan Y. and Z. Wu (1991) "Algorithm of fuzzy dynamic programming in AGV scheduling," *Proceedings of International Conference on Computer Integrated Manufacturing—ICCIM '91*, 405–408.

Zadeh L. A. (1968) "Probability measures of fuzzy events," *J. Math. Anal. and Appl.* **23**, 421–427.

Zadeh L. A. (1983) "A computational approach to fuzzy quantifiers in natural languages," *Computers and Math. with Appl.* **9**, 149–184.

6

ON TRANSITIVITY OF FUZZY PREFERENCE RELATIONS IN RANKING FUZZY NUMBERS

Xuzhu Wang and Da Ruan

Fuel Research Unit
Nuclear Research Centre (SCK•CEN)
Boeretang 200
B-2400 Mol, Belgium

ABSTRACT

In this paper, we first carry out further investigations into relations between several existing transitivity-related notions in ranking alternatives or fuzzy numbers. Then we pay attention to the discussion of transitivity properties of some practical fuzzy preference relations introduced by Dubois and Prade. Finally, we develop an ordering procedure of alternatives under the assumption of some certain weak transitivity.

1 INTRODUCTION

In decision making analysis, the ordering of alternatives frequently leads to that of fuzzy sets or fuzzy numbers when fuzzy information is concerned and processed (see, e.g., [2, 5, 11, 12, 27]). Therefore, how to rank fuzzy sets becomes a very important topic, and many efforts have been put into this research field [2]-[15]. For the review of these approaches, the reader is referred to [1, 21]. All the existing methods can be roughly divided into three classes, which we discuss hereafter. Let $A_1, A_2, \ldots, A_n$ be n fuzzy numbers which represent n alternatives. The first class of methods [9, 13, 15] maps each fuzzy number into a positive real number, i.e., a mapping $F : A \to \mathbb{R}^+$ where $A = \{A_1, A_2, \ldots, A_n\}$ and $\mathbb{R}^+$ is the set of positive real numbers; the alternatives are ranked according to the natural ordering of $F(A_i), i = 1, 2, \ldots, n$. The second class of methods [2, 3, 5, 6] is quite similar to the first one, but the numbers which reflect the ordering of alternatives are obtained by comparing A_i to $A_1, A_2, \ldots, A_n$. The last class of methods [4, 7, 8, 10, 11, 12, 14, 20] constructs the preference $P(A_i, A_j)$ for each pair (A_i, A_j). The order relations in the former two classes of methods are certainly transitive, due to the transitivity of the natural complete order relation that is defined on the real number. Concerning the last class of methods, it is often difficult for the decision-maker to obtain a final total order relation of alternatives due to the inconsistency existing in fuzzy preference relations. To overcome this difficulty, some researchers [4, 20] aggregate all $P(A_i, A_j)(j = 1, 2, \ldots, n)$ into a crisp number with the min operator for each alternative A_i. The order relation is eventually determined by the corresponding value for each alternative. The others impose some certain transitivity restriction on fuzzy preference to eliminate this inconsistency; different transitivity notions have been introduced in [8, 10, 14, 18]. In a previous paper [22], we have briefly investigated relations between different transitivity notions. In this paper, we focus our attention on the following transitivity-related definitions:

(1) max – min transitivity:
$(\forall i, j \in N = \{1, 2, \ldots, n\})\ (a_{ij} \geq \max_{k \in N} \min(a_{ik}, a_{kj}))$,
where a_{ij} denotes the degree to which alternative A_i is preferred or equal to alternative A_j.

(2) $W3$-transitivity: $a_{ik} > a_{ki}$ and $a_{kj} > a_{jk}$ imply $a_{ij} > a_{ji}$.

(3) consistency: $a_{ik} \geq a_{ki}$ and $a_{kj} \geq a_{jk}$ imply $a_{ij} \geq a_{ji}$.

The max – min transitivity condition is a very strong one. It entails many other transitivity properties including $W3$-transitivity. However, this condition does not necessarily imply a requirement of consistency. We will explicitly indicate the case in which max – min transitivity is satisfied while the consistency condition is not. Moreover, we conclude that the consistent fuzzy preference relation is a $W3$-transitive one. Then, we pay attention to some practical fuzzy preference relations for ordering fuzzy numbers, particularly to those introduced by Dubois and Prade [20], and discuss their transitivity properties. An interesting result is that all four indices used by Dubois and Prade are $W3$-transitive, but only one of them satisfies the max – min transitivity. Finally, we develop a method to obtain a total ordering of alternatives when the $W3$-transitivity is satisfied. Therefore, many fuzzy preference relations, including those proposed by Dubois and Prade, can be directly used to rank alternatives without aggregating them into a crisp number for each alternative in some subjective way.

2 FURTHER INVESTIGATIONS INTO RELATIONS BETWEEN SOME TRANSITIVITY-RELATED NOTIONS

Let $A = \{A_1, A_2, \ldots, A_n\}$ be the set of n alternatives (sometimes it is also used to denote the set of n fuzzy numbers representing these alternatives if there is no confusion in the context). A fuzzy preference relation on A is the mapping $P : A \times A \to [0,1]$ with $P(A_i, A_j)$ indicating the degree to which A_i is preferred or equal to A_j. The fuzzy preference relation is firstly employed by Orlovsky for the choice of nondominated alternatives [16] and studied by other authors [14] [17]-[19] in detail. In the following, we list some transitivity-related notions of the fuzzy preference relation P on A in the ordering of fuzzy numbers.

Definition 2.1.

(1) P is called max – min transitive iff $P(A_i, A_j) \geq \max_{A_k \in A} \min(P(A_i, A_k), P(A_k, A_j))$ for any three elements A_i, A_j, A_k of A.

(2) P is called $W1$-transitive iff $P(A_i, A_k) \geq 0.5$ and $P(A_k, A_j) \geq 0.5$ imply $P(A_i, A_j) \geq \min(P(A_i, A_j), P(A_j, A_k))$ for any three elements A_i, A_j, A_k of A.

(3) P is called $W2$-transitive iff $P(A_i, A_k) \geq 0.5$ and $P(A_k, A_j) \geq 0.5$ imply $P(A_i, A_j) \geq 0.5$ for any three elements A_i, A_j, A_k of A.

(4) P is called $W3$-transitive iff $P(A_i, A_j) > P(A_j, A_i)$ and $P(A_j, A_k) > P(A_k, A_j)$ imply $P(A_i, A_k) > P(A_k, A_i)$ for any three elements A_i, A_j, A_k of A.

(5) P is called consistent iff $P(A_i, A_j) \geq P(A_j, A_i)$ and $P(A_j, A_k) \geq P(A_k, A_j)$ imply $P(A_i, A_k) \geq P(A_k, A_i)$ for any three elements A_i, A_j, A_k of A.

(6) P is called reciprocal iff $(\forall (A_i, A_j) \in A^2)(P(A_i, A_j) + P(A_j, A_i) = 1)$.

Remark 2.1. The consistency is also a type of transitivity. We use the term 'consistency' in conformity with [26].

In [22], some relations between the notions in Definition 2.1 are investigated; they can be summarized as follows:

Proposition 2.1. Let P be a preference relation on $A = \{A_1, A_2, \ldots, A_n\}$.

(1) If P is max – min transitive, then P is both $W1$-transitive and $W2$-transitive.

(2) If P is max – min transitive, then P is $W3$-transitive.

(3) If P is $W1$-transitive, then P is $W2$-transitive.

(4) Let P be a reciprocal preference relation, then P is $W2$-transitive iff P is consistent.

By Proposition 2.1, the max – min transitivity is the strongest one. We can derive all others from this transitivity property if P is reciprocal. However, that P is max – min transitive does not necessarily imply that P is consistent without reciprocity, as shown by the following example.

Example 2.1. The fuzzy preference relation P on $A = \{A_1, A_2, A_3\}$ is defined as

$P(A_i, A_j)$	A_1	A_2	A_3
A_1	1	0.5	0.7
A_2	0.5	1	0.5
A_3	0.8	0.5	1

One can easily prove that P is max – min transitive. Meanwhile, $P(A_1, A_2) = P(A_2, A_1)$, $P(A_2, A_3) = P(A_3, A_2)$ and $P(A_1, A_3) < P(A_3, A_1)$, which means that P is not consistent.

It is worth noting that $P(A_1, A_2) = P(A_2, A_1) = P(A_2, A_3) = P(A_3, A_2)$ in the above example. This is not by accident at all as will be seen from the coming discussions. For convenience, $P(A_i, A_j)$ is abbreviated as a_{ij}.

Proposition 2.2. If P is $W3$-transitive, then $\forall (i, j, k) \in \{1, 2, \ldots, n\}^3$, we have:

(1) $a_{ij} = a_{ji}$ and $a_{jk} > a_{kj}$ imply $a_{ik} \geq a_{ki}$.

(2) $a_{ij} > a_{ji}$ and $a_{jk} = a_{kj}$ imply $a_{ik} \geq a_{ki}$.

Proof. (1) Suppose (1) is not true, i.e., there exists $(i, j, k) \in \{1, 2, \ldots, n\}^3$ satisfying $a_{ij} = a_{ji}$, $a_{jk} > a_{kj}$ and $a_{ik} < a_{ki}$. Since P is $W3$-transitive, we infer $a_{ij} < a_{ji}$ from $a_{jk} > a_{kj}$ and $a_{ik} < a_{ki}$, which is in contradiction with $a_{ij} = a_{ji}$. Therefore, (1) should be true.

The proof of (2) is similar to that of (1), hence omitted.

Remark 2.1. (1) As a matter of fact, it can be easily verified that (1) and (2) in Proposition 2.2 are equivalent.

(2) If P is max – min transitive, (1) and (2) in the above proposition are also true taking Proposition 2.1(2) into account.

Proposition 2.3. If P is max – min transitive and not consistent, then there must exist $(i,j,k) \in \{1,2,\ldots,n\}^3$ such that $a_{ij} = a_{ji} = a_{jk} = a_{kj}$ and $a_{ik} < a_{ki}$.

Proof. Based on Proposition 2.1(2) and 2.2, we can conclude that if max – min transitivity is satisfied and consistency is not, then there must exist $(i,j,k) \in \{1,2,\ldots,n\}^3$ such that $a_{ij} = a_{ji}$, $a_{jk} = a_{kj}$ and $a_{ik} < a_{ki}$. Now it suffices to prove $a_{ij} = a_{jk}$. For this purpose, let us consider the following two cases:

If $a_{ij} > a_{jk}$, then $a_{ik} \geq \min(a_{ij}, a_{jk}) = a_{jk}$ and $a_{jk} \geq \min(a_{ji}, a_{ik})$, hence, $a_{ik} \leq a_{jk}$. Therefore, we have $a_{ik} = a_{jk}$. However, $a_{kj} \geq \min(a_{ki}, a_{ij}) > \min(a_{ik}, a_{jk})$, which is a contradiction.

If $a_{ij} < a_{jk}$, then $a_{ik} \geq \min(a_{ij}, a_{jk}) = a_{ij}$. By $a_{ij} \geq \min(a_{ik}, a_{kj})$ and $a_{jk} > a_{ij}$, we have $a_{ik} \leq a_{ij}$, hence, $a_{ij} = a_{ik}$. Therefore, $a_{ji} \geq (a_{jk}, a_{ki}) > \min(a_{ij}, a_{ik}) = a_{ij}$, which is also a contradiction.

Hence, $a_{ij} = a_{ji} = a_{jk} = a_{kj}$ and $a_{ik} < a_{ki}$.

Clearly, Example 2.1 is just a special case of Proposition 2.3. In the sequel, we discuss the relation between $W3$-transitivity and consistency.

Proposition 2.4. Consider three statements:

(i) $a_{ij} = a_{ji}$ $a_{jk} = a_{kj}$ imply $a_{ik} = a_{ki}$,

(ii) $a_{ij} > a_{ji}$ $a_{jk} = a_{kj}$ imply $a_{ik} > a_{ki}$,

(iii) $a_{ij} = a_{ji}$ $a_{jk} > a_{kj}$ imply $a_{ik} > a_{ki}$.

We have the following conclusions:

(1) If P is $W3$-transitive, then (i) $\Rightarrow$ (ii).

(2) (ii) $\Rightarrow$ (i).

(3) (ii) $\Leftrightarrow$ (iii).

Proof. (1) Suppose there exists $(i,j,k) \in \{1,2,\ldots,n\}^3$ such that $a_{ij} > a_{ji}$, $a_{jk} = a_{kj}$, and $a_{ik} \le a_{ki}$, then we have two cases:
case 1: $a_{ij} > a_{ji}$, $a_{jk} = a_{kj}$, and $a_{ik} < a_{ki}$.
case 2: $a_{ij} > a_{ji}$, $a_{jk} = a_{kj}$, and $a_{ik} = a_{ki}$.
For case 1, $a_{ki} > a_{ik}$, $a_{ij} > a_{ji}$, and $a_{jk} = a_{kj}$, which contradicts $W3$-transitive.
For case 2, $a_{ik} = a_{ki}$, $a_{kj} = a_{jk}$, and $a_{ij} > a_{ji}$, which contradicts (i).

(2) Suppose there exists $(i,j,k) \in \{1,2,\ldots,n\}^3$ such that $a_{ij} = a_{ji}$, $a_{jk} = a_{kj}$, and $a_{ik} \ne a_{ki}$, then it should be $a_{ij} = a_{ji}$, $a_{jk} = a_{kj}$, $a_{ik} < a_{ki}$ or $a_{ij} = a_{ji}$, $a_{jk} = a_{kj}$, $a_{ik} > a_{ki}$. We have a contradiction in either case.

(3) We only prove (ii) $\Rightarrow$ (iii) and the proof of the reverse is similar. If there exists $(i,j,k) \in \{1,2,\ldots,n\}^3$ such that $a_{ij} = a_{ji}$, $a_{jk} > a_{kj}$, and $a_{ik} \le a_{ki}$, then $a_{ij} = a_{ji}$, $a_{jk} > a_{kj}$, $a_{ik} < a_{ki}$ or $a_{ij} = a_{ji}$, $a_{jk} > a_{kj}$, $a_{ik} = a_{ki}$. Neither of them can be true, due to (ii).

Proposition 2.5. P is consistent iff P satisfies the following:

(1) $a_{ij} = a_{ji}$, $a_{jk} = a_{kj} \Rightarrow a_{ik} = a_{ki}$.

(2) P is $W3$-transitive.

Proof. First, it is impossible for a consistent fuzzy relation to satisfy $a_{ij} > a_{ji}$, $a_{jk} = a_{kj}$, and $a_{ik} = a_{ki}$, since it is equivalent to $a_{jk} = a_{kj}$, $a_{ki} = a_{ik}$, and $a_{ji} < a_{ij}$, which is a contradiction of consistency. Similarly, $a_{ij} = a_{ji}$, $a_{jk} > a_{kj}$, and $a_{ik} = a_{ki}$ is impossible. Therefore, P is consistent iff the following statements are true:

(i) $a_{ij} = a_{ji}$, $a_{jk} = a_{kj}$ imply $a_{ik} = a_{ki}$;

(ii) $a_{ij} > a_{ji}$, $a_{jk} = a_{kj}$ imply $a_{ik} > a_{ki}$;

(iii) $a_{ij} = a_{ji}$, $a_{jk} > a_{kj}$ imply $a_{ik} > a_{ki}$;

(iv) P is $W3$-transitive.

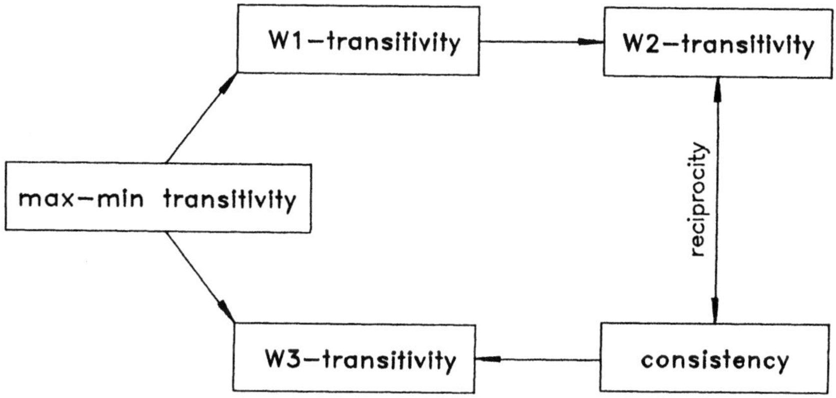

Figure 1 The relations between some transitivity-related notions.

Thanks to Proposition 2.4, we have (ii) ⇔ (iii) and (i)+(iv) ⇒ (ii). Therefore, these statements are true iff (i) and (iv) are true. That is just what this proposition indicates.

Corollary 2.1. Let P be $W3$-transitive. P is consistent iff $a_{ij} = a_{ji}$ and $a_{jk} = a_{kj}$ imply $a_{ik} = a_{ki}$.

Corollary 2.2. (1) If P is consistent, then P is $W3$-transitive.
(2) If P is $W2$-transitive and reciprocal, then P is $W3$-transitive.
(3) If P is $W1$-transitive and reciprocal, then P is $W3$-transitive.

Proof. (1) is obvious by Proposition 2.5. (2) can be derived from (1) and Proposition 2.1(4), while (3) holds from (2) and Proposition 2.1(3).

Figure 1 represents the relations between some transitivity notions in the form of diagram.

Remark 2.2. Reference [25] introduces transitive-2 notion. In our notations, it can be written as the following version:

P is called transitive-2 iff $(\forall(i,j,k) \in \{1,2,\ldots,n\}^3)$, $a_{ij} \geq a_{ji}$, and $a_{jk} \geq a_{kj}$ (with strict inequality holding at least once) imply $a_{ik} > a_{ki}$.

By Proposition 2.4(2) and Proposition 2.5, transitive-2 implies consistency. By the proof of Proposition 2.5, consistency implies transitive-2. Hence, they are identical concepts.

3 THE TRANSITIVITY OF INDICES INTRODUCED BY DUBOIS AND PRADE

In [20], Dubois and Prade proposed four indices to locate the relative position of two fuzzy numbers A and B reasonably (for the definition of a fuzzy number, see [23]). These indices have the following expressions:

$$\begin{aligned} PD(A,B) &= \sup_{a\geq b} \min(A(a), B(b)); \\ PSD(A,B) &= \sup_{a} \inf_{b\geq a} \min(A(a), 1-B(b)); \\ ND(A,B) &= \inf_{a} \sup_{b\geq a} \min(1-A(a), B(b)); \\ NSD(A,B) &= 1 - PD(B,A). \end{aligned}$$

In this section, we discuss these indices and their transitivity properties. We assume that the fuzzy sets involved are fuzzy numbers. Let $kerA = \{x|A(x) = 1\} = [a_1, b_1]$ and $kerB = \{x|B(x) = 1\} = [a_2, b_2]$; we can rewrite these indices in the following version.

Proposition 3.1.

(1) $PD(A,B) = \begin{cases} 1 & ; \text{ if } b_1 \geq b_2, \\ hgt(A \cap B) & ; \text{ otherwise}; \end{cases}$

(2) $PSD(A,B) = \sup_{x>b_2} \min(A(x), 1-B(x));$

(3) $ND(A,B) = \inf_{x<a_2} \max(1-A(x), B(x));$

(4) $NSD(A,B) = \begin{cases} 0 & ; \text{ if } b_2 \geq b_1, \\ 1-hgt(A \cap B) & ; \text{ otherwise}. \end{cases}$

Proof.
(1) The proof can be accomplished in an identical way as in [6].

(2) $PSD(A,B) = \max(\sup_{a \le b_2} \inf_{b \ge a} \min(A(a), 1 - B(b)), \sup_{a > b_2} \inf_{b \ge a} \min(A(a), 1 - B(b))$.

If $a \le b_2$, then we have

$$\inf_{b \ge a} \min(A(a), 1 - B(b)) \le \min(A(a), 1 - B(b_2)) = 0$$

and hence, $\sup_{a \le b_2} \inf_{b \ge a} \min(A(a), 1 - B(b)) = 0$.

Therefore,

$$\begin{aligned} PSD(A,B) &= \sup_{a > b_2} \inf_{b \ge a} \min(A(a), 1 - B(b)) = \sup_{a > b_2} \min(A(a), \inf_{b \ge a} 1 - B(b)) \\ &= \sup_{a > b_2} \min(A(a), 1 - B(a)). \end{aligned}$$

That is $PSD(A,B) = \sup_{x > b_2} \min(A(x), 1 - B(x))$.

(3) The proof is similar to that of (2).

(4) Obvious considering (1).

Proposition 3.2.

(1) $PD(A,B) + PD(B,A) \ge 1$;

(2) $PSD(A,B) + PSD(B,A) \le 1$;

(3) $ND(A,B) + ND(B,A) = 1$;

(4) $NSD(A,B) + NSD(B,A) \le 1$.

Proof. The proof of (1) and (4) is immediate from Proposition 3.1(1) and (4). For the proof of (2) and (3), see [20].

Proposition 3.3. $\max(PSD(A,B), PSD(B,A)) \ge 0.5$.

Proof. Suppose that there exist fuzzy numbers A and B satisfying

$$PSD(A,B) < 0.5 \text{ and } PSD(B,A) < 0.5,$$

or $\sup_{x>b_2} \min(A(x), 1 - B(x)) < 0.5$ and $\sup_{x>b_1} \min(1 - A(x), B(x)) < 0.5$.

Then

$$(\forall x > b_2)\ (\min(A(x), 1 - B(x)) < 0.5) \tag{3.1}$$

and

$$(\forall x > b_1)\ (\min(1 - A(x), B(x)) < 0.5). \tag{3.2}$$

Without loss of generality, we assume $b_2 \geq b_1$. Let $A(x_1) = B(x_0) = 0.5$ $(x_0 > b_2, x_1 > b_1)$. We have the following three cases:

(1) $x_0 = x_1$. $\min(A(x_0), 1 - B(x_0)) = 0.5$, which contradicts (3.1).

(2) $x_0 < x_1$. Let $x \in]x_0, x_1[$, i.e., $x_1 > x > x_0 > b_2 \geq b_1$. Hence, $A(x) > A(x_1) = 0.5$ and $B(x) < B(x_0) = 0.5$, which gives $\min(A(x), 1 - B(x)) > 0.5$. This is in contradiction with (3.1).

(3) $x_1 < x_0$. Let $x \in]\max(x_1, b_2), x_0[$, then $x > x_1 > b_1$ and $x_0 > x > b_2$. Hence, $A(x) < A(x_1)$ and $B(x) > B(x_0) = 0.5$. We obtain $\min(B(x), 1 - A(x)) > 0.5$, which contradicts (3.2). Therefore, we have contradiction in each case, which verifies our conclusion.

Proposition 3.4. Let $A_i, A_j, A_k \in A$ be fuzzy numbers:

(1) $PSD(A_i, A_j) > 0.5$ and $PSD(A_j, A_k) > 0.5$ imply $PSD(A_i, A_k) > 0.5$.

(2) $ND(A_i, A_j) > 0.5$ and $ND(A_j, A_k) > 0.5$ imply $ND(A_i, A_k) > 0.5$.

Proof.

(1) We give a proof by contradiction. Suppose there exists $(A_1, A_2, A_3) \in A^3$ such that $PSD(A_1, A_2) > 0.5$, $PSD(A_2, A_3) > 0.5$, and $PSD(A_1, A_3) \leq 0.5$. That is,

$$\sup_{x>b_2} \min(A_1(x), 1 - A_2(x)) > 0.5, \quad \sup_{x>b_3} \min(A_2(x), 1 - A_3(x)) > 0.5$$

and

$$\sup_{x>b_3} \min(A_1(x), 1 - A_3(x)) \leq 0.5,$$

where $b_i = \max\{x | A_i(x) = 1\}$, $i = 1, 2, 3$.

Equivalently, we have

$$\min(A_1(x_0), 1 - A_2(x_0)) > 0.5 \text{ for some } x_0 > b_2; \tag{3.3}$$

$$\min(1 - A_3(x_1), A_2(x_1)) > 0.5 \text{ for some } x_1 > b_3; \tag{3.4}$$

$$\min(A_1(x), 1 - A_3(x)) \leq 0.5 \text{ for every } x > b_3. \tag{3.5}$$

We consider two cases:
(i) If $b_3 \leq b_2$, then $x_0 > b_2 \geq b_3$.
By (3.5), $\min(1 - A_3(x_0), A_1(x_0)) \leq 0.5$.
By (3.3), $A_1(x_0) > 0.5$, hence $A_3(x_0) \geq 0.5$.
By (3.4), $A_3(x_1) < 0.5$ and $x_1 > b_3$, hence $x_0 < x_1$.
We have $x_1 > x_0 > b_2$ and thus $A_2(x_0) > A_2(x_1)$.
But $A_2(x_0) < 0.5$ by (3.3) and $A_2(x_1) > 0.5$ by (3.4), which is a contradiction.

(ii) If $b_3 > b_2$, then $x_1 > b_3 > b_2$.
From $A_2(x_0) < 0.5$ and $A_2(x_1) > 0.5$, we derive $x_0 > x_1 > b_3$.
Hence $A_3(x_0) < A_3(x_1)$, and thus $1 - A_3(x_0) > 1 - A_3(x_1) > 0.5$ because of (3.4). From (3.3), we get $A_1(x_0) > 0.5$, and hence $\min(1 - A_3(x_0), A_1(x_0)) > 0.5$, which contradicts (3.5) since $x_0 > x_3$ holds. This contradiction completes the proof of (1).

(2) Let $a_m = \min ker A_m$, $m = i, j, k$. Then $ND(A_i, A_j) > 0.5$ implies

$$(\forall x < a_j)\ (\max(1 - A_i(x), A_j(x)) > 0.5). \tag{3.6}$$

Similarly, $ND(A_j, A_k) > 0.5$ implies

$$(\forall x < a_k)\ (\max(A_k(x), 1 - A_j(x)) > 0.5). \tag{3.7}$$

Again, we consider two cases:
(i) If $a_k \leq a_j$, then both (3.6) and (3.7) hold when $x < a_k$.
Therefore, $\max(1 - A_i(x), A_k(x)) > 0.5$, and hence $ND(A_i, A_k) > 0.5$.
(ii) If $a_k > a_j$, similarly we obtain $\max(1 - A_i(x), A_k(x)) > 0.5$, for $x < a_j$.
As for $x = a_j$, $A_k(a_j) > 0.5$ by $a_j < a_k$ and (3.7).
Thus $\max(1 - A_i(a_j), A_k(a_j)) > 0.5$.
In the following, we consider the case $a_j < x < a_k$.
Let $A_k(x_0) = 0.5$ $(x_0 < a_k)$. Then $x_0 \leq a_j$ (otherwise, let $x_1 \in]a_j, x_0[$ such that $A_j(x_1) > 0.5$. From $x_1 < x_0 < a_k$, we have $A_k(x_1) < A_k(x_0) = 0.5$, which contradicts (3.7)).
Therefore, $x_0 < x < a_k$, and hence $A_k(x) > A_k(x_0) = 0.5$. This leads to

$\max(1 - A_i(x), A_k(x)) > 0.5.$
$ND(A_i, A_k) = \min(\inf_{x \le a_j} \max(1 - A_i(x), A_k(x)),$
$\inf_{a_j < x < a_k} \max(1 - A_i(x), A_k(x))) > 0.5.$

Proposition 3.5.

(1) PD, PSD, and ND are $W3$-transitive.

(2) NSD is max – min transitive.

Proof. (1) The $W3$-transitivity property of PSD and ND is obvious considering Proposition 3.2(2)(3) and Proposition 3.4.
Now we give the proof of $W3$-transitivity property of PD.
From $PD(A_i, A_j) > PD(A_j, A_i)$, we deduce $b_j < a_i$. Analogously, we deduce from $PD(A_j, A_k) > PD(A_k, A_j)$ that $b_k < a_j$. Hence, $b_k < a_j \le b_j < a_i$, which means $PD(A_i, A_k) > PD(A_k, A_i)$, i.e., PD is $W3$-transitive.

(2) The general version of this conclusion can be found in [14], therefore, the proof is omitted here.

Remark 3.1. From Proposition 3.5 and Proposition 2.1(2), we conclude that the four indices of Dubois and Prade are $W3$-transitive. But the max – min transitivity property is a stronger requirement to meet. In Figure 2, we give some examples in which PD, PSD, and ND are not max – min transitive.

Remark 3.2. The construction of $W3$-transitive fuzzy preference relations is relatively easy. Besides PD, PSD, ND, and NSD, there are many other ordering indices satisfying the $W3$-transitivity condition, e.g., those proposed in [7, 8, 10, 18]. Therefore, the $W3$-transitivity notion is very useful for the practical ordering of fuzzy numbers.

4 ORDERING METHOD UNDER $W3$-TRANSITIVITY

In this section, we give an approach to the ranking of alternatives when the $W3$-transitivity is satisfied. The method is based on the following proposition.

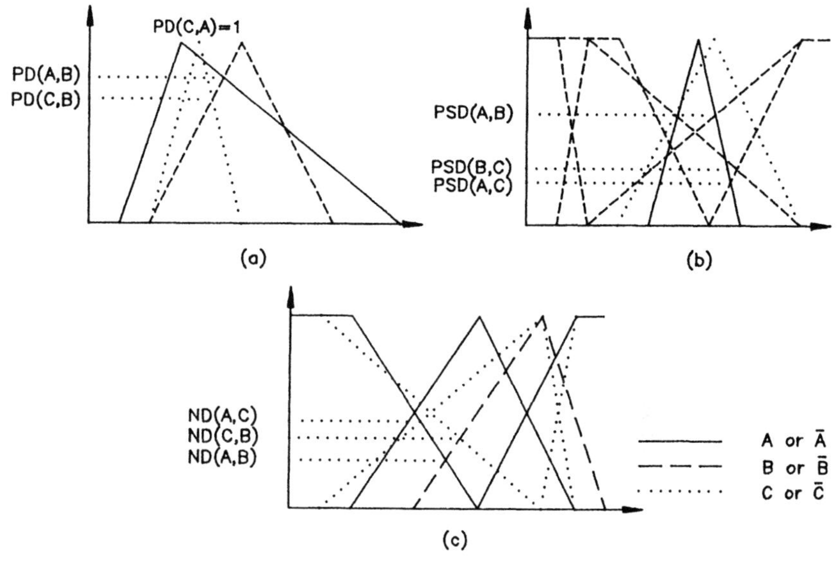

Figure 2 *PD*, *PSD*, and *ND* are not max – min transitive.

Proposition 4.1. Let P be a $W3$-transitive preference relation on A, then

(1) P is a $W3$-transitive preference relation on each subset A' of A;

(2) there exists A_i satisfying $(\forall A_j \in A)\ (P(A_i, A_j) \geq P(A_j, A_i))$.

Proof. (1) Obvious.

(2) Let us suppose that this statement is not true. Then for any $A_i \in A$, there exists $A_j \in A$ such that $P(A_i, A_j) < P(A_j, A_i)$. In particular, for $i = 1$, there exists $A_{i_1} \in A$ such that $P(A_1, A_{i_1}) < P(A_{i_1}, A_1)$. Similarly, for $i = i_1$, there should exist A_{i_2} such that $P(A_{i_1}, A_{i_2}) < P(A_{i_2}, A_{i_1})$. Following the same reasoning, we have the following inequalities:

$$P(A_{i_2}, A_{i_3}) < P(A_{i_3}, A_{i_2})\ ,\ \ldots\ ,\ P(A_{i_{n-1}}, A_{i_n}) < P(A_{i_n}, A_{i_{n-1}}).$$

Note that $A_{i_0} = A_1, A_{i_1}, \ldots, A_{i_n}$ are $n+1$ alternatives. Therefore, there must exist two A_{i_j}, A_{i_k} ($i_k > i_j$ without loss of generality) such that $A_{i_j} = A_{i_k}$.

From $P(A_{i_j}, A_{i_{j+1}}) < P(A_{i_{j+1}}, A_{i_j})$, ... , $P(A_{i_{k-1}}, A_{i_k}) < P(A_k, A_{i_{k-1}})$, we derive $P(A_{i_j}, A_{i_k}) < P(A_{i_k}, A_{i_j})$ due to the $W3$-transitivity. This is impossible considering $A_{i_k} = A_{i_j}$. Hence, (2) should be true.

We state the ordering procedure of alternatives based on $W3$-transitivity.

First, we construct $\mathcal{H}_1 = \{A_i \in A | P(A_i, A_j) \geq P(A_j, A_i)\ \forall A_j \in A\}$. If $\mathcal{B}_1 = A \setminus \mathcal{H}_1 \neq \emptyset$, then $\mathcal{H}_2 = \{A_i \in \mathcal{B}_1 | P(A_i, A_j) \geq P(A_j, A_i)\ \forall A_j \in \mathcal{B}_1\}$. In a similar way, $\mathcal{H}_3$ can be constructed if $\mathcal{B}_2 = \mathcal{B}_1 \setminus \mathcal{H}_2 \neq \emptyset$. Keep constructing $\mathcal{H}_4, \mathcal{H}_5, \ldots, \mathcal{H}_m$ until $\mathcal{B}_m = \mathcal{B}_{m-1} \setminus \mathcal{H}_m = \emptyset$ for some m. Then, we construct the order relations '>', '≥', and '=' on $A = \{A_1, A_2, \ldots, A_n\}$ as:
$A_i > A_j \Leftrightarrow \exists \mathcal{H}_s, \mathcal{H}_t$, such that $A_i \in \mathcal{H}_s, A_j \in \mathcal{H}_t$ and $s > t$.
$A_i = A_j \Leftrightarrow \exists \mathcal{H}_s$ such that $A_i \in \mathcal{H}_s$ and $A_j \in \mathcal{H}_s$.
$A_i \geq A_j \Leftrightarrow A_i > A_j$ or $A_i = A_j$.

Obviously, the above-defined order relations have the following properties:

(1) '≥' and '=' are reflexive.

(2) '=' is symmetric and '≥' is antisymmetric.

(3) '>' , '=', and '≥' are transitive.

(4) Given any two alternatives A_i and A_j, then one of the following holds: $A_i > A_j$, $A_i < A_j$, and $A_i = A_j$. Therefore, '≥' is a total order.

(5) If P is consistent, then $A_i > A_j \Leftrightarrow P(A_i, A_j) > P(A_j, A_i)$ and $A_i = A_j \Leftrightarrow P(A_i, A_j) = P(A_j, A_i)$.

Remark 4.1. Let P be a $W3$-transitive relation on A. By Proposition 4.1, we infer that for any subset A' of A, there must exist $A_i \in A'$ such that $P(A_i, A_j) \geq P(A_j, A_i)\ \forall A_j \in A'$. The reverse is not true as can be seen in the following example.

Example 4.1. $A = \{A_1, A_2, A_3\}$ and the fuzzy preference P is defined as

P	A_1	A_2	A_3
A_1	1	0.8	0.5
A_2	0.4	1	0.7
A_3	0.5	0.2	1

It can be easily seen that for any subset A' of A, there exists $A_i \in A'$ such that $P(A_i, A_j) \geq P(A_j, A_i)\ \forall A_j \in A'$. But $P(A_1, A_2) > P(A_2, A_1)$,

$P(A_2, A_3) > P(A_3, A_2)$, and $P(A_1, A_3) = P(A_3, A_1)$, which indicate that P is not $W3$-transitive on A.

Remark 4.2. If P is reciprocal and $W3$-transitive, an ordering approach is developed in [24]. However, it is often difficult for a preference relation to meet the reciprocity requirement, as can be seen from some existing ordering indices [7, 20]. Our method can be regarded as an extension of this approach.

Finally, we give some examples to illustrate the ordering procedure.
Example 4.2. $A = \{A_1, A_2, A_3, A_4, A_5, A_6\}$

P	A_1	A_2	A_3	A_4	A_5	A_6
A_1	1	0.5	0.3	0.1	0.5	0
A_2	0.5	1	0.7	0.1	0.6	0.2
A_3	0.4	0.9	1	0.2	0.4	0.4
A_4	0.9	0.9	0.7	1	0.4	1
A_5	0.5	0.9	0.9	0.6	1	0.6
A_6	0.9	0.4	0.8	1	0.6	1

The preference relation P is $W3$-transitive.
For $i = 1, 2, \ldots, 6$, $P(A_6, A_i) \geq P(A_i, A_6)$ and $P(A_5, A_i) \geq P(A_i, A_5)$.
Therefore, $H_1 = \{A_5, A_6\}$. Similarly $H_2 = \{A_4\}$, $H_3 = \{A_3\}, H_4 = \{A_1, A_2\}$.
The obtained order relation is: $A_6 = A_5 > A_4 > A_3 > A_2 = A_1$.

In the next example, we use PD, PSD, ND, NSD, and their $W3$-transitivity to rank fuzzy numbers.

Example 4.3. Consider four fuzzy numbers A_1, A_2, A_3, and A_4 as depicted in Figure 3. The corresponding indices PD, PSD, ND, and NSD are calculated and listed below.

PSD	A_1	A_2	A_3	A_4
A_1	0.5	0.5	0.75	0.83
A_2	0.5	0.5	0.75	0.83
A_3	0.25	0.25	0.5	0.62
A_4	0.17	0.17	0.37	0.5

PD	A_1	A_2	A_3	A_4
A_1	1	1	1	1
A_2	1	1	1	1
A_3	0.75	1	1	1
A_4	0.6	0.63	0.63	1

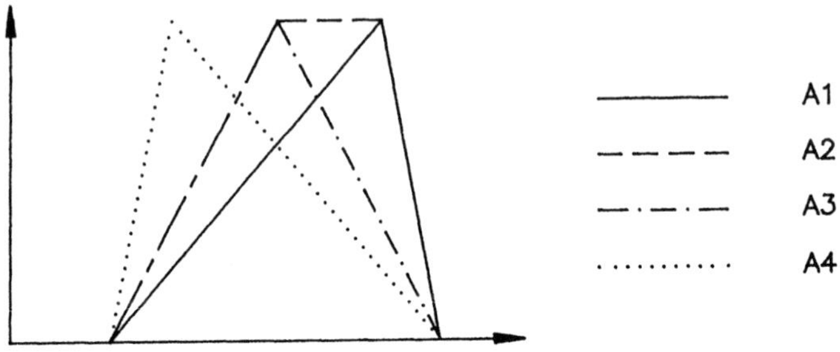

Figure 3 Fuzzy numbers A_1, A_2, A_3, and A_4.

NSD	A_1	A_2	A_3	A_4
A_1	0	0	0.25	0.4
A_2	0	0	0	0.38
A_3	0	0	0	0.38
A_4	0	0	0	0

ND	A_1	A_2	A_3	A_4
A_1	0.5	0.63	0.63	0.83
A_2	0.37	0.5	0.5	0.75
A_3	0.17	0.5	0.5	0.75
A_4	0.17	0.25	0.25	0.5

By using these preference relations, we obtain the following orderings:
for PD: $A_1 = A_2 > A_3 > A_4$;
for PSD: $A_1 = A_2 > A_3 > A_4$;
for ND: $A_1 > A_2 = A_3 > A_4$;
for NSD: $A_1 = A_2 > A_3 > A_4$.

In this example, the order relation due to the use of ND is different from the others. It is easy to understand this result taking into account that ND evaluates the relative locations of the left sides of two fuzzy numbers. The left sides of A_2 and A_3 are identical, hence, they have the same ordering by ND.

5 CONCLUDING REMARKS

In this paper, we discussed in depth some relations between transitivity-related notions, particularly max – min transitivity, $W3$-transitivity, and consistency. The max – min transitivity is the most frequently used one. However, it is also a very strong restriction on fuzzy preference relations. Hence, it is often difficult to satisfy it in the construction of ordering indices for practical decision-making. The $W3$-transitivity condition is a relatively weaker one. As pointed out in Section 2, many ordering indices satisfy this transitivity property. Some researchers construct fuzzy preference relations with $W1$ or $W2$-transitivity properties. These properties are always accompanied by reciprocity in the ordering process. In this situation, it is $W3$-transitive. Therefore, the ordering procedure developed in Section 3 is applicable again.

Acknowledgements

We would like to dedicate this paper to Professor Kerre's 50th birthday for his kind guidance during our research period at the University of Ghent and his further cooperation with our present FLINS research project at the Belgian Nuclear Research Centre (SCK•CEN). One of the authors (Wang) was sent to Belgium by Taiyuan University of Technology, Shanxi, China, as a visiting scholar in November 1993. This paper was made possible due to the support of SCK•CEN.

REFERENCES

[1] Bortolan, G. and Degani, R., "A review of some methods for ranking fuzzy subsets," *Fuzzy Sets and Systems* **15** (1985), pp. 1–19.

[2] Baas, S. M. and Kwakernaak, H., " Rating and ranking of multiple-aspect alternatives using fuzzy sets," *Automatica* **13** (1977), pp. 47–58.

[3] Jain, R., "Decision-making in the presence of fuzzy variables," *IEEE Trans. Systems Man Cybernet* **6** (1976), pp. 698–703.

[4] Baldwin, J. F. and Guild, N. C. F., "Comparison of fuzzy sets on the same decision space," *Fuzzy Sets and Systems* **2** (1979), pp. 213–233.

[5] Kerre, E. E., "The use of fuzzy set theory in electrocardiological diagnostics," in: *Approximate Reasoning in Decision-analysis* (North-Holland Publishing Company, 1982), pp. 277–282.

[6] Gu, J. and Wang, X., "A class of method of ranking alternatives," *Journal of Systems Engineering* **3** (1988), pp. 14–23.

[7] Saade, J. J. and Schwarzlander, H., "Ordering fuzzy sets over the real line: an approach based on decision making under uncertainty," *Fuzzy Sets and Systems* **50** (1992), pp. 237–246.

[8] Nakamura, K., "Preference relations on a set of fuzzy utilities as a basis for decision making," *Fuzzy Sets and Systems* **20** (1986), pp. 147–162.

[9] Yager, R. R., "A procedure for ordering fuzzy sets of the unit interval," *Information Sciences* **24** (1981), pp. 143–161.

[10] Yuan, Y., "Criteria for evaluating fuzzy ranking methods," *Fuzzy Sets and Systems* **43** (1991), pp. 139–157.

[11] Buckley, J. J., "Ranking alternatives using fuzzy numbers," *Fuzzy Sets and Systems* **15** (1985), pp. 21–31.

[12] Watson, S. R., Weiss, J. J. and Donnell, M. L., "Fuzzy decision analysis," *IEEE Transactions on Systems, Man, Cybernetics* **9** (1979), pp. 1–9.

[13] Gonzalez, A., "A study of the ranking function approach through mean value," *Fuzzy Sets and Systems* **35** (1990), pp. 29–42.

[14] Ovchinnikov, S. V., "Transitive fuzzy ordering of fuzzy numbers," *Fuzzy Sets and Systems* **30** (1989), pp. 283–295.

[15] Choobineh, F. and Li, H., "An index for ordering fuzzy numbers," *Fuzzy Sets and Systems* **54** (1993), pp. 287–294.

[16] Orlovsky, S. A., "Decision-making with a fuzzy preference relation," *Fuzzy Sets and Systems* **1** (1978), pp. 155–167.

[17] Barrett, C. R., Pattanaik, P.K. and Salles, M., "On choosing rationally when preferences are fuzzy," *Fuzzy Sets and Systems* **34** (1990), pp. 197–212.

[18] Kolodziejczyk, W., "Orlovsky's concept of decision-making with fuzzy preference relation—further results," *Fuzzy Sets and Systems* **19** (1986), pp. 11–20.

[19] Roubens, M., "Some properties of choice functions based on valued binary relations," *European Journal of Operational Research* **40** (1989), pp. 309–321.

[20] Dubois, D. and Prade, H., "Ranking fuzzy numbers in the setting of possibility theory," *Information Sciences* **30** (1983), pp. 183–224.

[21] Chen, S. and Hwang, C., *Fuzzy multiple attribute decision making*, (Springer-Verlag, 1992), pp. 101–486.

[22] Wang, X., Kerre, E. E., Cappelle, B. and Ruan, D., "Transitivity of fuzzy orderings based on pairwise comparisons," In: Ruan, D. *et al.* eds., *Fuzzy Logic and Intelligent Technologies in Nuclear Science—Proceedings of the 1st international FLINS workshop* (World Scientific, Singapore, 1994), pp. 38–43.

[23] Dubois, D. and Prade, H., "Operations on fuzzy numbers," *International Journal of System Sciences* **9** (1978), pp. 613–626.

[24] Basile, L., "Ranking alternatives by weak transitivity relations," In: Kacprzyk, J. and Fedrizzi, M. eds., *Multiperson Decision Making using Fuzzy Sets and Possibility Theory* (Kluwer Academic Publishers, 1990), pp. 105–112.

[25] Banerjee, A., "Rational choice under fuzzy preferences: The Orlovsky choice function," *Fuzzy Sets and Systems* **53** (1993), pp. 295–299.

[26] Ok, E. A., "On the approximation of preferences by exact relations," *Fuzzy Sets and Systems* **67** (1994), pp. 173–179.

[27] Van Laarhoven, P. J. M. and Pedrycz, W., "A fuzzy extension of Saaty's priority theory," *Fuzzy Sets and Systems* **11** (1983), pp. 229–241.

7

INFORMATION AGGREGATION: ETHICAL AND COMPUTATIONAL ISSUES

Vincenzo Cutello and Javier Montero*

Dept. of Mathematics
University of Catania
Catania, Italy

** Dept. of Statistics and OR*
Complutense University
Madrid, Spain

ABSTRACT

In this paper we survey on aggregation operators and in particular hierarchies of them. As a case study, we will analyze the problems of the aggregation of truth values of fuzzy predicates and the aggregation of individual opinions into a single group opinion, based upon hierarchical intensity aggregation rules. We will see that hierarchical amalgamations are supported from an ethical and rational point of view. Two different hierarchical procedures will be recalled: cover-based hierarchical aggregations and ordered hierarchical aggregations.

Finally we will see that when we deal with ordered hierarchical aggregations of OWA operators, some interesting computational problems appear quite naturally. Such problem admit polynomial time solutions.

1 INTRODUCTION AND PRELIMINARIES

The topic of our discussion is intuitively very simple yet at the basis of many applications. Controllers, expert and decision making systems, theorem provers, learning algorithms, to mention a few, are all (computer aided) applications in which it is of primary importance to have mechanisms able to analyze the gathered information and produce from it some aggregated value(s) to be (possibly) used in some next evaluation steps.

Common hypothesis to several applications of such kind is that information is passed to an aggregation operator as an ordered sequence of real numbers, which without loss of generality can be supposed to belong to the unit interval. Formally, an aggregation rule of dimension n, will be a mapping

$$\phi^{(n)} : [0,1]^n \to [0,1].$$

To capture the essential meaning of aggregation, $\phi^{(n)}$ must satisfy at least the following two conditions (see [15]):

- $\phi^{(n)}(0,0,\ldots,0) = 0, \quad ,\phi^{(n)}(1,1,\ldots,1) = 1$ and
- monotonicity, i.e. $\phi^{(n)}(a_1,a_2,\ldots,a_n) \le \phi^{(n)}(b_1,b_2,\ldots,b_n)$ if $a_i \le b_i$ for all i.

If we think of $\phi^{(n)}$ as an algorithm that takes as inputs n numbers and then it outputs the aggregated value of the n numbers, we can obtain aggregation rules that are independent from the number of aggregands by generalizing the above definition as follows. Put

$$\mathcal{I} = \bigcup_{n\ge 2} [0,1]^n.$$

For every $L \in \mathcal{I}$ there exists one and only one $k \ge 2$ such that $L \in [0,1]^k$; such an index k is obviously the "dimension" of L, which we will denote by $|L|$.

Then ϕ on $\mathcal{I}$ can be defined as

$$\phi(L) = \phi^{(|L|)}(L).$$

For simplicity, we will not specify the dimension of the operator whenever it will be clear from the context.

In what follows we will concentrate on two particular applications

1. the aggregation of truth values of fuzzy predicates in a most general and comprehensive way.
2. the aggregation of preference intensities expressed by individual/ groups over finite set of alternatives, so to obtain a *rational* and *democratic* social opinion.

Notice that when we say that we want to aggregate group opinions to obtain a social opinion we are basically saying that we want to aggregate values which are themselves an aggregation of values. Therefore, a further generalization of the above definition of aggregation rules is obtained by introducing the notion of hierarchical aggregations, that is to say aggregations of chunks of information which in turn represent aggregated information.

In general, hierarchical aggregation procedures for individual preferences are defined by means of a basic classification of the individuals. The set of individual is divided into groups, in such a way that each individual is present in at least one of these groups. An immediate example is provided by national Parliaments. Each of the representative of the Parliament has been elected by a particular group of people and therefore he/she represents the aggregated opinion of such people. When the Parliament express its opinion then what we have is an hierarchical aggregation of the people opinions. Since voters are generally divided into electoral districts, we have a fixed "cover" of the set of individuals, and the social opinion is obtained by using such cover. Hierarchical aggregations of this kind will be called "cover-based".

Alternatively, basic classification can be made on the basis of a natural ordering on all possible intensity values, and then we can talk about "ordered" hierarchical aggregation rules. This second type of hierarchical aggregations are obviously more natural when dealing with fuzzy predicates truth aggregations.

2 TRUTH VALUES AGGREGATION

One of the main issues in Fuzzy Logic is the choice of operators to generalize the classical logic operators *and, or.* Many proposal have been made and the most general formalizations are represented by T-norms, T-conorms and OWA operators (see [23, 24]).

A T-norm is a map $T : [0,1] \times [0,1] \to [0,1]$ such that

(T1) $T(a,b) = T(b,a)$

(T2) $T(a,T(b,c)) = T(T(a,b),c)$

(T3) $T(a,b) \geq T(c,d)$ if $a \geq c$ and $b \geq d$

(T4) $T(a,1) = a$

A T-conorm is a map $S : [0,1] \times [0,1] \to [0,1]$ such that

(S1) $S(a,b) = S(b,a)$

(S2) $S(a, S(b,c)) = S(S(a,b),c)$

(S3) $S(a,b) \geq S(c,d)$ if $a \geq c$ and $b \geq d$

(S4) $S(a,0) = a$

T-norms and T-conorms represent aggregation operators that generalize the notion of conjunction and disjunction of classical logic. The min operator is the maximal T-norm and the max operator is the minimal T-conorm, in the sense that for all a, b we have that $T(a,b) \leq \min\{a,b\}$ and $S(a,b) \geq \max\{a,b\}$ hold for any T-norm T and any T-conorm S (see [15]).

Ordered Weighted Averaging (OWA) operators fill the gap between min and max, in such a way that by means of these OWA operators we can go from conjunction (intersection) to disjunction (union) in a continuous way. OWA operators were initially introduced by Yager in [24], and they have been used and applied to many fields, such as Neural Networks, Database systems, Learning systems and Fuzzy Logic Controllers (see [25] for a comprehensive review on the subject). In order to get the aggregated values, OWA operators make use of the relative order within intensity values.

Formally, an OWA operator of dimension n is an aggregation operator ϕ that has an associated list of weights $W = [w_1, \ldots, w_n]$ such that

1. $w_i \in [0,1]$ for all $1 \leq i \leq n$
2. $\sum_{i=1}^{n} w_i = 1$
3. for any $L = [a_1, a_1, \ldots, a_n] \in [0,1]^n$

$$\phi(L) = \sum_{i=1}^{n} w_i b_i.$$

   where $b_1 \geq \cdots \geq b_n$, is the sequence obtained when we sort in non decreasing order the inputs $a_1, a_2, \ldots, a_n$.

Therefore, OWA operators are commutative, monotone and idempotent, but in general not associative. As a consequence, the semantic problems of using the *same* OWA operator on inputs of different dimensions, i.e.

the problem of defining the OWA operator $\phi : \mathcal{I} \to [0,1]$ given the OWA operators $\phi^{(n)}$ is quite challenging. An answer to such a problem has been given in [10] where it is introduced the notion of recursive families of OWA operators. For the time being, we will ignore such a problem and we will only deal with a particular type of hierarchical aggregation which we will call *ordered* hierarchical aggregation and that will allow us to deal in a very natural way with OWA operators.

3 AGGREGATION OF PREFERENCE INTENSITIES

The classical Arrow's paradox in group decision making (cf. [1]) when translated into a fuzzy context (see [2, 13, 14, 17, 20]), can be avoided in several ways, according to axiomatics which are similar to those proposed in the crisp context. As we mentioned above it is very common to obtain aggregated preferences of large groups of people by means of rules which allow the successive aggregation of degrees of preferences. The set of individuals is divided into smaller subsets of individuals -not necessarily disjoint- and the global aggregation will be the aggregation of all partial aggregations within each one of these subsets of individuals. We follow the model proposed in [18, 19] and subsequently characterized in [5], and the formalization given in [7].

At the basis of such a model are non absolutely irrational (in the sense of [18, 19]) complete fuzzy preference relations. In particular, it is assumed that each individual is able to express her/his opinion about any possible set of alternatives through some complete fuzzy binary preference relation, as formalized below.

Let $\mu : X \times X \to [0,1]$ be a fuzzy preference relation over an arbitrary finite set of alternatives X. $\mu(x,y)$ represents the degree to which the relation x *not worse than* y holds. The completeness hypothesis is expressed by

$$\mu(x,y) + \mu(y,x) \geq 1 \quad \forall x,y \in X\,. \tag{7.1}$$

Following [4], completeness is required in order to assure that all individuals consider the set of alternatives on which they are expressing their opinions, feasible and comprehensive.

The values

$$\mu_I(x,y) \quad = \quad \mu(x,y) + \mu(y,x) - 1 \tag{7.2}$$

$$\mu_B(x,y) = \mu(x,y) - \mu_I(x,y)$$
$$\mu_W(x,y) = \mu(y,x) - \mu_I(x,y)$$

can be understood, respectively, as the degree to which the two alternatives are indifferent (xIy), the degree of strict preference of x over y, $(xBy$, x *is better than* $y)$ and the degree of strict preference of y over x $(xWy$, x *is worse than* $y)$. We clearly have that

$$\mu_B(x,y) + \mu_I(x,y) + \mu_W(x,y) = 1 \ \ \forall x,y\ . \tag{7.3}$$

We want to define rationality as a fuzzy property. We then consider cycles of preferences over chains $G = (x_1 - x_2 - \cdots - x_k - x_1)$ of k distinct alternatives, defined as

$$x_1P_1x_2P_2\cdots x_kP_kx_1$$

where $P_h \in \{W, I, B\}$ for all $h = 1, 2, \ldots, k$. A cycle $x_1P_1x_2P_2\cdots x_kP_kx_1$ is *irrational* if either

- $P_h \in \{B, I\}$ for all $h = 1, 2, \ldots, k$ and $B \in \{P_h : h = 1, 2, \ldots, k\}$; or
- $P_h \in \{W, I\}$ for all $h = 1, 2, \ldots, k$ and $W \in \{P_h : h = 1, 2, \ldots, k\}$.

We say that a cycle is rational if it is not irrational. Then, given any fuzzy preference μ over a fixed set of alternatives and a chain of alternatives, we can look for all possible rational cycles of preferences, weigh them in some way and assign to the chain a degree of rationality (see [5, 18]). Specifically, this is done as follows. Given a a cycle $C \equiv x_1P_1\cdots x_kP_kx_1$ where $P_h \in \{B, I, W\}$ for all $h = 1, 2, \ldots, k$, a quite natural weight associated to C and denoted by $\Delta(C)$ is

$$\Delta(C) = \Pi_{h=1}^{k}\mu_{P_h}(x_h, x_{h+1})$$

where $x_{k+1} = x_1$ for convenience.

Therefore, given a chain $G = (x_1 - x_2 - \cdots - x_k - x_1)$ a degree of rationality associated to G and denoted by $A_\mu(G)$ can be defined as

$$A_\mu(G) = \sum_{C\in \text{rat.cycles}} \Delta(C).$$

As proven in [5, 18], $A_\mu(G)$ verifies

$$1 - A_\mu(G) = \Pi_{h=1}^{k}\mu(x_h, x_{h+1}) + \Pi_{h=1}^{k}\mu(x_{h+1}, x_h) - 2\Pi_{h=1}^{k}\mu_I(x_h, x_{h+1}). \tag{7.4}$$

In view of (7.4), once a finite set of alternatives X has been fixed, rationality can be defined as a fuzzy property $A : \mathcal{P}(X) \to [0,1]$ with

$$A(\mu) = \min_G A_\mu(G) \tag{7.5}$$

and where $\mathcal{P}(X)$ is the set of all complete fuzzy preferences. Needless to say, the above is just a particular rationality measure, which may or may not seem "rational". A formal characterization of rationality measures, was given in [8], following the approach described in [6]. The above described rationality measure as well as the most commonly found in literature, are particular instance of such general definition. It would then be desirable to extend the results that we will describe below to "generic" fuzzy rationality measures.

Once a group of $n \geq 2$ individuals is fixed, we should be able to aggregate their opinions about any set of alternatives in a coherent way. Therefore, in [5] were defined aggregation operations that can take into account any extra alternative x so to properly extend any previous aggregated opinion relative to a collection of alternatives not containing x. The key properties are the standard conditions

(IIA) *Independence of Irrelevant Alternatives*: each aggregated preference relation $\mu(x,y)$ depends solely on the values $\mu^i(x,y)$, i.e. on the individual preference intensities of x over y.

(UD) *Unrestricted Domain*: the aggregation rule is defined over all possible profiles of fuzzy preferences.

Given an aggregation rule $\phi^{(n)}$, if we also assume

(N) *Neutrality*: given any permutation of the set of alternatives π, if we have $v^i(x,y) = \mu^i(\pi(x),\pi(y))$ for all $i = 1,2,\ldots,n$ and any pair of alternatives x,y, then

$$\phi^{(n)}(v^1(x,y),\cdots,v^n(x,y)) = \phi^{(n)}(\mu^1(x,y),\cdots,\mu^n(x,y))$$

It is clear then that the same aggregation $\phi^{(n)}$ will be associated to any pair of alternatives and therefore each possible aggregation procedure is characterized by one of these aggregation rules.

For the time being, we will suppose that conditions IIA, UD, N hold.

Given n individuals expressing their opinion on the set of alternatives X and $\phi^{(n)}$, the aggregated preference μ defined on $X \times X$ associated to $\phi^{(n)}$

is defined as

$$\mu(x,y) = \phi^{(n)}(\mu^1(x,y),\cdots,\mu^n(x,y)) \ \forall x,y \in X.$$

Standard *ethical* conditions may also be imposed on the intensity aggregation rules, among them:

(PR) *Positive Responsiveness*:

$$\phi^{(n)}(a_1,a_2,\ldots,a_n) > \phi^{(n)}(b_1,b_2,\ldots,b_n)$$

if $a_i \geq b_i$ for all $i = 1,2,\ldots,n$ and there exists $1 \leq j \leq n$ such that $a_j > b_j$.

(A) *Anonymity*: given any permutation $\pi : \{1,\ldots,n\} \to \{1,\ldots,n\}$, we have

$$\phi^{(n)}(a_1,a_2,\ldots,a_n) = \phi^{(n)}(a_{\pi(1)},\ldots,a_{\pi(n)}).$$

(U) *Unanimity*: if $a_i = a$ for all $i = 1,2,\ldots,n$, then

$$\phi^{(n)}(a_1,a_2,\ldots,a_n) = a.$$

(CS) *Citizen Sovereign*: for any given $a \in [0,1]$ there exists a profile $(a_1,a_2,\ldots,a_n) \in [0,1]^n$ such that $\phi^{(n)}(a_1,a_2,\ldots,a_n) = a$.

(ND) *Non Dictatorship*: there is no individual i such that

$$\phi^{(n)}(a_1,a_2,\ldots,a_n) = a_i$$

for any $(a_1,\ldots,a_{i-1},a_{i+1},\ldots,a_n) \in [0,1]^{n-1}$.

We will say that an intensity aggregation rule is *complete* if and only if the associated aggregated fuzzy preference is complete for any profile of complete individual preferences.

A characterization of complete aggregations rules is given by the following lemma proven in [5].

LEMMA 1 *An intensity aggregation rule* $\phi^{(n)} : [0,1]^n \to [0,1]$ *is complete if and only if*

$$\phi^{(n)}(a_1,\ldots,a_n) + \phi^{(n)}(b_1,\ldots,b_n) \geq 1$$

whenever $a_i + b_i \geq 1$ *for all* $i = 1,2,\ldots,n$. *Moreover, since* $\phi^{(n)}$ *is monotone non decreasing then it is complete if and only if*

$$\phi^{(n)}(a_1,\ldots,a_n) + \phi^{(n)}(1-a_1,\ldots,1-a_n) \geq 1$$

for all $(a_1, \ldots, a_n) \in [0,1]^n$. ■

The above given fuzzy property of rationality is extended to aggregation rules in the following way.

DEFINITION 1 *Given* n *individuals, an aggregation rule* $\phi^{(n)} : [0,1]^n \to [0,1]$ *is* non absolutely irrational *(NAI), or simply* non irrational, *if for any arbitrary finite set of alternatives* X*, the associated aggregated preference* $\mu : X \times X \to [0,1]$ *is complete and non absolutely irrational, i.e.* $A(\mu) > 0$*, whenever all individuals are complete and non absolutely irrational themselves, i.e.* $A(\mu^i) > 0$ *for all* $i = 1, 2, \ldots, n$*, with* $\mu^i : X \times X \to [0,1]$ *for all* i*.*
□

It is clear that in this way both individual and social opinions are required to belong to the set of Non-Absolutely Irrational (NAI) complete fuzzy preference relations. Therefore, we are in fact modifying the Unrestricted Domain condition.

The main result proven in [5] is the following.

THEOREM 1 *Let* $\phi : [0,1]^n \to [0,1]$ *be a complete intensity aggregation rule verifying condition A. Then* ϕ *is NAI if and only if the following conditions hold:*

(i) if $a_i + b_i > 1$ *for all* $i = 1, 2, \ldots, n$*, then* $\phi(a_1, \ldots, a_n) + \phi(b_1, \ldots, b_n) > 1$;

(ii) $\phi(a_1, \ldots, a_n) = 1$ *implies* $a_i = 1$ *for all* $i = 1, 2, \ldots, n$. ■

Moreover, from the proof of theorem 1 it can be concluded that in order for a complete intensity aggregation rule to be NAI, conditions (i) and (ii) are sufficient.

Two immediate corollaries of theorem 1 are

COROLLARY 1 *A complete intensity aggregation rule* $\phi : [0,1]^n \to [0,1]$ *verifying conditions CS and A is NAI if and only if conditions (i)-(ii) of theorem 1 hold.* ■

COROLLARY 2 *Let* $\phi : [0,1]^n \to [0,1]$ *be a complete intensity aggregation rule verifying condition PR. Then* ϕ *is NAI.* ■

Example 1 By applying Corollary 2 we can easily prove that the following intensity aggregation rule (see [15], page 60) is NAI:

- *Weighted Generalized Mean:* given $w_1, \ldots, w_n$, positive real numbers and $r \geq 1$, then

$$WGM(a_1, \ldots, a_n) = \left(\sum_{i=1}^{n} \frac{w_i (a_i)^r}{\sum_{j=1}^{n} w_j} \right)^{\frac{1}{r}}$$

where the assumption $r \geq 1$ has been made in order to assure completeness.

4 COVER-BASED HIERARCHICAL AGGREGATION RULES

Cover-based hierarchical aggregation rules appear in practice when the whole set of individuals is previously divided into groups (not necessarily disjoint), in such a way that the amalgamation is obtained by means of those subsets of individuals. Sometimes the society under study is naturally classified into those smaller groups, and sometimes such a classification is just claimed because of the large dimension of the group. Hence, final aggregation will be obtained as an amalgamation of all those partial aggregations.

Let us now introduce some notations which will be useful for our future discussions.

By $[i_1, i_1, \ldots, i_n]$ we will denote the ordered list whose first element is i_1, second element is i_2 and so on. $[\,]$ will denote the empty list and given a list L, $|L|$ will denote the length of the list. Moreover, let $\cdot$ be the classical list composition operator. Finally, given a list L we define the operator $\star$ that produces the set of elements of the list L, i.e. $\star L = \{j : j \text{ occurs in } L\}$. Thus, given a list L the notation $j \in L$ has the meaning of $j \in \star L$. $| \star L|$ will denote the cardinality of the set $\star L$ and since sets have no repetition of elements it is clear that $| \star L| \leq |L|$ for any list L.

So, if

$$L_1 = [1, 2, 3] \;\; L_2 = [3, 2, 4, 3].$$

We have,

$$\star L_1 = \{1, 2, 3\} \ , \ \star L_2 = \{3, 2, 4\}.$$

Given a list I and m lists $I_1, \dots, I_m$ we say that the list $\mathcal{I} = [I_1, \dots, I_m]$ is a *cover* of I if the following conditions are verified:

- $I_k \neq []$ for all $k = 1, 2, \dots, m$;
- for all $k = 1, 2, \dots, m$, we have $| \star I| = |I|$ i.e. all the elements in I_k are different;
- $\star I = \bigcup_{k=1}^{m} \star I_k$.

Given a list of indices $I = [i_1, \dots, i_m]$ and an aggregation rule ϕ, we introduce the following notation

$$\phi(a_h | h \in I) \equiv \phi(a_{i_1}, \dots, a_{i_m}).$$

Then we can define the concept of cover-based hierarchical aggregation.

DEFINITION 2 *Let I be a finite list of indices and let $[I_1, \dots, I_{c_0}]$ be a fixed cover of I. Let $m = |I|$ and $c_k = |I_k|$ for all $k = 1, 2, \dots, c_0$. A* cover-based hierarchical aggregation *ϕ is characterized by a collection $\phi_0, \phi_1, \phi_2, \dots, \phi_{c_0}$ of aggregation rules with $c_0 \geq 2$ and with $m_k > 1$ for some $1 \leq k \leq c_0$, such that $\phi_k : [0,1]^{c_k} \to [0,1]$ for all $k = 0, 1, 2, \dots, c_0$, in such a way that the composition*

$$\phi \equiv \phi_0(\phi_1, \phi_2, \dots, \phi_{c_0}) : [0,1]^m \to [0,1]$$

is defined as

$$\phi(a_k | k \in I) = \phi_0(\phi_1(a_h | h \in I_1), \dots, \phi_{c_0}(a_h | h \in I_{c_0}))$$

□

To clarify definition 2 consider the following example. Let $I = [1, 2, 3, 4]$ and for $c_0 = 3$ consider the cover $[I_1, I_2, I_3]$ with $I_1 = [1, 2]$, $I_2 = [2, 3, 1]$, $I_3 = [4, 1, 3]$. So given any input (a_1, a_2, a_3, a_4) we have

$$\phi(a_1, a_2, a_3, a_4) = \phi_0(\phi_1(a_1, a_2), \phi_2(a_2, a_3, a_1), \phi_3(a_4, a_1, a_3)).$$

Assuming an a priori fixed cover $[I_1, \dots, I_{c_0}]$ of I, each hierarchical aggregation is therefore characterized by the composition $\phi_0(\phi_1, \phi_2, \dots, \phi_{c_0})$.

DEFINITION 3 *If a property P that holds for $\phi_0, \phi_1, \phi_2, \ldots, \phi_{c_0}$ holds for $\phi_0(\phi_1, \ldots, \phi_{c_0})$ as well, with respect to any fixed cover, we will say that P* propagates under cover-based hierarchical aggregations. □

One important property that propagates under cover-based hierarchical aggregation is *stability* as defined and analyzed in [7].

DEFINITION 4 *An intensity aggregation rule ϕ is* stable *if there exists a constant K (called stability constant) such that for all $\epsilon > 0$ and for all $i = 1, \ldots, m$,*

$$|\phi(a_1, \ldots, a_{i-1}, a_i', a_{i+1}, \ldots, a_m) - \phi(a_1, \ldots, a_m)| \le K\epsilon$$

for all $a_1, \ldots, a_m$ and for $|a_i' - a_i| < \epsilon$. □

As a simple consequence of functional composition many other properties propagate under cover-based hierarchical aggregations, among them completeness, positive responsiveness (PR), unanimity (U) and conditions (i) and (ii) of theorem 1. Being NAI also propagates, as shown in the following theorem.

THEOREM 2 *If $\phi_0, \phi_1, \phi_2, \ldots, \phi_{c_0}$ are NAI aggregation rules then $\phi \equiv \phi_0(\phi_1, \phi_2, \ldots, \phi_{c_0})$ is also NAI, i.e. rationality propagates under cover-based hierarchical aggregations.*

Proof. Let X be a set of alternatives and let $\mu^1, \ldots, \mu^n$ be n NAI individuals. Let us denote by ν the aggregation function associated to ϕ and by ν_k the aggregation function associated to ϕ_k for all $k = 0, 1, 2, \ldots, c_0$. Since all the individuals $\mu^1, \ldots, \mu^n$ are NAI so are $\nu_1, \ldots, \nu_{c_0}$. On the other hand, since ϕ_0 is NAI and in view of definition 1, we have that its aggregation function ν_0 defined as

$$\nu_0(x, y) = \phi_0(\nu_1(x, y), \ldots, \nu_{c_0}(x, y))$$

is NAI. Therefore, since

$$\nu(x, y) = \nu_0(x, y)$$

we have that ϕ is NAI. ■

It is easy to observe that anonymity (A) and citizen sovereign (CS) do not in general propagate under cover-based hierarchical aggregations. However, it can be proved they propagate in specific but important cases (see [7] for details).

5 OWA OPERATORS AND ORDERED HIERARCHIES

We will now introduce ordered hierarchical aggregations. Such hierarchical aggregations are related very naturally to the *Ordered Weighted Averaging (OWA)* operators.

As previously commented, for any OWA operator ϕ we have

$$\min_i(a_1,\ldots,a_n) \le \phi(a_1,\ldots,a_n) \le \max_i(a_1,\ldots,a_n) \tag{7.6}$$

Two significative measures are associated with OWA operators (see Yager [24]). Given ϕ of dimension n with associated weights $w_1,\ldots,w_n$ these measures are the following.

(m1) *degree of orness,* i.e. how close an OWA operator is to the max operator. It is defined as

$$orness(\phi) = \frac{1}{n-1}\sum_{i=1}^{n}(n-i)w_i.$$

Dual to the measure of *orness* is the measure of *andness* defined as

$$andness(\phi) = 1 - orness(\phi)$$

which therefore measures how close an OWA operator is to the min operator.

(m2) The second measure called *dispersion* gives us the degree to which all aggregates are used equally, i.e. the degree to which an OWA operator is close to the simple average operator. The dispersion measure is defined as

$$Disp(\phi) = -\sum_{i=1}^{n} w_i \ln w_i.$$

As shown in [24], $orness(\phi) \le orness(\phi')$ when their associated weights $(w_1,\ldots,w_n)$ and $(w'_1,\ldots,w'_n)$ verify that

$$\sum_{j=1}^{i} w_j \le \sum_{j=1}^{i} w'_j \quad \forall i = 1,\ldots,n \tag{7.7}$$

This property gives in fact the intuitive idea of when an OWA operator must be more an "or" than other OWA operator. An important class of OWA operator is the class of *buoyancy* measures defined as follows (see [24]).

DEFINITION 5 *An OWA operator ϕ is a* buoyancy *measure if the associated weights $w_1, w_2, \dots, w_n$ verify the condition $w_i \geq w_j$ for all $i \leq j$.* □

Buoyancy measures verify the property of being *orlike* measures, i.e. if ϕ is a buoyancy measure then $orness(\phi) \geq \frac{1}{2}$. Buoyancy measures verify as well an important property for our future discussions (see [11])

THEOREM 3 *Let ϕ be an OWA operator of dimension n with associated weights $w_1, \dots, w_n$. Then ϕ is complete if and only if for all integer $k \leq \frac{n}{2}$ we have*

$$\sum_{i=1}^{k} w_i \geq \sum_{i=1}^{k} w_{n-i+1}. \tag{7.8}$$

■

Therefore, since buoyancy measures verify condition (7.8) they are complete.

A characterization of OWA operators in terms of "fuzzy rationality" is given by the following theorem.

THEOREM 4 *Let ϕ be a complete OWA operator with associated weights $w_1, \dots, w_n$. Then ϕ is NAI if and only if $w_n > 0$.*

Proof. We remark that OWA operators verify the pre-conditions of theorem 1. We will now show that any complete OWA operator verifies condition (i) of theorem 1.

Suppose then that $a_i + b_i > 1$ for all $i = 1, 2, \dots, n$. Therefore $a_{[n]} > 0$ and $b_{[n]} > 0$ from which we also have that there exists $\epsilon > 0$ such that for all $i = 1, 2, \dots, n$

$$\begin{aligned} a_i + b_i - \epsilon &\geq 1 \\ b_i - \epsilon &\geq 0 \end{aligned}$$

We then have

$$\begin{aligned}\phi(a_1,\ldots,a_n)+\phi(b_1,\ldots,b_n) &= \sum_{i=1}^{n} w_i a_{[i]} + \sum_{i=1}^{n} w_i b_{[i]} > \\ \sum_{i=1}^{n} w_i a_{[i]} + \sum_{i=1}^{n} w_i (b_{[i]}-\epsilon) &= \phi(a_1,\ldots,a_n)+\phi(b_1-\epsilon,\ldots,b_n-\epsilon)\end{aligned}$$

and the latter is not smaller than 1 for the completeness hypothesis.

In order to prove our theorem it is enough to prove that $w_n > 0$ if and only if condition (ii) of theorem 1 holds, i.e. if and only if $\phi(a_1,\ldots,a_n) = 1$ implies $a_1 = \ldots = a_n = 1$. Suppose that $a_1 \geq a_2 \geq \ldots \geq a_n$. We have

$$\phi(a_1,\ldots,a_n) = \sum_{i=1}^{n} w_i a_i = 1 \leftrightarrow w_i = 0 \;\; \forall \; a_i < 1.$$

In turn we have:

- if $\phi(a_1,\ldots,a_n) = 1$ and $a_n < 1$ then $w_n = 0$;
- if $w_n = 0$ then $\phi(1,\ldots,1,a_n) = 1$ for all $a_n \in [0,1]$

and the theorem is proven. ■

As a simple consequence we have

COROLLARY 3 *A buoyancy measure is NAI if and only if all its associated weights are non null.* ■

5.1 Ordered hierarchies of OWA operators

It can easily be observed that cover-based hierarchical aggregations of OWA operators do not in general produce OWA operators since cover-based hierarchical aggregations are characterized by a fixed cover of individuals, independently of their opinions. We will now give a characterization of hierarchical aggregations of OWA operators that produce OWA operators.

Let $\phi_0, \phi_1, \ldots, \phi_c$ be $c+1$ OWA operators such that

- ϕ_0 has dimension c;

- ϕ_i has dimension h_i for any $i = 1, 2, \ldots, c$
- $\sum_{i=1}^{c} h_i = n$

Let $w_{0,1}, \ldots, w_{0,c}$ be the weights associated to ϕ_0, and for all $i = 1, \ldots, c$ let $w_{i,1}, \ldots, w_{i,h_i}$ be the weights associated to ϕ_i.

DEFINITION 6 *The* ordered hierarchical composition *of $\phi_0, \phi_1, \ldots, \phi_c$ is defined by*

$$\phi_0(\phi_1, \ldots, \phi_c)(a_1, \ldots, a_n) = \phi_0(\phi_1(a_{[1]}, \ldots, a_{[h_1]}), \ldots, \phi_c(a_{[n-h_c+1]}, \ldots, a_{[n]}))$$

for all n-uples $(a_1, \ldots, a_n)$. □

DEFINITION 7 *If a property P that holds for $\phi_0, \phi_1, \phi_2, \ldots, \phi_c$ holds for the ordered hierarchical aggregation $\phi_0(\phi_1, \ldots, \phi_c)$ as well, we will say that P* propagates under ordered hierarchical aggregations. □

The result we were looking for follows.

THEOREM 5 *The property of being an OWA operator propagates under ordered hierarchical aggregations.*

Proof. Given any n-uple $a_1, \ldots, a_n$ such that $a_1 \geq a_2 \geq \ldots \geq a_n$, because of property 7.6 for the OWA operators, we have

$$\phi_1(a_1, \ldots, a_{h_1}) \geq \cdots \geq \phi_c(a_{n-h_c+1}, \ldots, a_n).$$

Let $\alpha_j = \sum_{i=1}^{j} h_i$ for any $j = 1, 2, \ldots, c$.

We have

$$\phi_0(\phi_1, \ldots, \phi_c)(a_1, \ldots, a_n) = \phi_0(\sum_{i=1}^{h_1} w_{1,i}a_i, \ldots, \sum_{i=\alpha_{c-1}+1}^{n} w_{c,i-\alpha_{c-1}}a_i) =$$

$$\sum_{j=1}^{c} w_{0,j} \sum_{i=\alpha_{j-1}+1}^{\alpha_j} w_{j,i-\alpha_{j-1}}a_i = \sum_{i=1}^{h_1} w_{0,1}w_{1,i}a_i + \ldots + \sum_{i=\alpha_{c-1}+1}^{n} w_{0,c}w_{c,i-\alpha_{c-1}}a_i$$

Finally, since

$$\sum_{i=1}^{h_1} w_{0,1}w_{1,i} + \ldots + \sum_{i=\alpha_{c-1}+1}^{n} w_{0,c}w_{c,i-\alpha_{c-1}} = 1$$

we can claim that the ordered hierarchical aggregation produces an OWA operator. ■

Thus, the above defined ordered hierarchical aggregations are such that the property of being an OWA operator propagates. However, as it is easy to see, completeness does not propagate in general for ordered hierarchical aggregations. Moreover -in view of theorem 5- Monotonicity, Idempotency, Commutativity, Stability and Non Absolute Irrationality (whenever completeness propagates) propagate under ordered hierarchical aggregations of OWA operators.

5.2 Orness and Dispersion

Given a fixed ordered hierarchical aggregation $\phi \equiv \phi_0(\phi_1, \dots, \phi_c)$ based upon $c+1$ OWA operators, the degree of orness is given by

$$orness(\phi) = \phi(\frac{n-1}{n-1}, \frac{n-2}{n-1}, \dots, \frac{1}{n-1}, 0) = \phi_0(z_1, \dots, z_i, \dots, z_c)$$

where

$$z_i = \phi_i(\frac{n - (\sum_{j=1}^{i-1} h_j + 1)}{n-1}, \dots, \frac{n - \sum_{j=1}^{i} h_j}{n-1}) = \sum_{m=1}^{h_i} \frac{n - (\sum_{j=1}^{i-1} h_j + m)}{n-1} w_{i,m} =$$

$$\frac{n - \sum_{j=1}^{i} h_j}{n-1} + \frac{\sum_{m=1}^{h_i} (h_i - m) w_{i,m}}{n-1} = \frac{n - \sum_{j=1}^{i} h_j}{n-1} + (h_i - 1)\frac{orness(\phi_i)}{n-1}$$

for all $i = 1, \dots, c$, with h_i the dimension of each ϕ_i and $(w_{i,1}, \dots, w_{i,h_i})$ its associated weights. Hence,

$$orness(\phi) = \sum_{i=1}^{c} \frac{(h_i - 1) orness(\phi_i) + n - \sum_{j=1}^{i} h_j}{n-1} w_{0,i}$$

Obviously, changes in ϕ_j's weights for $j = 1, \dots, c$ such that $orness(\phi_j)$ does not decrease for any $j = 1, \dots, c_0$ will never make $orness(\phi)$ decrease. However, increasing $orness(\phi_0)$ does not necessarily increase $orness(\phi)$ as shown by the following example.

Example 2 Let ϕ_0, ϕ_0' be two OWA operators of dimension c such that their associated weights are, respectively, $w_{0,2} = 1, w_{0,i} = 0$ for all $i \neq 2$

and $w'_{0,1} = w'_{0,3} = 1/2, w'_{0,i} = 0$ for all $i \neq 1,3$. Then, $orness(\phi_0) = orness(\phi'_0)$. Let now ϕ_i be the minimum rule (i.e., $orness(\phi_i) = 0$ for all $i = 1,\dots,c$). Then the difference between $orness(\phi_0(\phi_1,\dots,\phi_c))$ and $orness(\phi'_0(\phi_1,\dots,\phi_c))$ will still depend on the relative sizes h_2 and h_3 of OWA operators ϕ_2, ϕ_3. □

The dispersion of the hierarchical aggregation can be computed as follows.

$$
\begin{aligned}
Disp(\phi) &= -\sum_{j=1}^{c}\sum_{i=1}^{h_j} w_{0,j}w_{j,i}\ln[w_{0,j}w_{j,i}] \\
&= -\sum_{j=1}^{c}\sum_{i=1}^{h_j}(w_{0,j}w_{j,i}\ln w_{0,j} + w_{0,j}w_{j,i}\ln w_{j,i}) \\
&= -\sum_{j=1}^{c} w_{0,j}\ln w_{0,j} - \sum_{j=1}^{c} w_{0,j}Disp(\phi_j) \\
&= Disp(\phi_0) + \sum_{j=1}^{c} w_{0,j}Disp(\phi_j)
\end{aligned}
$$

Therefore, the dispersion of the hierarchical aggregations depends directly upon the dispersions of the given OWA operators.

In both cases it is however clear that once the operators ϕ_0 and $\phi_1,\dots,\phi_c$ are fixed then the orness and dispersion values of the hierarchical aggregation depend upon the particular ordering of the c OWA operators $\phi_1,\dots,\phi_c$.

It is then natural to ask the following questions:

(1) How can one *quickly* find an ordering of those c OWA aggregated operators which maximizes [resp. minimizes] the dispersion value of the hierarchical aggregation ?

(2) How can one *quickly* find an ordering of those c OWA aggregated operators which maximizes [resp. minimizes] the orness value of the hierarchical aggregation ?

In the next section we will provide complete solutions for them following the results in [9, 12].

6 MAXIMIZING AND MINIMIZING DISPERSION AND ORNESS

Let ϕ_0 be an OWA operator of dimension c and let $\Phi = \{\phi_i : 1 \le i \le c\}$ be a given set of OWA operators and let $h_1, \ldots, h_c$ their respective dimensions.

6.1 Dispersion

We start by providing polynomial algorithms to maximize [resp. minimize] the dispersion of the hierarchical aggregation.

Let π be a permutation of the indices $\{1, 2, \ldots, c\}$ such that

$$i \le j \text{ if and only if } w_{0,\pi(i)} \ge w_{0,\pi(j)}.$$

Intuitively, for $i = 1, 2, \ldots, c$, $w_{0,\pi(i)}$ is the i-th weight in non-increasing order.

Let $\phi_1, \ldots, \phi_c$ be a given ordering of the OWA operators in Φ.

We will now prove that the following lemmas hold.

LEMMA 2 *The value* $Disp(\phi) = Disp(\phi_0) + \sum_{j=1}^{c} w_{0,j} Disp(\phi_j)$ *is maximum if and only if the ordering* $\phi_1, \ldots, \phi_c$ *verifies*

$$i \le j \textit{ if and only if } Disp(\phi_{\pi(i)}) \ge Disp(\phi_{\pi(j)}).$$

■

LEMMA 3 *The value* $Disp(\phi) = Disp(\phi_0) + \sum_{j=1}^{c} w_{0,j} Disp(\phi_j)$ *is minimum if and only if the ordering* $\phi_1, \ldots, \phi_c$ *verifies*

$$i \le j \textit{ if and only if } Disp(\phi_{\pi(i)}) \le Disp(\phi_{\pi(j)})$$

■

In order to prove the two lemmas above, we make use of the elementary result that if $a \ge b \ge 0$ and $c \ge d \ge 0$ then $ac + bd \ge ad + bc$

(L1) Let us then prove Lemma 2.

Suppose first that $Disp(\phi)$ is maximum and suppose by contradiction that there exist i, j with $i < j$ such that $Disp(\phi_{\pi(i)}) < Disp(\phi_{\pi(j)})$. Then since

$$w_{0,\pi(i)}Disp(\phi_{\pi(j)}) + w_{0,\pi(j)}Disp(\phi_{\pi(i)}) > \\ w_{0,\pi(i)}Disp(\phi_{\pi(i)}) + w_{0,\pi(j)}Disp(\phi_{\pi(j)})$$

by switching ϕ_i and ϕ_j we would obtain an ordering which would cause the dispersion of the hierarchical aggregation to increase, contradicting the initial hypothesis.

Conversely, suppose that the ordering $\phi_1, \ldots, \phi_c$ verifies the condition of the Lemma. Let $\phi'_1, \ldots, \phi'_c$ be an ordering which maximizes $Disp(\phi)$. From what has been proven above, this ordering verifies the condition of the Lemma as well.

Let us prove that

$$\sum_{j=1}^{c} w_{0,j}Disp(\phi_j) = \sum_{j=1}^{c} w_{0,j}Disp(\phi'_j).$$

If $\phi_{\pi(1)} \neq \phi'_{\pi(1)}$ in view of the condition of the Lemma it must be

$$Disp(\phi_{\pi(1)}) = Disp(\phi'_{\pi(1)}).$$

Thus, we are left to prove that

$$\sum_{j \neq \pi(1)} w_{0,j}Disp(\phi_j) = \sum_{j \neq \pi(1)} w_{0,j}Disp(\phi'_j).$$

which can be shown easily to hold, by applying a recursive argument.

(L2) Lemma 3 can be proven analogously.

In view of the above Lemmas and from the well-known result (see, e.g., [3]) that any given n numbers can be sorted in $\mathcal{O}(n \log n)$ time, we can claim that the following theorem is true.

THEOREM 6 *Given an OWA operator ϕ_0 of dimension c and a set of c OWA operators Φ and given their dispersions, there exists a $\mathcal{O}(c \log c)$ algorithm which produces an ordering of Φ maximizing (or minimizing) the dispersion of the hierarchical aggregation.* □

6.2 Orness

The problem of maximizing or minimizing the orness of the hierarchical aggregation at a first sight appears to be quite more difficult from a computational point of view and its (polynomial) solution definitely more subtle. However we can see that we can easily reduce such a problem to the classical combinatorial *assignment problem* for which there exists a $\mathcal{O}(c^3)$ algorithm.

Let us go back to the algebraic expression for computing the orness of the hierarchical aggregation once an ordering $\phi_1, \dots, \phi_c$ of the OWA operators in Φ has been fixed:

$$orness(\phi) = \sum_{i=1}^{c} \frac{(h_i - 1)orness(\phi_i) + n - \sum_{j=1}^{i} h_j}{n-1} w_{0,i}$$

Thus

$$orness(\phi) = \sum_{i=1}^{c} \frac{(h_i - 1)orness(\phi_i)w_{0,i} + n - \sum_{j=i}^{c} w_{0,j}h_i}{n-1}.$$

We can therefore conclude that the contribution to the orness of ϕ of an OWA operator $\bar{\phi} \in \Phi$ with dimension h will be

$$\begin{aligned}
\delta(\bar{\phi},1) &= \tfrac{(h-1)orness(\bar{\phi})w_{0,1}+n-h}{n-1} && \text{if } \bar{\phi} \text{ is the first}\\
\delta(\bar{\phi},2) &= \tfrac{(h-1)orness(\bar{\phi})w_{0,2}+n-(1-w_{0,1})h}{n-1} && \text{if } \bar{\phi} \text{ is the second}\\
&\vdots && \vdots\\
\delta(\bar{\phi},i) &= \tfrac{(h-1)orness(\bar{\phi})w_{0,i}+n-\sum_{j=i}^{c} w_{0,j}h}{n-1} && \text{if } \bar{\phi} \text{ is the } i\text{-th}\\
&\vdots && \vdots\\
\delta(\bar{\phi},c-1) &= \tfrac{(h-1)orness(\bar{\phi})w_{0,c-1}+n-(w_{0,c-1}+w_{0,c}h)}{n-1} && \text{if } \bar{\phi} \text{ is next to the last}\\
\delta(\bar{\phi},c) &= \tfrac{(h-1)orness(\bar{\phi})w_{0,c}+n-w_{0,c}h}{n-1} && \text{if } \bar{\phi} \text{ is the last}
\end{aligned}$$

Summing up we can say that:

- denoted with $\phi_1, \phi_2, \dots, \phi_c$ the OWA operators in Φ, and
- denoted for simplicity with δ_{ij} the contribution to the orness of ϕ of the operator ϕ_i once it is chosen as the j-th in the hierarchical aggregation, then

- the problem of maximizing [resp. minimizing] the orness of the hierarchical aggregation is equivalent to the problem of finding a permutation π of the rows of the matrix

$$M = \begin{pmatrix} \delta_{11} & \delta_{12} & \dots & \delta_{1c} \\ \delta_{21} & \delta_{22} & \dots & \delta_{2c} \\ \vdots & \vdots & \ddots & \vdots \\ \delta_{c1} & \delta_{c2} & \dots & \delta_{cc} \end{pmatrix}$$

such that the sum of the elements of the main diagonal of

$$M_\pi = \begin{pmatrix} \delta_{\pi(1)1} & \delta_{\pi(1)2} & \dots & \delta_{\pi(1)c} \\ \delta_{\pi(2)1} & \delta_{\pi(2)2} & \dots & \delta_{\pi(2)c} \\ \vdots & \vdots & \ddots & \vdots \\ \delta_{\pi(c)1} & \delta_{\pi(c)2} & \dots & \delta_{\pi(c)c} \end{pmatrix}$$

is maximum [resp. minimum], i.e. such that

$$\sum_{i=1}^{c} \delta_{\pi(i)i}$$

is maximum [resp. minimum].

The above is the classical *assignment problem* (also known as the *bipartite weighted matching problem*) (see [21], chapter 11) and can be solved by the so-called *hungarian method* (see [16]) in time $\mathcal{O}(c^3)$. More recently in [22] it has been proven that by using special data structures it is possible to improve the above bound to $\mathcal{O}(c^2 \log c)$.

6.3 Some particular cases

$\mathcal{O}(c \log c)$ algorithms can instead be given for the most significative cases in practice. Since each one of the given c OWA operators represents partial information to be aggregated into only one index, it is common to define them in such a way that either

- they contain the same amount of information. In this case, all OWA operators in Φ will have the same dimension h such that $n = hc$; or
- they treat their inputs with the same degree of optimism (pessimism). In this case, all OWA operators will have the same degree of orness.

Let us consider the above two cases in more details.

(case 1) In case all OWA operators in Φ have the same dimension $h = n/c$, we have to find an ordering $\phi_1, \ldots, \phi_c$ such that

$$orness(\phi) = \sum_{i=1}^{c} \frac{(h-1)orness(\phi_i) + n - \sum_{j=1}^{i} h}{n-1} w_{0,i}$$

is maximum. Therefore,

$$orness(\phi) = \frac{(h-1)}{n-1} \sum_{i=1}^{c} orness(\phi_i) w_{0,i} + \frac{\sum_{i=1}^{c}(c-i)h}{n-1} w_{0,i}.$$

It follows that the sought ordering can be obtained as in Lemma 2, i.e. by producing any ordering which verifies

$$i \leq j \text{ if and only if } orness(\phi_{\pi(i)}) \leq orness(\phi_{\pi(j)}).$$

Analogously, if we want to minimize the orness of the hierarchical aggregation we act as in Lemma 3.

(case 2) In case all OWA operators in Φ have the same orness r, we have

$$orness(\phi) = \frac{1}{n-1} \sum_{i=1}^{c} (rh_i + \sum_{j=i+1}^{c} h_j) w_{0,i} - \frac{r}{n-1}.$$

As a consequence, any ordering which sorts the elements of Φ in increasing [resp. decreasing] order with respect to their dimensions maximizes [resp. minimizes] the orness.

7 FINAL COMMENTS

In this paper, we surveyed on some research work that has been done to obtain a full characterization of rational aggregation operators and hierarchies of them. In particular we focused on the propagation of key ethical conditions, rationality and stability properties of cover-based hierarchies and ordered hierarchies. In this last case, we have analyzed Yager's OWA operators and seen that the property of being an OWA operator propagates under ordered hierarchical aggregations. As by product, we observed interesting theoretical and computational results on the propagation of OWA operators significant measures such as *orness* and *dispersion*,

We conclude by again pointing out the practical importance of such hierarchical aggregation rules. Preference intensity of large groups of people are in common life usually aggregated by means of hierarchical procedures. The fact that standard ethical conditions as well as non absolute irrationality, together with stability, propagate under general conditions legitimize hierarchical aggregations from both theoretical and applicative point of view.

Acknowledgment: This research has been partially supported by Dirección General de Investigación Científica y Técnica (Spain).

REFERENCES

[1] K.J. Arrow. *Social Choice and Individual Values.* John Wiley & Sons, Inc., New York, 1964.

[2] W. Cholewa. Aggregation of fuzzy opinions: an axiomatic approach. *Fuzzy sets and systems*, 17:249–258, 1985.

[3] T.H. Cormen, C.E. Leiserson and R.R. Rivest. *Introduction to Algorithms.* MIT Press, Cambridge, MA, 1990.

[4] V. Cutello and J. Montero. A model for amalgamation in group decision making. In *NAFIPS 1992*, Puerto Vallarta, Mexico, 1992. Proceedings of the North American Fuzzy Information Processing Society International Conference on Fuzzy set theory and applications. NASA conference publications 10112, Vol. 1, 215–223.

[5] V. Cutello and J. Montero. A characterization of rational amalgamation operations. *International Journal of Approximate Reasoning*, vol. 8 (4), 325–344, June 1993.

[6] V. Cutello and J. Montero. An axiomatic approach to fuzzy rationality *Proceedings of IFSA'93*, Seoul, Korea, 634–636, 1993.

[7] V. Cutello and J. Montero. Hierarchies of intensity preference aggregations. *International Journal of Approximate Reasoning*, 10:123–133, 1994.

[8] V. Cutello and J. Montero. Fuzzy Rationality Measures *Fuzzy Sets and Systems*, 62:39–54, 1994.

[9] V. Cutello and J. Montero. Computational problems of the hierarchical aggregation of OWA operators In *Proceedings of IPMU'94*, Paris, July 1994. B. Bouchon-Meunier and R.R. Yager Eds, 407–411.

[10] V. Cutello and J. Montero. Recursive Families of OWA operators In *Proceedings of IEEE World Congress on Computational Intelligence, FUZZ-IEEE 94*, Orlando, Florida, June 1994. P. Bonissone Ed, IEEE Press, 2046–2049.

[11] V. Cutello and J. Montero. Hierarchies of aggregation operators. *International Journal of Intelligent Systems*, 9:1025–1045, 1994.

[12] V. Cutello and J. Montero. Hierarchical aggregation of OWA operators: basic measures and related computational problems. *Submitted*

[13] D. Dubois and J.L. Koning. Social choice axioms for fuzzy set aggregation. *Fuzzy sets and systems*, 43:257–274, 1991.

[14] L.W. Fung and K.S. Fu. An axiomatic approach to rational decision making in a fuzzy environment. In L.A. Zadeh, K.S. Fu, K. Tanaka, and M. Shimura, editors, *Fuzzy sets and their applications to Cognitive and decision processes*, 227–256. Academic Press, 1975.

[15] G.J. Klir and T.A. Folger. *Fuzzy sets, uncertainty and information*. Prentice Hall, Englewood Cliffs, NJ, 1988.

[16] H.W Kuhn. The Hungarian Method for the assignment Problem. *Naval Research Logistics Quarterly*, 2:83–97, 1955.

[17] J. Montero. A note on Fung-Fu's theorem. *Fuzzy sets and systems*, 17:259–269, 1985.

[18] J. Montero. Arrow's theorem under fuzzy rationality. *Behavioral Science*, pages 267–273, 1987.

[19] J. Montero. Social welfare functions in a fuzzy environment. *Kybernetes*, 16:241–245, 1987.

[20] S. Ovchinnikov. Means and social welfare function in fuzzy binary relation spaces. In J. Kacprzyk and M. Fedrizzi, editors, *Multiperson decision making using fuzzy sets and possibility theory*, 143–154. Kluwer, 1990.

[21] C.H. Papadimitriou and K. Steiglitz. *Combinatorial Optimization. Algorithms and Complexity*. Prentice Hall, Englewood Cliffs, NJ, 1982.

[22] M.L. Fredman and R.E. Tarjan. Fibonacci heaps and their uses in improved network optimization algorithms. *Journal of A.C.M.*, 34(3):596-615, 1987.

[23] R.R. Yager. Connectives and quantifiers in fuzzy sets. *Fuzzy sets and Systems*, 40:39-75, 1991.

[24] R.R. Yager. On ordered weighted averaging aggregation operators in multi-criteria decision making. *IEEE transactions on systems, man and cybernetics*, 18:183-190, 1988.

[25] R.R. Yager. Families of OWA operators. *Fuzzy sets and systems*, 59:125-148, 1993.

8

ON THE NOTION STATE IN MULTISTATE STRUCTURE FUNCTION THEORY

Bart Cappelle

Department of Applied Mathematics and Computer Science
Universiteit Gent
Krijgslaan 281-S9,
B-9000 Gent, Belgium

ABSTRACT

Binary structure function theory has been developed to model the reliability aspects of both physical and nonphysical systems. Each component and system assumes either one of two possible states, *failure* or *work*, and the notion reliability has been defined in a probabilistic way. There are, however, some important deficiencies that cannot be overcome without extending the state space of both components and systems and, hence, allowing more than two possible states. Although many authors developed and described several models that extend the binary model, some fundamental notions like *state* and *state space* are almost never discussed or described. In this paper, however, we introduce some new ideas about the notion state in reliability theory and we construct a general framework to describe the reliability aspects of multistate systems and their components.

1 INTRODUCTION

German researchers who were developing the V1- and V2-missiles, have formulated the first foundations of reliability theory, and shortly after the Second World War, many scientists, especially in the United States, developed a complete new theory to study the *reliability of complex systems*. Till the beginning of the sixties, reliability theory was considered as a subtheory of applied probability theory, since the notion *reliability of a system* was defined in a probabilistic way. In 1961 Birnbaum, Esary and Saunders published an important paper on binary structure function theory [6]. A binary system, i.e., a system that assumes one of only two possible states, was represented by a binary structure function. From their point of view, a system is made up of a finite number of components and, hence, the determination of the system reliability, when the component reliabilities are known, is the main problem in classical—i.e. binary and probabilistic—reliability theory. Since then, many reliability specialists consider a binary structure function as the mathematical representation of the reliability aspects of a physical system. Hence, studying this mathematical representation of the system must give us more insight into the nature, the strength and the weaknesses of the physical system. A binary structure function is a boolean polynomial [15] and there is an important link between binary structure function theory and binary graph theory [7, 16]. This link allows a graphical representation of the reliability aspects of the physical systems under study [23]. Since the introduction of the notion *software reliability*, that describes the reliability aspects of software, binary structure functions are also applied to model the reliability aspects of both physical and nonphysical systems. Let us have a closer look at this classical, binary theory.

Birnbaum, Esary and Saunders—as well as many other reliability scientists—assume that a system is made up of a finite number of components, and the binary structure function actually represents the interaction between on the one hand the system and on the other hand its components. In their binary model, each component and system has one of the following properties:

- the component or system is working or functioning,
- the component or system fails, or is not able to work or to function.

These two properties of components and systems are called *possible states*, hence, a state is a kind of label indicating the ability to work. Classically, the *reliability* of a system or a component is defined as the probability that the component or system assumes the label *is working*, in a time interval $[t_1, t_2]$. In other words, the reliability of a component or system is the probability that the state of a component or system equals *is working* or, equivalently, *is functioning*, during a time interval. The axiomatic approach to classical reliability theory, then, is made up of two parts:

- the study of the relationship between components and the systems they belong to, which is the topic of binary structure function theory;
- the study of probabilistic problems like the mean time until the first system failure, the availability of components, subsystems and the system itself, the mean time between two failures, etc.

Due to the relationship between boolean polynomials and graph theory, the reliability theoreticians and engineers mainly devoted their attention to the probabilistic aspects of reliability theory. The main results of this research field have been collected in two excellent books by Barlow and Proschan [2, 3], which are till now basic works for everyone studying the reliability aspects of systems.

Although the classical theory has been very successful in the past, there are some severe shortcomings. Let us summarize the two main shortcomings. On the one hand, each system and component assumes either one of two possible states, hence, each system must be dichotomous in nature. Obviously, this is not always the case, many systems are not binary in nature and the application of binary structure function theory is an oversimplification of reality. The application of binary structure function theory implies that we idealize the systems and, hence, study some *ideal systems and components*. On the other hand, the confidence in the ability to function is measured by a probability measure on the state spaces of the components and the systems they are part of. Since the development of alternative, nonprobabilistic measures of uncertainty and vagueness, that allow an ordinal approach to uncertainty (see, e.g. [18, 20]), probability theory no longer is considered as the only tool to model the uncertainty of events. Several alternative models based upon fuzzy set theory and possibility theory have been proposed. For some examples we refer to [8, 11, 12, 27, 28, 29].

We shall not dwell upon the nonprobabilistic approaches to the notion *reliability* of a component or system, but in the sequel we shall introduce some new ideas and further develop some existing ideas about the foundations of the notion *state* of a component or system. In the next section, we focus our attention to two important types of multistateness of systems and their components, and we illustrate them with some concrete examples.

2 FROM BINARY TO MULTISTATE MODELS: A SUMMARY

Until 1978, most papers that discuss multistate components and systems, mainly are devoted to the extension of the notion reliability. Moreover, the classical notion *reliability* was extended straightforwardly and only applied to components or systems that are subject to either degraded states or n different types of failure, and assume one of a finite number of states.

From 1978 on, however, two important American research teams developed multistate structure function theory in an axiomatic way [4, 21]. These teams tried to overcome the main shortcomings of the binary model, by introducing the notion *intermediate state* and presented two models that, nevertheless, are closely related to the binary model, since the conditions that they imposed on a mapping to be a *multistate structure function* highly resembled the conditions imposed on a mapping to be a binary structure function (see, e.g., [3]). Hence, the proposed models are only able to cope with degraded states and the imposed conditions restrict the application of these models considerably. In 1988, Montero, Tejada and Yáñez [25] have introduced lattice theory to deal with multistateness in reliability theory. Their model is able to cope with components that are subject to both degraded states and different types of failure. Lattice theory turns out to be a valuable tool to study many aspects of multistate systems and the lattice approach, e.g., is able to define minimal path and cut elements properly (see e.g., [10, 11, 25]). The foundations of the notion *state*, however, are hardly discussed from a reliability point of view.

Several authors, however, mentioned the problem of defining the notion state and the related notion state space. One of the oldest papers on this matter, has been written in 1963 by Dubes [19], who discussed some basic ideas about the working and failing of components and systems that are subject to two or more possible failures. Like in most reliability papers, Dubes focussed on the solution of probabilistic problems. The structural aspects and the possible representations of this kind of systems were less important for him and the notion state was not really explained and was based upon some very intuitive ideas. Therefore, the papers written by Barlow and Wu [4], and El-Neweihi, Proschan and Sethuraman [21] are highly relevant from our point of view: they tried to develop an axiomatic approach to multistate reliability theory and they tried to answer some important questions about the origin of multistateness.

The previous summary clearly indicates the need for a solid theory to study the multistate behaviour of both components and systems in reliability theory. Therefore, we must not only define the notion *state*, properly, but we also must formulate clear concepts about the nature of a system and component in reliability theory. This will be the topic of the next sections.

In the previous paragraphs, we have mentioned already two possible types of failure that often appear in reliability papers:

- components subject to n different failures,

- components that degrade and therefore, their working aspects can hardly be represented by working or failing.

Let us have a closer look at two examples.

Example 1 Consider a valve that is open and that cannot be closed on the one hand and a valve that is closed but that cannot be opened on the other hand. The two types of failure are basically different and therefore, we may say that the valve is subject to two different types of failure. Some reliability specialists call this specific type of failure of the example *dual* (see, e.g., [24]). Electronic equipment often is subject to dual failures, therefore many reliability specialists studied them intensively. We, however, shall consider components subject to dual failures as components that are subject to two different types of failure.

Example 2 Consider a pump with a flow that varies between 0 and 20 t/h (tons per hour) and assume that the actual setpoint of the pump is 20 t/h. The pump obviously fails when its flow is 0 t/h and works when its flow is 20 t/h. Assume that the actual flow is 15 t/h and that it cannot be raised to the setpoint. Does this pump fail or not? We only may say that the pump assumes an intermediate state that seems to be closer to work than to fail.

3 SOME SEMINAL IDEAS ABOUT THE NOTION STATE

Let us consider some remarkable notes by Gnedenko, Beliaev and Soloviev [22]. The main problem in reliability theory is to find the answer to the question *what is the reliability of a system.* Hence, the notion *reliability* must be described somehow. What is the *reliability* of a system? Reliability is a measure of the confidence that the system will work sufficiently. How can we measure whether the performance of a system is sufficient or not? For some systems it is quite obvious whether their performance is sufficient or not. A light bulb, e.g., functions until it suddenly does not emit any light. Obviously, the bulb functions sufficiently until it suddenly stops emitting any light. Does the pump in example 2 functions sufficiently or not? The performance of the pump is surely not optimal but it is much better than a complete failure. Hence, it is not very realistic to assume that the pump works perfectly only when its actual flow is 20 t/h and that it fails otherwise. Therefore, we must introduce something in between. From similar considerations, Gnedenko, Beliaev and Soloviev [22] introduced two different types of failure:

- an *immediate failure*, i.e., a sudden change of the values of the parameters that characterize the quality of the system or its components;
- a *gradual failure*, i.e., a gradual change of the values assumed by the parameters that characterize the quality of the system; the possible values of the parameters can thus be divided into several classes, that can be marked by a label which indicates the quality of the system or the component.

Their observations about the nature of a failure indicate that the binary model applies only when the failures, that the components and the system are subject to, are immediate failures. When we model gradual failures by the binary model, we must divide the values of the parameters, that characterize the quality of the system or component, into two classes. Let us, therefore, have a closer look at example 2 and assume that we apply the binary model, which implies that there are only two possible quality labels determined from the parameter *flow*. Can we define some number L satisfying the following conditions:

- when the actual flow is smaller than L the quality label of the pump is *fail*,
- when the actual flow is greater than L the quality label of the pump is *work*.

Any number L, $L \in [0, 20]$ introduces a very artificial split up of the set of possible values of the parameter *flow*, hence, the binary model cannot be applied without loss of essential information. One notices, that the same considerations lead to the introduction of fuzzy sets by Zadeh in 1965 [30] and the introduction of evaluation sets by De Cooman in 1991 [17].

The considerations of Gnedenko, Beliaev and Soloviev do not explain the possibility that components may be subject to n different kinds of failure. From example 1, however, it is quite clear that components, systems respectively, are subject to n different kinds of failure, when n totally different causes lead to a failure and when the state must refer to the cause of the component failure, respectively system failure.

The considerations of Gnedenko, Beliaev and Soloviev, and the ideas of the previous section do answer some very important questions about the nature of states of components and the systems they belong to:

- a system is completely characterized by a set of parameters that may take numerical or linguistic values,
- a component obviously is a part of a system, and, hence, must be characterized by a subset of the set of parameters that characterize the system the component belongs to,
- a state is a kind of quality label that transforms the information of the parameters and reduces the amount of available information; the binary case allows only two possibilities *work* or *fail*,
- intermediate states and the multistateness caused by these intermediate states stem from the vagueness of the notion *well functioning* when the components or systems are subject to gradual failures, and, hence, cannot be modeled by binary structure functions without loss of essential information;
- components and systems are subject to n different failures when there are different causes of a component or system failure, and the component or system state must refer to the cause of the component or system failure;
- in most cases, the binary approach is not able to represent the available knowledge without loss of essential information; only immediate failures can be represented by the binary model.

Obviously, a component may belong to more than one system. In section 4, we introduce some new ideas about the notion *system, component* and *state* that are based upon the ideas explained in this section.

4 SYSTEMS, COMPONENTS AND THEIR POSSIBLE STATES

The ideas of Gnedenko, Beliaev and Soloviev explain the nature and the origin of degraded states partially and they have indicated some important ideas behind the basic concept *state* in reliability theory. From their point of view, a state is some interpreted information but, unfortunately, they have focussed their attention to the probabilistic aspects of reliability theory and did not dwell upon the foundations of these notions. Dubes' comments on the notion *state* and *good* component [19] are more or less similar to the ideas of Gnedenko, Beliaev and Soloviev. Based upon their work, we shall introduce a more formal and uniform approach to the notions *state*, *state mapping* and *state space* from a reliability point of view.

Reliability theory deals with the reliability aspects of systems and their components. The reliability of a system is determined from the component reliabilities. But what is a component or a system in the reliability context? First, we introduce the notion *entity*. An entity may be physical or not, and is observable to the reliability engineer. To each entity we associate a set of parameters, that may take numerical or linguistic values and this set characterizes the entity completely. Let us have a closer look at some examples.

Example 3 A car is an entity and the set of parameters that characterize the car include, e.g., the price, the colour, the type of engine, the number of doors, its maximum speed, etc. A computer programme may be another entity. The language the application is written in, the speed of the programme, the type of application (spreadsheet, database, etc.) are possible parameters that characterize this entity. The pump of example 2 is another entity. This entity can be characterized, e.g., by the maximal flow of the pump, the type (electrical, steam), etc.

Hence, from the previous examples, it is clear that an entity refers to an object in the *real world* that may be either physical or not, and that is characterized by a set of parameters. The notion system that we introduce in this section is the mathematical representation of the reliability aspects of an entity, taking into account the conditions imposed by the *context* of the entity. Hence, we must define the notion *context*.

Definition 1 *A context is a set of boundaries and requirements that are imposed on an entity.*

We explain this notion by some examples.

Example 4 Assume that a company wants to buy a car for its sales managers. The price of the cars must not exceed 30,000 USD and they must be blue, according to the company instructions. These conditions define a context for the set of cars and make it possible to determine whether a car can be bought or not. Assume that the same company wants to buy a computer programme and must choose one out of a number of possible programmes. The automation engineer introduces a list of minimal requirements for these programmes like the minimal speed of the programmes, the availability of a user friendly interface, the possibility to link the programme to an optical disk system, etc. The list of minimal requirements is a context. The requirements that we imposed on the pump in example 2 is a context.

Example 5 Consider an army that wants to buy some missiles to protect its armoured ground forces against tank attacks. A possible context for the missiles could be:

- the missiles must strike a moving tank at a distance of 2 kilometer with a certainty of 99%,
- the missile must not explode within 400 m of the launching place in order to protect the own troops,
- the speed of the missile when launched is at least 1,400 m/s.

The set of conditions that are imposed on the missiles is a context of the entity ground-ground missile.

Many reliability theoreticians call an entity a system, but the application of this intuitive definition cause some fundamental problems when defining a mathematical model of the reliability aspects of a system (see, e.g., the problems that arise when the classical intuitive definition of system, component and state is applied to electronic equipment [1, 9]). These problems can be overcome when we notice that the reliability aspects of the entity depend upon the context we are working in. Hence, under one set of conditions an entity may be working but when other conditions are imposed, the entity may not work. Therefore, a system refers to an entity and its nonempty, finite set of parameters, determined by the context we are working in.

Example 6 Consider the pump of example 2. The set of parameters that describe the pump of the problem under consideration is a singleton, {flow}. The flow characterizes the pump completely for the problem under study, since the context only discusses the values of the actual flow of the pump. Consider a computer programme that must perform some accountant calculations without errors. The set of parameters that characterize this system may be

$$\{\text{the number of wrong calculations per million calculations}\}.$$

One notices the importance of the context. Without a context we are not able to describe the entity by means of a set of parameters. In the sequel, we assume that a system refers to an entity and its context. Obviously, a component is a part of a system, hence, we can define a component as a subset of a system.

Definition 2 *A system is a nonempty, finite set of parameters that characterize the behaviour of the entity in a context. A component of a system is a nonempty subset of a system.*

A system is *atomic* when it is a singleton and, hence, is not made up of more than one component. The pump of example 6 is atomic. Obviously, most systems are not atomic and are characterized by many, many parameters.

From the previous definitions, it is clear that the definition of a system and its components highly depend upon the context of the entity the system and components refer to. Without a well-defined context we are not able to describe a system or its components properly. Components are defined by the reliability engineer who models the system related to the entity. Therefore, they are not uniquely defined. They may refer to some physical subentity of the entity that is modeled—e.g., a pump in a feedwater system—or they may not refer to a physical subentity—e.g., the set of all parameters that are related to the pressure in a steam grid can be a component of the representation of the physical entity steam grid. We implicitly assume that a system is characterized by a finite number of parameters, hence, we assume that each system is made up of a finite number of components.

Definition 3 *A system S that is related to an entity in a context is completely represented by a set of components C_i, $1 \leq i \leq n$, if and only if*

- *every parameter of the system is included in at least one component, i.e.,*

$$S = \bigcup_{i=1}^{n} C_i,$$

- *a component never includes another component, i.e.,*

$$(\forall (i,j) \in \{1,\ldots,n\}^n)(i \neq j \Rightarrow C_i \nsubseteq C_j \text{ and } C_j \nsubseteq C_i).$$

Example 7 Consider a pump that is represented by the following set of parameters S in a context:

$$S = \{\text{maximal flow, age, electrical or not}\}.$$

The system S is not completely represented by its subsets C_1 and C_2,

$$C_1 = \{\text{maximal flow}\} \text{ and } C_2 = \{\text{maximal flow, electrical or not}\},$$

while the system is completely represented by the subsets D_1 and D_2,

$$D_1 = \{\text{maximal flow, age}\} \text{ and } D_2 = \{\text{age, electrical or not}\}.$$

Remark that we do not demand that the intersection of two different components is empty, since components may have common parameters. In the sequel, we implicitly assume that a system is completely represented by its set of components.

Since we have defined systems and their possible components, we are now able to define the *state* of a component and a system. First, we have a closer look at the binary approach. In the binary approach to each combination of values that the parameters may take, one of two possible values is assigned: *fail* or *work*. These two values are often represented by 0, 1 respectively. Hence, the information that is included in the values that the parameters take, is transformed into a more feasible, easier to handle bit of information. We have assigned to each set of values that the parameters may take, one and exactly one new value. This value is called the *state* of the system or component, the mapping that transforms the values of the parameters into a state is called a *state mapping*. We resume these ideas in definition 4.

Definition 4 *Let e be an entity and $S = \{p_1, \ldots, p_n\}$ be its set of parameters. Assume that each parameter p_i, $1 \leq i \leq n$, assumes values in the set P_i and let T be an arbitrary set. A mapping f,*

$$f : P_1 \times \cdots \times P_n \to T : (x_1, \ldots, x_n) \mapsto f(x_1, \ldots, x_n),$$

is a system state mapping. When the parameters assume the values $x_1, \ldots,$ x_n, then $f(x_1, \ldots, x_n)$ is the system state, defined by the context of the entity e.

Analogously, to each component we can assign a *component state*.

Definition 5 *Let e be an entity and* $S = \{p_1, \ldots, p_n\}$ *be its set of parameters. Let C be a component of S,* $C = \{p_{i_1}, \ldots, p_{i_m}\}$*, and* T_C *be an arbitrary set. A mapping f,*

$$f : P_{i_1} \times \cdots \times P_{i_m} \to T_C : (x_{i_1}, \ldots, x_{i_m}) \mapsto f(x_{i_1}, \ldots, x_{i_m}),$$

is a component state mapping. When the parameters assume the values $x_{i_1}, \ldots, x_{i_m}$ *then* $f(x_{i_1}, \ldots, x_{i_m})$ *is the component state.*

One notices that basically, there is no difference between a component state or a system state. We exemplify definition 4 and 5.

Example 8 Consider the pump of example 2 and consider the set $T = \{t_1,, t_2\}$. The mapping f,

$$f : [0, 20] \to \{t_1, t_2\} : x \mapsto \begin{cases} t_1 & ; \quad x = 20 \\ t_2 & ; \quad \text{elsewhere,} \end{cases}$$

is a state mapping.

Now, we are able to determine the component or system state, but the information included in the set of states is not sufficient, since there is no relation amongst the states of the set of possible states and we are not able to *compare* states. In the next section, we have a closer look at this relation and we define the notion state space.

5 ON THE NOTION STATE SPACE

In the previous section, we have pointed out that a state mapping transforms the information about a component or system into new information that is easier to handle. Although a component or a system state is a kind of quality label that indicates the ability to function and, hence, any set can be a state space, the structure *set* is not sufficient to solve some basic reliability problems. Therefore, we must introduce the notion *state space.*

When covering a reliability problem—whether we apply the binary model or a multistate model—we must be able to answer basic questions like "*the state of component i is better than the state of component j,*" or "*the state of component k is degraded.*" To answer these questions, we must define a binary relation R on the set of possible states T. Hence, a state space consists of two parts: a set of possible states T and a binary relation R on

T. Taking into account the nature of the statement *"the state of component i is better than the state of component j,"* the relation R mathematically represents the linguistic property *" state x is at most as good as state y"* or, dually, *"state y is at least as good as state x"*, for any $(x, y) \in R$. The evaluation of these linguistic statements depend upon the context. Hence, the relation R also depends upon the context of the entity.

Definition 6 *The set of possible states of a component, system respectively, provided with the relation R is called the state space of the component, system respectively.*

Obviously, R cannot be any binary relation on the set T. In fact, R must satisfy three basic properties:

- let x be an arbitrary element of T; since *"state x is at most as good as state x"* is meaningful, the relation R must be reflexive or, equivalently, the equality relation "=" on the set T is a subset of R, i.e.,

$$(\forall x \in T)((x, x) \in R);$$

- consider two arbitrary but different elements of the set T, denoted by x and y, and assume that (x, y) is an element of R; taking into account the interpretation of $(x, y) \in R$, i.e., *"state x is at most as good as state y,"* it would be quite unusual to accept that x and y also satisfy the property *"state y is at most as good as state x,"*; hence, the relation R is antisymmetric, i.e.

$$(\forall (x, y) \in T^2)(x \neq y \text{ and } (x, y) \in R \Rightarrow (y, x) \notin R);$$

- R is also transitive; let x, y and z be three arbitrary elements of T, and assume that $(x, y) \in R$ and $(y, z) \in R$ hold, i.e., *"state x is at most as good as state y"* and *"state y is at most as good as state z"*, obviously, *"state x is at most as good as state z"* must hold, i.e., $(x, z) \in R$; hence the relation R must be transitive, i.e.,

$$(\forall (x, y, z) \in T^3)((x, y) \in R \text{ and } (y, z) \in R \Rightarrow (x, z) \in R).$$

Taking into account the previous properties, the relation R on T must be a partial order relation and, hence, a state space must be a partially ordered set. Assumption 1 resumes the previous remarks.

Assumption 1 *The binary relation on the set of possible states of a component or system is a partial order relation that mimics the linguistic property* *"state ... is at most as good as state ..."A state space is determined by the set of possible states and the partial order relation on the set of possible states.*

Since the relation R is a partial order relation, we shall denote it by $\leq$. The linguistic property *"state ... is at least as good as state ..."* is denoted by $\geq$. By $(T, \leq)$, we denote the state space determined by the set T and $\leq$. The state spaces $(T_1, \leq_1)$ and $(T_2, \leq_2)$ are *equivalent* if and only if $(T_1, \leq_1)$ and $(T_2, \leq_2)$ are order isomorphic [5], and, hence, both structures are basically the same.

Example 9 Consider the set of possible states T, $T = \{t_1, t_2\}$, and let R be a binary relation on R,

$$R = \{(t_1, t_1), (t_1, t_2), (t_2, t_2)\}.$$

Since R is a partial order relation on T, (T, R) is a state space. One easily verifies that (R, T) and $(\{0, 1\}, \leq)$ are order isomorphic. Hence, (R, T) is an equivalent representation of the well-known binary state space.

Example 10 The partially ordered set $(\{0, \ldots, m\}, \leq)$, $m \geq 2$, is the state space for both components and systems in the El-Neweihi, Proschan and Sethuraman approach [21], and the Barlow and Wu approach [4].

Since the relation on the set of possible states is a partially ordered set, two states x and y, $(x, y) \in T^2$, may be incomparable, i.e.,

$$x \not\leq y \text{ and } y \not\leq x.$$

When x and y are incomparable, we denote this by $x \parallel y$. This situation is rather unusual, considering the classical multistate models. In this paper, however, we shall not dwell upon the interpretation of incomparable states. For some considerations about incomparability in reliability theory, we refer to [26]. We may conclude, however, that for two arbitrary states x and y

- $x \leq y$ mimics that y is a better state than x, i.e., x is more degraded than y;
- $x \parallel y$ mimics that x and y are two different types of failures, that is failures that stem from different causes.

The structure of a partially ordered set, however, is too poor to model the state space of components and systems. Montero, Tejada and Yáñez [25], and Montero, Cappelle and Kerre [26] already indicated that the minimal structure of a state space must be a complete lattice, i.e., for any subset its least upper bound (supremum) and its greatest lower bound (infimum) must exist. From a practical point of view, there are no reasons to impose this rather strong condition, but there are many theoretical reasons. Amongst others, we mention the possibility to deduce important approximation theorems and the ability to apply them to concrete systems—see, e.g., [13, 14] where several approximation theorems are developed and applied to an emergency coolant system in a Belgian nuclear power plant—the ability to define minimal path and cut sets from an order theoretic point of view (see, e.g., [11]) and the possibility to study the duality problem in reliability theory in a much broader, lattice-based context (see, e.g., [11, 26]).

In most cases, however, the minimal state space, i.e., any state of the state space can be assumed by the component or system and there are no other states the component or system may assume, for a component or a system must not be a complete lattice. Therefore, we must extend the minimal state space to a complete lattice, and, preferably, in such a way that all existing suprema and infima are not affected by the extension. This problem is well-known in lattice theory and the procedure to extend an arbitrary partially ordered set to a complete lattice, satisfying the imposed conditions is known as the Dedekind-McNeille completion [5]. The definition of the reals out of the set of rational numbers, e.g., is a very famous application of the Dedekind-McNeille completion of partially ordered sets. Moreover, one easily verifies that the Dedekind-McNeille completion is *minimal*, i.e., it is the smallest complete lattice that includes the initial partially ordered set, preserving all existing infima and extrema. The added states have no real meaning in a practical problem but are very useful for theoretical and approximation purposes, i.e., determining upper and lower bounds of both component and system performances. Therefore, we introduce the second assumption about the notion state space.

Assumption 2 *The state space of a component or a system is a complete lattice.*

Example 11 Consider the pump of example 2 and consider the context: the actual flow of the pump, must equal its maximal flow, i.e., 20 t/h. The unit interval provided with its natural order relation $\leq$ is a complete lattice, even a complete chain, i.e., a complete lattice for which each pair of elements is comparable. The pump is an entity and its context is: *the actual flow of the pump is 20 t/h.* We have indicated, previously, that the pump is subject to a gradual failure and that it functions perfectly when its flow is 20 t/h and fails when its flow is 0 t/h. The mapping f,

$$f : [0, 20] \to [0, 1] : x \mapsto \frac{x}{20},$$

is a state mapping. When the actual flow of the pump, e.g., equals 15 t/h, its state is 0.75. The state of the pump indicates that the pump does not function perfectly but that its performance is closer to perfect functioning than to complete failure. When the actual flow of the pump is 10 t/h, the state of the pump equals 0.50. Hence, in the latter case, the pump is closer to a failure than in the former case.

Example 12 Consider a system that is subject to three different types of failure and that the system performance does not degrade gradually to one of the possible failures. The set of states T, that are assumed by the system, equals $\{w, f_1, f_2, f_3\}$, where w means working and f_i, $1 \leq i \leq 3$, are three possible, but different failures. The partial order relation $\leq$,

$$\leq = \{(f_1, w), (f_2, w), (f_3, w)\},$$

does not define a complete lattice on the set of possible states T. Hence, we must extend it by means of the Dedekind-McNeille completion of partially ordered sets. One easily proves that one element, denoted by f, must be added to T. The extended set of states T_e,

$$T_e = \{w, f_1, f_2, f_3, f\},$$

is a complete lattice when the partial order relation $\leq$ is extended to $\leq_e$,

$$\leq_e = \{(f, w), (f, f_1), (f, f_2), (f, f_3), (f_1, w), (f_2, w), (f_3, w)\}.$$

This section completes our search to the foundations of the notion state in reliability theory. However, we did neither define the notion structure function in the developed lattice-based framework nor explain its meaning. In the previous sections we have explained that a structure function relates the component states to the system states. Formally, a structure function is an isotone mapping from the product state space to the system state space, that satisfies two boundary conditions. We shall not dwell upon this topic but we refer to, e.g., [11, 14, 25, 26], where the authors develop a mathematical theory of structure functions, starting from the definition of a structure function.

6 CONCLUSION

Most papers in reliability theory do not discuss the meaning of the notion *state* from a reliability point of view, and most authors assume that it is clear to the reader what a state is and, hence, it does not need further explanation. This assumption, however, is dangerous and often leads to disagreement among scientists (see, e.g., the critics of Amendola and Contini [1] and Caldarola [9] on the classical definitions of multistate structure functions). In this paper, we started our search to the foundations of the notion state and the related notions, and in a sense the foundations of reliability theory, from the notion entity. An entity may be physical (e.g., a steam grid) or nonphysical (e.g., a computer programme) and is subject to at least one failure. An entity must be modeled and can be subject to two basically different types of failure: gradual failures and failures caused by different deficiencies. A state mimics the quality of the entity—in most cases called a system—or a smaller part of an, entity—which then is called a component. Hence, a state is a quality label, an indication of the ability to function, that refers to the cause of malfunctioning, when a failure occurs. A state space, then, is a set of possible states that has been provided with a binary relation that is the mathematical transformation of the linguistic property *"state ... is at most as good as state ..."* The binary relation on the set of possible states is a partial order relation and the different types of failure are reflected by the comparable and incomparable elements:

- gradual failures are modeled by the comparable elements of the state space,
- different types of failure are represented by the incomparable elements of the state space.

The structure partially ordered set, however, is too poor. Therefore, we have assumed that a state space is a complete lattice. The role of a structure function, now, is obvious. A structure function links the component states and the system states in a special way and allows to study the reliability aspects of an entity. There is a one-to-one relationship between the entity and the notion system, that represents the reliability aspects of the entity, when there is a context, i.e., a set of requirements that the entity must satisfy and that allow to measure the quality or performance of the entity. The need to define components, and, hence, study the relationship between component states and the system state stem from the complexity of most systems. It is easier to define components and to model them first, and afterwards to determine the relation between the component states and the system states.

The link between multistate structure function theory and fuzzy set theory is rather obvious now. Fuzzy sets have been introduced to overcome the shortcomings of classical set theory when, e.g., linguistic variables must be modeled. Fuzzy sets often model the vagueness of these linguistic terms and the uncertainty of some statements when probability theory cannot be applied. Especially ordinal problems can be modeled rather easily by fuzzy sets. Applying classical sets to these cases is very artificial and eventually leads to an unacceptable loss of essential information. The same considerations lead to the introduction of multistate structure function theory, several years ago. Although, this theory has been developed separately from fuzzy set theory, many ideas resemble the basic ideas of fuzzy set theory. It is only recently that the relationship between fuzzy sets and multistate structure function theory has been explored by Montero, Tejada and Yáñez [25], and by Kerre and Cappelle [12, 13, 14]. The combination of these two theories is extremely rewarding, although not well enough propagated in the fuzzy and the reliability community.

Acknowledgements

The author likes to thank Prof. Dr. Etienne E. Kerre for his support and their fine cooperation since 1988. His enthusiasm and openness for new and promising theories made the author's work possible. Therefore, the author would like to dedicate this paper to Prof. Dr. Etienne E. Kerre for his fiftieth birthday.

REFERENCES

[1] Amendola, A. and Contini, S., "About the Definition of Coherency in Binary Systems Reliability Analysis," in Synthesis and Analysis for Safety and Reliability Studies, eds. Apostolakis, G., Garriba, S., and Volta, G. (Plenum Press, New York 1980) 79–84.

[2] Barlow, R.E. and Proschan, F., "Mathematical Theory of Reliability" (John Wiley and Sons Inc., New York, 1965).

[3] Barlow, R.E., and Proschan, F., "Statistical Theory of Reliability and Life Testing" (Holt, Rinehart and Winston, New York, 1975).

[4] Barlow, R.E., and Wu, R., "Coherent Systems with Multistate Components," Mathematics in Operations Research 3 (1978) 275–281.

[5] Birkhoff, G., "Lattice Theory," AMS Colloquium Publication Volume 24, Providence, Rhode Island, 1967.

[6] Birnbaum, Z.W., Esary, J.D., and Saunders, S.C., "Multicomponent Systems and Structures, and their Reliability," Technometrics 3 (1961) 55–77.

[7] Bollobaś, B., "Extremal Graph Theory" (Academic Press Inc. Ltd., London 1978).

[8] Cai, K.-Y., and Wen, C.-Y., "Street-lighting Lamps Replacement: a Fuzzy Viewpoint," Fuzzy Sets and Systems 37 (1990) 161–172.

[9] Caldarola, L., "Fault-tree Analysis with Multistate Components," in Synthesis and Analysis for Safety and Reliability Studies, eds. Apostolakis, G., Garriba, S., and Volta, G., (Plenum Press, New York 1980) 199–248.

[10] Cappelle, B., "Multistate Structure Functions and Possibility Theory: an Alternative Approach to Reliability Theory," in Introduction to the Basic Principles of Fuzzy Set Theory and Some of its Applications, ed. Kerre, E.E., (Communication and Cognition, Gent 1991) 252–293.

[11] Cappelle, B., "Structure Functions and Reliability Mappings, a Lattice Theoretic Approach to Reliability Theory" (in Dutch, Doctoral Dissertation, Universiteit Gent, Gent, 1994).

[12] Cappelle, B., and Kerre, E.E., "Issues in Possibilistic Reliability Theory," accepted for publication in Fuzzy Sets and Possibility Theory in Reliability and Safety Analysis, eds. Onisawa, T., and Kacprzyk, J., 20 pages.

[13] Cappelle, B., and Kerre, E.E., "The CARA System: An Intelligent Tool to Support Qualitative Reliability Analysis," in Proceedings of the 7th International AMSE Symposium on Fuzzy Systems, Neural Networking and Artificial Intelligence, Lyon (France), July 1994, 25–35.

[14] Cappelle, B., and Kerre E.E., "Computer Assisted Reliability Analysis: an Application of Possibilistic Reliability Theory to a Subsystem of a Nuclear Power Plant," in Proceedings of the FLINS '94 workshop, Mol (Belgium) September 1994, 260–265.

[15] Carvallo, M., "Principes et applications de l'analyse booléenne" (Gauthier-Villars, Paris, 1965).

[16] Christofides, N., "Graph Theory, an Algorithmic Approach" (Academic Press Inc., London) 1975.

[17] De Cooman, G., "Evaluation Sets: An Order Theoretic Approach to Fuzzy Sets," The Journal for the Integrated Studies of Artificial Intelligence, Cognitive Science and Applied Epistemology 8 (1991) 81–101.

[18] De Cooman, G., "Evaluation Sets and Mappings: an Order Theoretic Approach to Vagueness and Uncertainty" (in Dutch, Doctoral Dissertation, Universiteit Gent, Gent, 1993).

[19] Dubes, R.C., "Two Algorithms for Computing Reliability," IEEE Transactions on Reliability 12 (1963) 55–63.

[20] Dubois, D., and Prade, H., "Théorie des Possibilités" (Masson, Paris) 1985.

[21] El-Neweihi, E., Proschan, F., and Sethuraman, J., "Multistate Coherent Systems," Journal of Applied Probability 15 (1978) 675–688.

[22] Gnedenko, B., Beliaev, Y., and Soloviev, A., "Méthodes mathématiques en théorie de la fiabilité" (Mir, Moscow, 1972).

[23] Kauffman, A., Grouchko D., and Cruon, R., "Mathematical Models for the Study of the Reliability of Systems" (Academic Press, New York, 1977).

[24] Lesanovský, A., "Asymptotic Behaviour of Reliabilities of the Optimal Systems with two Dual Modes of Failure," Proceedings of the Eighth Symposium on Reliability in Electronics—Relectronic '91 Volume 1, Budapest (1991) 371–380.

[25] Montero, J., Tejada, J., and Yáñez, J., "General Structure Functions," in Proceedings Workshop on Knowledge-Based Systems and Models of Logical Reasoning December 1988 Cairo, Egypt.

[26] Montero, J., Cappelle, B., and Kerre, E.E., "The Usefulness of Complete Lattices in Reliability Theory," accepted for publication in Fuzzy Sets and Possibility Theory in Reliability and Safety Analysis, eds. T. Onisawa and J. Kacprzyk, 16 pages.

[27] Onisawa, T., "An Approach to Human Reliability in Man-Machine Systems using Error Possibility," Fuzzy Sets and Systems 27 (1988) 87–103.

[28] Singer, D., "A Fuzzy Set Approach to Fault Tree and Reliability Analysis," Fuzzy Sets and Systems 34 (1990) 145–155.

[29] Tanaka, H., Fan, L.T., Lai, F.S., and Toguchi, K., "Fault-Tree Analysis by Fuzzy Probability," IEEE Transactions on Reliability 32 (1983) 453–457.

[30] Zadeh, L.A., "Fuzzy Sets," Information and Control 8 (1965) 338–353.

Part 3

FUZZY LOGIC, FUZZY CONTROL, AND NEURO-FUZZY CONTROL

9

FUZZY LOGIC AND NEURAL SYSTEMS

Madan M. Gupta

Intelligent Systems Research Laboratory
Centre of Excellence on Neuro-Vision Research
College of Engineering
University of Saskatchewan
Saskatoon, Sask., S7N OWO, Canada

ABSTRACT

The path that leads to scientific discovery very often begins when one of us takes an adventurous step into the world of endless possibilities. Scientists intrigued by a mere glimpse of a subtle variation may uncover a clue or link, and from that fragment emerges an idea to be developed and worked into shape.

1 INTRODUCTION

Man has always dreamed of creating a portrait of himself, a machine with humanlike attributes, such as locomotion, speech, vision, and cognition (memory, learning, thinking, adaptation, intelligence). Through his creative actions, he has been able to realize some of his dreams. In today's technological society, there are machines that have some of the human attributes which do emulate several human functions with tremendous capacity and capabilities. Some examples of these are transportation systems to human locomotion, communications systems to human speech and vision, and computing systems to human low-level cognition. No doubt the machines which are an extension of human muscular power (cars, tractors, airplane, trains, robots, etc.), have brought luxury to human life. But who provides a control to all these mighty machines—the human intelligence.

When a computer appeared first, almost five decades ago, we admired it is an *artificial brain*, and we thought that we were successful in creating a low-level decision-making cognitive machine. We called it "artificial intelligence" and waited for many potential applications to evolve. However, after several years we realized that the so-called artificial intelligence (AI) was indeed very artificial in nature. In AI, about half of what we hear is *not true* and the other half is *not possible.*

Now, we are moving into a new era of *information technology*: from the technology of processing statistical information only to the new mathematics and technology for processing both statistical and cognitive information. The cognitive information embodies the human thought process—the process which originates from human cognitive faculty—the brain. New computing theories with a sound biological understanding have been evolving. This new category of computing falls under a general field of *neural computing systems.* The new computing technology has been evolving with new hardwares such as optical computing, opto-electronics, and molecular computing. This new technology seems to have the potential of surpassing the micro-, nano-, and pico-technologies. Also, the neural computing systems have been proven (theoretically) to supplement the enormous processing power of the von Neuman digital computer[1]. Hopefully, these new computing methods, which embody the neural architecture as a basis, will be able to provide a thinking machine—a low-level cognitive machine, of course, for which scientists have been striving for so long.

Today, we are in the process of designing neural-computing-based information processing systems using the biological neural systems as a basis. The highly *parallel* processing and layered morphology with learning attributes of the human cognitive faculty—the brain—provides us with a new tool for designing a cognitive machine that can learn and recognize complicated patterns—like human faces and Japanese characters. The theory of fuzzy logic, on the other hand, provides a mathematical power for the emulation of the higher-order cognitive functions—the thought and perception processes. A marriage between these two disciplines—the neural computing and fuzzy logic—may provide a new class of computing systems—the neural-fuzzy systems—for the emulation of higher-order cognitive functions. The chaotic behavior inherent in biological systems, heart and brain, for example, and the natural phenomena, and the genetic algorithms are some of the other important subjects which promise to provide robustness to the newly developing field of neural computing methods.

[1]In terms of memory and processing power, the carbon-based cognitive computer possessed by a 'fly' is much superior to the combined power of many silicon-based super computers.

"Biology has many interesting mathematical functions to offer to scientists and engineers in their advanced research. We feel strongly that a basic course in the mathematical physiology will provide a sound basis for the development of robust sensors, control mechanisms, and fascinating paradigms for new computing systems—the neural computing systems."

Uncertainty is an inherent phenomenon in our universe and in our cognitive lives, and it stands continuously open to our gaze. To some, it may become a cause of anxiety, but to scientists, it becomes a chapter full of challenges. Scientists are attempting to comprehend the language of this uncertainty through mathematical tools, but these mathematical tools are still incomplete.

2 COGNITION, UNCERTAINTY, AND INTELLIGENCE

It was a cold New Year's eve in Saskatoon—only −30°C. After cleaning the clutters of 1987 from my office and rearranging certain tasks for 1988, I was driving home. Light snow was falling and the roads were icy, but not too busy. Suddenly, my hands steered the car towards the side lane and my foot worked the brake rendering the car motionless. I soon realized that a speeding car, chased by the police, was coming towards me in my lane and would have collided with my car had I not steered it to the side lane. I took a long breath of relief and thanked the Almighty for giving us such a wonderful *sensory* system and control mechanisms and a *cognitive* computational organ, the *brain*, which can perform many urgent tasks, such as this collision avoidance, at an ultra high speed in such an unexpected and uncertain environment.

The *natural sensors* (in this case my eyes and ears) were acquiring the information continuously and relaying it through the sensory neurons to the cognitive processor (my central nervous system) which perceived the driving environment and the danger associated with it almost instantly. The *cognitive processor* then synthesized the control signals and transmitted them through the motor neurons to my hands and foot, which activated the steering and braking mechanisms instantly to avoid the collision. I was left alone on the road and the experience created a number of emotional feelings in my mind: thoughts of drunk drivers on the road, on modern technology, on a government which encourages the selling of liquor in order to collect the revenue and then spends more revenue to curb drunk drivers, and so on.

This incident upset me so greatly that a magnitude of thoughts appeared momentarily keeping the *cognitive processor* busy in various types of *reasoning* and *perception.* After having supper and sitting down in my study, I recalled a thought: the Almighty has given many wonderful attributes to humans. The attributes of the $3h$, which I have described in one of the earlier papers and needs special mention here.

The attributes of $3h$ are associated with the *hand*, the *head* and the *heart.* We have successfully mimicked the function of the *hand* (muscular power), and have created many mighty machines. Now we are working on the emulation of some of the important attributes of the *head* (brain), such as learning, thinking, reasoning, or in general cognition and perception, on a silicon chip for the creation of an intelligent machine. The *heart* is associated with the attribute "emotions", and this is the attribute which is mainly responsible for the "survival" of humanity. Perhaps this is the subject which must be considered very seriously in future research, because a *mighty thinking* machine without *emotions* (*heart*) may become the cause of destruction of humanity.

Intelligence implies the ability to think, reason, learn, and memorize, or, in general, it refers to the process of human mentation and cognition. One of the last frontiers of science, perhaps its ultimate challenge, is understanding the biological basis of mentation and cognition: how we think, reason, learn, remember, perceive, and act. I still cannot understand how the brain perceived the dangerous driving situation and acted instantaneously while it might take us several seconds to multiply two three-digit numbers ($365 \times 461 = ?$). How do the genes contributes to the process of mentation and cognition and how do they develop with the environment?

Here we have two computational tools: the carbon-based organic brain which has existed in humans and animals for several billions of years, and the silicon-based modern computers which have evolved only over the last three decades. Recent technological advances in micro-, nano-, and pico-technologies have made it possible to carry in a briefcase a very powerful computer (a logarithmic machine) which is ultrafast and efficient for numerical computations. The *cognitive information* acquired by our natural sensors and its perception are *non-numerical.* But the mentation process (a non-algorithmic process) can process such cognitive information almost instantaneously. It also can store and retrieve such non-numerical cognitive information and take the appropriate decisions. The modern-day computers are incapable of processing such cognitive information.

The fact that the human mentation and cognition process is so marvelously efficient and effective poses a question to scientists and engineers. *Can some of the functions and attributes of the human sensory system, mentation and the cognitive processor, and motor neurons be emulated in a robotic system?*

For such an emulation process, it is necessary to understand the biological basis and physiological functions of the brain. It is a difficult question to answer. However, it is thought that, if we reexamine some of the "mathematical aspects" of our thinking process and "hardware aspects" of the "neurons", the principle element of the brain, we may succeed to some extent in our emulation process.

3 COGNITION AND RELATIVE GRADES

The mentation and cognitive activities of the brain, unlike the computational functions of the binary computer, are based upon the *relative grades* of the information acquired by the natural sensory system. The conventional mathematical tools, whether deterministic or probabilistic, are based upon some absolute measure of the information. Our natural sensors acquire information in the form of *relative grades* rather than in absolute numbers. While driving on an icy road, for example, we perceive the driving environment in a *relatively graded* sense and act accordingly.

The mentation and cognitive process thus acts upon the graded information. Information may appear in a numerical form (temperature of the body is 38.4°C), however, during the process of cognition we perceive this temperature as *near normal*, in the form of *relative grades*. Thus, the cognitive process acts upon the different forms of information and this leads to "formless" uncertainty. During diagnosis, the cognitive process of a doctor perceives the patient's temperature as *near normal*. This "formless uncertainty" provides us with a new scientific challenge. Recently, scientists have started thinking of giving a morphology to this amorphous uncertainty. In the past, mathematics have disdained this challenge and have chosen to avoid natural mentation by devising theories unrelated to human perception and cognition.

4 INTRODUCTION TO FUZZY LOGIC

It was in 1965 that Lotfi A. Zadeh published his first celebrated paper on *fuzzy sets.* It is now almost two and a half decades since he first introduced to me this new type of information and uncertainty, and showed me the path which leads to somewhat beautiful gardens full of immortal and ever increasing fragrance. Though I was taught the notion of cognition and perception in schools, I was so ignorant about the mathematical morphology of this uncertainty and its pervasiveness around us.

No one had seen the beauty of *fuzzy sets* before Lotfi Zadeh, and it was he who first showed the promise of consolidating this beauty into an organized field of *fuzzy logic,* full of rich theories and promising applications.

The theory of fuzzy logic is based upon the notion of *relative graded membership* and so is the function of the mentation and cognitive process. In the past, studies of cognitive uncertainty and its cognate, the cognitive information, were hindered by the lack of suitable mathematical tools for modeling such information. However, with the introduction of the theory of fuzzy logic, it is possible now to expand the studies in the important fields of cognitive information, neural networks, and cognitive-neural computing systems.

My own laboratory is heavily committed to studies in the field of cognitive information processing, cognitive vision fields, vision perception, cognitive-neural computing systems, and cognitive feedback controllers with promising applications to intelligent robotic systems in manufacturing, space, and medical image processing.

5 PERSPECTIVES

Recent progress in information-based technology has significantly broadened the capabilities and applications of computers. Today's computers are merely being used for the storage and processing of numerical data (information). Should we not reexamine the functions of these computing tools in view of the increasing interests in subjects such as knowledge-based systems, expert systems, and intelligent robotic systems, and for solving problems related to decision and control? Human mentation acts upon cognitive information and the cognitive information is characterized by using relative grades: "Although it is snowing, driving conditions are not so bad." Human mentation and cognition function through the *confluence* of the fresh information (acquired from the environment by our natural sensors) and the information (experience, knowledge-base) stored in the memory.

Shannon's definition of "information" was based upon certain physical measurements of random activities in systems, in particular, in communication channels. This definition of information was restricted only to a class of information arising from physical systems.

If we wish to mimic some of the cognitive functions (learning, remembering, reasoning, intelligence and perception, etc.) of humans in a machine, we have to generalize the definition of information and develop new mathematical tools and hardware. These new mathematical tools and hardware must deal with the simulation and processing of cognitive information. Many new notions, although still at a primitive stage, are springing up around the mathematics of neural-fuzzy logic and genetic paradigms and, hopefully, we will be able to nurture some interesting studies in the not too distant future.

It will be interesting to study the field of *cognition* and *uncertainty*, and research further on the notion of cognitive uncertainty—the uncertainty which arises from human thought processes and provide a base for the challenging field of *cognitive information.*

The human cognition and mentation process is marvelously so efficient and effective that I can almost instantly recognize faces and speech and can take appropriate actions in the case of a dangerous situation. Now, it poses a question for scientists and engineers. *Can we emulate some of the mathematical functions and attributes of the human sensory systems, mentation and cognitive process, and motor neurons, and create a virtual cognitive machine?*

6 VIRTUAL INTELLIGENCE

Over the last decade or so, several parallel advances have been made in the two distinct disciplines: *fuzzy logic* and *neural networks.* As the names imply, fuzzy logic provides a mathematical strength to the process of emulation of certain perceptual and linguistics attributes associated with human cognition, whereas the science of neural networks provides new computing morphologies with learning and adaptive capabilities.

A marriage between these two distinct disciplines has the potential of producing some off-springs with the capability of generating a new discipline—virtual intelligence (VI) with a new generation of computing systems—virtual cognitive systems (VCS). Such virtual cognitive systems will, hopefully, capture certain aspects of human cognition. The integration of these two fields, neural networks and fuzzy logic, has a potential of producing robust sensors and robust control mechanisms. Scientists and engineers are exploring the possibility of creating new paradigms under this new discipline which seem to have the potential for creating a new class of intelligence—virtual intelligence.

7 HUMAN COGNITION

Modern neural science has evolved from the research studies of many individuals working in many disciplines over a relatively large time span. These studies have included some philosophical dialogue on the *cognitive* (thought processes) and *affective* (emotional processes) aspects of the brain to the basic understanding of the neuron, the elementary building block of our central nervous system (CNS). Indeed, the neuronal morphology of the human central nervous system is too complex to analyze, though we always present certain speculative analogies, and based upon this superficial understanding, we produce some computational networks—known in the literature as *artificial neural networks* (ANN).

Given the basic knowledge of digital components such as OR, AND, NOT, and NAND gates, the human mind is capable of synthesizing a super digital computing system. However, if we are presented with a supercomputer with a superficial knowledge of its computing functions, we can only speculate on the morphology and the organizational aspects of its behavior. If this problem is given to specialists in different fields, they will come up with different speculative responses based upon their observations and their fields of specialization. A synthesis of these speculative responses, hopefully, may provide some clues as to the working of digital machines. It reminds me of the story of six blind people who were asked to examine the morphology of an elephant. Based upon their own individual experiences and observations, each blind person gave a correct narration of his findings. However, a synthesis of these findings did not make much sense to a person examining the elephant with full vision.

Nature has given us a beautiful and robust cognitive and affective faculty—the brain—and we do not have any choice but to examine its working from various angles embodied in various disciplines, and provide a synthesis. As system scientists, our ultimate objective is that, based upon this superficial understanding of the brain, we create a virtual cognitive system (VCS) which can aid humans in their various decision-making tasks[2].

The discipline of neural science has evolved over the centuries through the understanding of the basic neuron and the neuronal morphology of the central nervous system. Essentially, the central nervous systems acquires information internally from the external environment through the natural sensors, and stores previous experiences. It also processes the new information in conjunction with the past experiences, and provides new decision signals. Although various aspects of biological plausibilities of neuronal morphologies are still not well understood, on a speculative basis these biological plausibilities have led to various morphologies of artificial neural networks with learning and adaptive capabilities. A host of these neural networks have been proven to provide efficient computational tools for many information processing and decision-making tasks.

8 THE NEURON

Nature has developed a very complex neuronal morphology in most biological species. The biological neurons, over one hundred billion, in the central nervous systems (CNS) of humans play a very important role in various complex sensory, control, affective, and cognitive aspects of information processing and decision making. In neuronal information processing, there are a variety of complex mathematical operations and mapping functions involved which synergically act in a parallel-cascade structure forming a complex pattern of neuronal layers evolving into a sort of pyramidal pattern. The information flows from one neuronal layer to another in

[2]Man has always dreamed of creating machines with humanlike attributes. In this technological world, there are machines which have provided an aid to several human functions with tremendous capacity and capabilities. Some examples of these are transportation systems to human locomotion, communication systems to human speech and vision, and computers to human cognition (very low level of cognition).

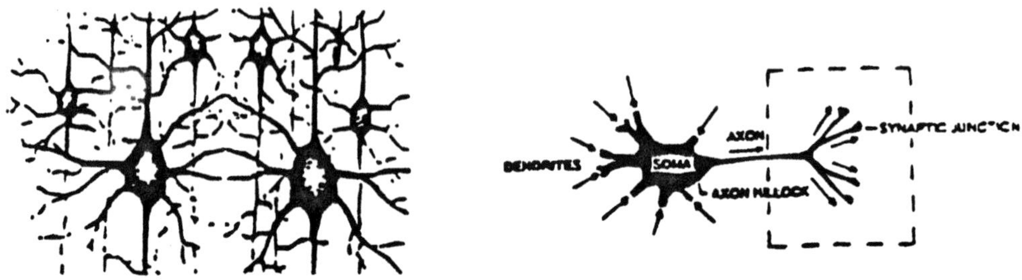

(a) Biological Neuronal Field (b) A Biological Neuron.

Figure 1 Morphology of Biological Neuron.

the forward direction with a continuous feedback evolving into a dynamic pyramidal structure. The pyramidal structure is in the sense of the extraction and convergence of information at each point in the forward direction. The understanding of biological neuronal morphology (structure) provides not only a clue, but also a challenge, in the design of a realistic cognitive computing machine—a virtual cognitive system (VCM).

From the neurobiological, as well as the neuromathematical point of view, we do identify two key neuronal elements in a biological neuron which are responsible for providing neuronal attributes such as learning, adoption, knowledge (storage or memory to the past experience), aggregation, and nonlinear mapping operations on neuronal information:

(i) *Synapse*: The synapse is a storage element of the past experience (knowledge). It learns from the neuronal environment and continuously adapts its strength. The past experience appears in the form of synaptic strength, what we call synaptic weight. There are over one hundred billion neurons in our central nervous systems and, on the average, there are over 1000 synapses per neuron. Mathematically, a synapse provides a confluence operation between the new neuronal inputs (information) and the past experience, and sends signals to the dendrite—the input node to the main body (soma) of the neuron. We will refer to the mathematical operation in a synapse as the "synaptic operation" or "synaptic confluence operation."

(ii) *Soma*: The soma (neural cell) refers to the main body of the neuron. It receives signals from a very few to over 1000 synapses through its dendrites and provides an aggregation operation. If the aggregated

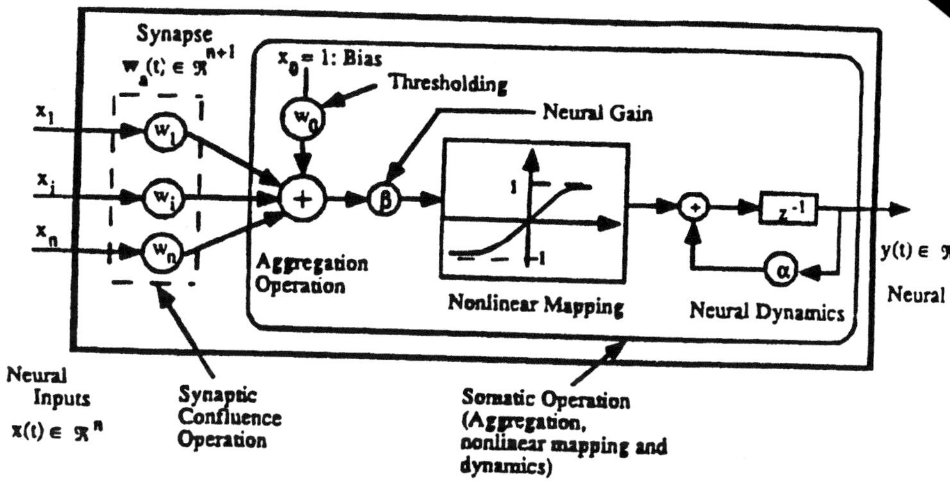

Figure 2 Mathematical Morphology of a Neuron.

value of the dendritic inputs exceeds a certain threshold, it fires, providing an axonal (output) signal which is a sort of nonlinear function of the aggregated value. We refer to this mathematical operation in the soma as "somatic operation." In the somatic operation, one can identify three distinct mathematical operations: (i) *aggregation*, (ii) *thresholding*, and (iii) *nonlinear transformation (mapping).* Figure 3.1 shows the morphology of a biological neuron, and Figure 3.2 depicts the mathematical morphology of the neuron.

9 NONLINEAR SYSTEMS

The biological neuronal processes generate some interesting mathematical mapping properties because of their nonlinear operations combined with a thresholding in the soma. As a matter of fact, if the neuron would carry out only linear operations, the mathematical attractiveness and the robustness of the neural circuits in most of the applications would disappear.

Systems theory is a mathematical discipline which deals with the understanding of dynamic behavior and the robust control of processes found in technological, socio-economic, and biological processes. There are pro-

ɩproximated to linear time-invariant systems over a erations, while such an approximation may result in applied to real-world applications.

Most of the time, … our undergraduate and graduate systems courses, we start with linear time-invariant systems and expose our students with beautiful results found in the time and frequency domains. These results usually embody some beautiful concepts such as stability, optimality, and the notion of poles and zeros. I find it frustrating (I am sure others must have experienced similar frustrations), however, that once we linearize the minds of our young students through our linear lectures and available linear textbooks, it is difficult to feed them the notions of nonlinear systems, the systems dealing with most of the real-world complex situations.

Perhaps, a better approach would be to expose the students, at the very outset, to the general concept of systems theory, with a host of examples from the nonlinear world supported by the appropriate mathematics (depending upon the level of the students), and then to present a linearized situation, keeping the students aware of the restrictions.

The pedagogical issue of nonlinear system theory is very important in today's technological world, and it needs a close reexamination in our educational institutions—through the reorganization of our lectures and rewriting of most of the textbooks.

Simplifications in our educational systems certainly destroy the robustness in the ability of our graduates who usually face the complex technological world full of nonlinearities.

10 NEURAL-FUZZY LOGIC

The theory of fuzzy logic has been developed in order to capture the uncertainties associated with human cognitive processes such as in thinking, reasoning, and perception. "Driving conditions in Saskatoon in January, when it is *extremely* cold, are *not so bad* compared to what they are in *late* February or *early* March, especially when there is a *large* temperature variation between day time (snow is melting) and night time (freezing cold)." In this statement, the words in italics express various vague notions, which can be modeled using the notion of graded membership inherent in the

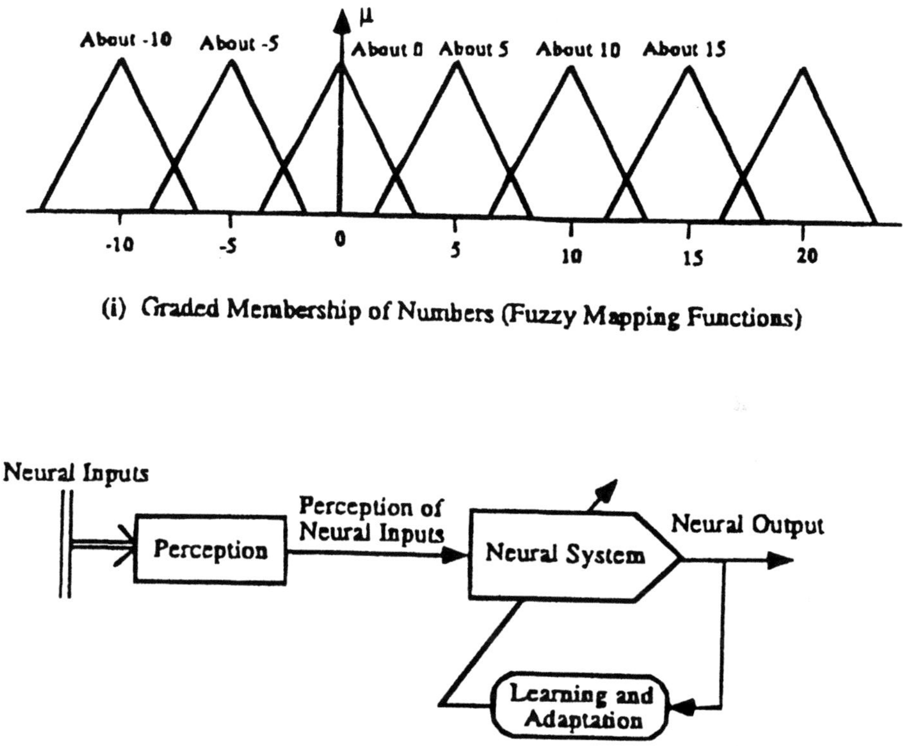

Figure 3 Fuzzy Neural System.

fuzzy set theory. The theory of fuzzy logic, unlike that of the binary logic, is developed using the notion of *graded membership* distributed over the closed unit interval $[0, 1]$.

The basic mathematical notion of the graded membership was introduced in 1965; however, the recent marriage between the mathematical fields of fuzzy logic and neural networks is responsible for the development of a new field: the field of neuronal-fuzzy logic. This field embodies the attributes of graded membership of the fuzzy logic and the learning and adaptation strength found in the neuronal morphology. This newly developing field promises to provide some emulation aspects of the cognitive faculty.

11 NEURONAL LEARNING

Biological species have adopted strategies in their processes which are based upon learning, adaptation, and self-organization in an uncertain environment. "Learn while functioning" is the most important attribute which makes the biological species so robust and flexible. It is natural that we adopt this strategy in our own design of the virtual cognitive systems.

Certainly, the process of neuronal learning and adaptation in biological species is enormously complex, and the progress made in the understanding of the field through experimental observations in such fields as neurophysiology and psychology during the last century is limited and crude compared to the achievements made in the physical sciences during the same period. It is, nevertheless, true that the neurophysiological and psychological understanding of the biological process has provided a tremendous impetus to the emulation of certain neurological morphologies and their learning behavior through the fields of mathematics and systems sciences. We have a long way to go before we can speak of understanding the principles of cognition (learning, thinking, reasoning and perception) and thus of the virtual cognitive computing to the degree that we understand the principles of the electrochemical, biochemical and ionic behaviors of neuronal populations in biological species[3].

We have realized the difficulties in the understanding of neurophysiology. Nevertheless, biology has inspired the work of system scientists during the past and at an accelerated pace during the more recent years. There has been an exponentially increasing number of attempts to develop neuronal paradigms for applications to problems such as machine vision and control systems [11, 12]. Indeed, the neuronal learning paradigms developed during the recent years combined with the cognitive strength of the notion of graded membership promises to provide robust solutions to the problems in pattern recognition, decision making, control of complex dynamic flexible structures in space, and general robotic systems for applications to manufacturing and medical sciences.

[3]A understanding of the functional morphology of the brain needs a real brain.

12 VIRTUAL COGNITION

One of the tenets of recent research in robotics and system science is that intelligence can be cast into a machine. The thought of creating virtual intelligence on a silicon chip (machine), unlike the thought of creating a mighty (powerful) machine, brings some strange feelings into our minds.

Intelligence implies the ability to think, reason, learn, and memorize, or, in general, it refers to the human mentation and cognition processes. One of the last frontiers of science, perhaps its ultimate challenge, is understanding the biological basis of mentation and cognition. Can this process be described by known physical laws, or does it lie within the domain of metaphysical principles? What is the process of thinking, reasoning, learning and remembering, and perceiving? We still cannot understand how the brain can perceive a dangerous situation and act instantaneously while it usually takes several seconds to multiply two three-digit numbers. How do genes contribute to the process of mentation and cognition, and how do they adapt to the environment?

The recent technological advances in computer hardware (opto-electronics, programmable lasers, molecular computing, etc.) have made it possible to carry a very powerful and ultrafast computer in a briefcase. Such computers are very efficient for numerical computations. However, the *cognitive information*, the information that our natural sensors acquire and process, is not numerical. However, the cognitive process is very efficient and fast in the processing of such non-numerical cognitive information. The logarithmic modern day computers fail to process such cognitive information. The question arises whether the mathematical functions and attributes of the human sensory systems, mentation and cognitive processes can be emulated for the creation of a virtual cognitive machine. Do we have the mathematics to emulate such cognitive functions, and do we have the hardware to implement such processes?

For such an emulation process, it is necessary to understand the biological, physiological, and metaphysical functions of the brain and develop a new class of mathematics. The mathematics that we know today was developed for the understanding of physical processes, but the process of cognition does not necessarily follow such mathematical laws. Then, what are the *cognitive mathematics*? It is a difficult and challenging question to answer. However, scientists have realized that if we reexamine some of the "mathematical aspects" of our thinking process and "hardware aspects" of the "neurons," the principle element of the brain, we may succeed to some extent in the emulation process.

The mentation and cognitive functions of the brain, unlike the computation functions of the binary computer, are based upon the *relative grades* of the information acquired by the natural sensory systems. The conventional mathematical tools, whether deterministic or probabilistic, are based upon some absolute measure of information. Our natural sensors acquire information in the form of *relative grades* rather than in absolute numbers. The "perceptions" and "actions" of the cognitive process also appear in the form of relative grades. While driving on an icy road, for example, we perceive the driving environment in a relatively graded sense and take action accordingly. The attributes of elasticity and robustness are inherent in our cognitive functions.

The process of mentation and cognition thus acts upon the graded information. Information may appear in a numerical form (temperature of the body is 39.012°C); however, during the process of cognition, a physician perceives this temperature as near normal, in the form of a relative grade. Thus, the cognitive process acts upon the different forms of information and this leads to "formless" uncertainty: *temperature is near normal.*

The theory of fuzzy logic (soft logic) founded by Lotfi A. Zadeh in 1965 is based upon the notion of graded membership and so are the functions of the cognition and mentation processes. In the past, studies of *cognitive uncertainty* and its cognate, the *cognitive information*, were hindered by the lack of suitable tools for modeling such information. However, with the introduction of theory of fuzzy logic, it seems to be possible to expand these studies in the important field of cognitive information. A marriage between the cognitive information aspects and the neuronal morphology in the design of information systems may provide better computational tools in the years to come.

The emerging paradigms based upon cognitive-neural computing have potential uses in the development of new research studies such as cognitive information processing, cognitive feedback controllers and neuro-vision systems with promising applications to intelligent robotic systems and health sciences.

13 SOME REMARKS

Recent progress in information-based technology has significantly broadened the capabilities and application of computers. Today's computers, however, are merely being used for the storage and processing of numerical data (hard uncertainty and hard information). Should we not reexamine the functions of these computing tools in view of the increasing interest in subjects such as knowledge-based systems, expert systems, and intelligent robotic systems, and for solving problems related to decision and control? Human mentation acts upon cognitive information and cognitive information is characterized by using relative grades. Human mentation and cognition functions use fresh information (acquired from the environment by our natural sensors) and the information (experience, knowledge-base) stored in the biological memory.

Shannon's definition of "information" was based upon certain physical measurements of random activities in physical systems, in particular, in communication channels. This definition of information was restricted only to a class of information arising from physical systems.

If we wish to emulate some of the cognitive functions (learning, remembering, reasoning, intelligence, perceiving, etc.) of human in a machine, we have to generalize the definition of information and develop new mathematical tools and hardwares. These new mathematical tools and hardwares must deal with the simulation and processing of cognitive information and soft logic. Many new notions, although still in primitive stages, are emerging around the mathematics of fuzzy neural logic and, hopefully, we will be able to nurture some interesting studies in the not too distant future. Indeed, biological processes have much to offer to engineers, system scientists and mathematicians to solve many practical problems in the world we live in today.

After I finish some initial studies on an intelligent machine, the next natural stage for me would be to embark upon the design of a robotic machine which could play a game of ping-pong with me with the same degree of emotion, enthusiasm, and pleasure that I receive when I am playing with my students, friends, and family members. Such studies would need, of course, a basic understanding of the theory of meta-physics.

Acknowledgements

This paper is dedicated to Professor Etienne E. Kerre, who introduced many novel concepts in the field of fuzzy logic and neural systems over the last twenty-five years.

REFERENCES

[1] Amari, S.I. and Arbib, M.A., Eds., (1982), "Competition and Cooperation in Neural Nets," *Lecture Notes in Biomathematics*, vol. 45, Springer-Verlag, New York.

[2] Amit, D.J., Gutfreund, H. and Sompolinsky, Y., (1985), "Spin-Glass Model of Neural Networks," Physical Review A, vol. 32, 1007–1018.

[3] Anderson, J.A., (1983), "Cognitive and Psychological Computation with Neural Models," *IEEE Trans. Systems, Man, and Cybernetics*, SMC-13, 799–815.

[4] Black, M., (1937), "Vagueness: An Exercise in Logical Analysis," *Philosophy of Science*, vol. 4, 427–455.

[5] Brillouin, L., (1956), "Science and Information Theory," Academic Press, New York.

[6] Cajal, S. Ramon y, (1934), "Les preuves objectives de l'unite anatomique de cellules nerveuses," *Trob. Lab. Inest. Biol. Univ. Madrid*, vol. 29, 1–37, (Translation: Purkiss, M.V. and Fox, C.A., Madrid: Instituto "Ramon y Cajal," 54).

[7] Conant, R.C., (1976), "Law of Information which Govern Systems," *IEEE Trans. Systems, Man and Cybernetics*, vol. 6, 334–338.

[8] Gupta, M.M., (1975), *Fuzzy Automata and Decision Processes: The First Decade*, Sixth Triennial World IFAC Congress, Boston, Cambridge.

[9] Gupta, M.M., (1988), "On the Cognitive Computing Perspectives," in Fuzzy Computing: Theory, Hardware and Applications, North Holland.

[10] Gupta, M.M., (1988), "Cognition, Perception and Uncertainty," in Fuzzy Computing and Theory, Hardware and Application, North Holland, 7–10.

[11] Gupta, M.M., (1989), "Biological Basis for Computer Vision: Some Perspective," SPIE Conf. on Intelligent Robots and Computer Vision, November 5-10, Philadelphia, Paper #1192-49, 811–823.

[12] Gupta, M.M., (1991), "Fusion of Fuzzy Logic and Neural Networks with Applications to Decision and Control Problems," Automatic Control Conference, Boston, June, 30–31.

[13] Gupta, M.M., (1991), "Uncertainty and Information: The Emerging Paradigms," *International Journal on Neuro and Mass-Parallel Computing and Information Systems*, vol. 2, 65–70.

[14] Gupta, M.M., and Knopf, G.K. (Editors) [1994], '*Neuro-Vision Systems: Principles and Applications*, A Volume of Selected Reprints, IEEE Neural Networks Council, IEEE-Press, New York, 555 pages.

[15] Gupta, M.M., and Rao, D.H., (Editors) [1994], *Neuro-Control Systems: Theory and Applications*, A Volume of Selected Reprints, IEEE Neural Networks Council, IEEE-Press, New York, 607 pages.

[16] Gupta, M.M., and Sinha, N.K., (editors) [1995], *Intelligent Control Systems: Theory and Applications*, A Volume of Invited Chapters, IEEE Neural Networks Council, IEEE-Press, New York, 700 pages [In Press].

[17] Gupta, M.M., [1994], "Fuzzy Sets and Systems," McGraw-Hill Encyclopedia of Science and Technology, [15 pages, In Press].

[18] Hebb, D.O., (1949), *The Organization of Behavior*, John Wiley and Sons, New York.

[19] Kaufmann, A., (1975), *Introduction to the Theory of Fuzzy Subsets*, vol. 1, Academic Press, New York.

[20] Kaufmann, A. and Gupta, M.M., (1985), *Introduction to Fuzzy Arithmetic: Theory and Applications*, Van Nostrand Reinhold, New York, Second Edition, 1991.

[21] Kaufmann, A. and Gupta, M.M., (1988), *Fuzzy Mathematical Models in Engineering and Management Science*, North Holland, Amsterdam, New York, (Also, Japanese Translation, Ohmsha Publication, Tokyo, 1991).

[22] Klir, G.J., (1977), "Where Do We Stand on Measure of Uncertainty, Ambiguity, Fuzziness and the Like," *Fuzzy Sets and Systems*, Special Issue on Measure of Uncertainty, vol. 24, No. 2, 141–160.

[23] McCulloch, W.S. and Pitts, W., (1943), "A Logical Calculus of the Ideas Immanent in Nervous Activity," *Bulletin of Mathematical Biophysics*, vol. 5, 115-133.

[24] McCulloch, W.S., (1965), "Embodiments of Minds," MIT Press, Cambridge, MA, p. 20 of Introduction by S. Papert.

[25] Wiener, N., (1948), *Cybernetics*, John Wiley and Sons, New York.

[26] Zadeh, L.A., (1965), "Fuzzy Sets," *Information and Control*, vol. 8, 338-353.

10

SOME MATHEMATICAL PROBLEMS OF FUZZY CONTROLLERS

László T. Kóczy

Department of Telecommunication and Telematics
Technical University of Budapest
Sztoczek u. 2
H-1111 Budapest, Hungary

ABSTRACT

I was very glad when Da Ruan asked me to contribute to the volume dedicated to my old fuzzy friend Etienne Kerre. We have known each other for rather a long time, we met so many times at various fuzzy conferences, and now we have even a cooperation project on teaching fuzzy logic and related areas.

I am not a mathematician myself, nevertheless I tried to write a brief paper on some mathematical problems of the rather engineering technique of rule based fuzzy control. I hope, Etienne will like it.

And happy birthday to you, Etienne!

1 INTRODUCTION

Fuzzy control as it was introduced by Zadeh [1] and then by Mamdani [2] has proved to be a useful tool for systems where the exact model is not known, or where it is too complex to be tractable in real time. The technique as it was proposed had very nice philosophical and general logical foundations, but essentially it was and has been until now a smart heuristic technology. Many appliers see its main advantage in its transparency, i.e. in the overviewability of what happens when a certain fuzzy rule base representing the (usually rather vague) knowledge and an observation are combined in order to get a conclusion that might be used as a control action. As such models are usually not very good at the beginning – that is a natural fact, considering that the real model is unknown –, it is often necessary to tune it, as long as it satisfies some criteria concerning the accuracy of fitting the behavior of the real system, that might be a very loose fitting under certain circumstances. Engineers using this technique are often happy for not being forced to study the sometimes very complex mathematics of even approximate models for the systems they deal with, and for having a tool that is easy to handle. The rule based approach combined with fuzzy logical terms representing linguistic values for the variables in the model offers especially good features in the area of modifying the model, while formal techniques with explicit or implicit formulae are very hard to interpret. Even the simplest polynomial transference functions have not the least transparency in the sense that nobody has any idea what will happen with the hyper-surface representing the function when a certain parameter in the polynomial is changed. In addition, the behavior of e.g. polynomial approximators is often surprising in the areas where there is no control point, and no specification for its behavior.

However the linguistic fuzzy logic based rules, and a subsequently used simple defuzzification method might be intuitive and heuristic, the behaviour of such systems can be discussed with the tools of exact mathematics. Recently, there are more and more approaches that connect this very plausible and simple method with approximation theory and other aspects of mathematics.

An alternative approach to fuzzy control is the Takagi-Sugeno model [3], where the consequents of the rules are given in the form of linear or, more generally, arbitrary functions. This seems to be "more mathematical" in its attitude, even though it preserves some of the transparency of the purely linguistic approach. An interesting question is what is the connection between these two approaches. An analysis of the explicit behaviour of the two shows that these models are not very different from each other, the type of functions realized is rather similar.

2 THE LINGUISTIC APPROACH TO FUZZY CONTROL

Fuzzy control in the linguistic rule based sense was first applied in 1974 by Mamdani [2], who constructed a computationally acceptable version of the CRI method of Zadeh [1].

In these approaches knowledge about a system is given as a set of rules in the form

R_i = "If x is A_i then y is B_i"
where x stands for the input state variable, usually a compound variable

$$x = (x_1, x_2, \ldots, x_k) ,$$

y is the output state variable (or action variable) which can always be considered as a scalar, A_i and B_i are linguistic terms (or combinations of them), like "positive big" or "rather tall" and they are represented by fuzzy sets of the universes of discourse $X = \{x\}$, and $Y = \{y\}$, resp. A rule is mathematically a fuzzy relation of the space $X \times Y$:

$$R_i : X \times Y \to [0, 1]$$

Mamdani's technique supposes that the relations are decomposable as

$$R = S_1 \times S_2 \times \ldots \times S_k \times S_{k+1} ,$$

where

$$S_i : X_i \to [0, 1] \text{ for } i = 1, \ldots k \text{ and } S_{k+1} : Y \to [0, 1] .$$

A crisp singleton or linguistic fuzzy observation on X in the form "x is x^*", or more generally, "x is A^*" is combined with the rules by relational composition. The result is a fuzzy conclusion on Y:

$$B^* = \mathbf{R} \circ A^* \text{ where } \mathbf{R} = \bigvee_{i=1}^{r} R_i$$

(there are r rules in the base).

From here the crisp conclusion/action is calculated by defuzzification as

$$y^* = defuzz(B^*)$$

where $defuzz()$ is some *averaging* operator like the *Center of Area* method

$$defuzz(S) = \frac{\int ySdy}{\int Sdy} ,$$

or the *First of Maxima* technique:

$$defuzz(S) = \inf\{Y_{max} | Y_{max} = \{y_j | \neg\exists y \in Y : S(y) > S(y_j)\}\} .$$

The other approach to fuzzy control the Takagi-Sugeno model [3] has rules like

"If x is A_i then $y = a_i x + b_i$"
or, in general, there might be any function in the consequent part. There is no need for defuzzification here, as the consequent is crisp.

If the function is a single constant b_i (a special case of both discussed models), the method is referred to as "Sugeno-controller". All three fuzzy control methods are discussed in detail in some recent monographs like [4, 5, 6, 7].

3 MATHEMATICAL FORMULAE OF FUZZY CONTROLLERS

The previous fuzzy controllers have a common structural model. Suppose that the input is a crisp singleton x. It goes to the Degree of Matching unit that determines the fuzzy degrees of all the rules matching with the given x, and these values are lead into the Inference Engine that produces a fuzzy conclusion as output. (Both use the Rule Base as reference.) The result is obtained by

$$w_{ij}^* = \max_{x_j}\{\min\{A^*(x_j), A_{ij}(x_j)\}\} ,$$

$$w_i^* = \min_j\{w_{ij}^*\}$$

$$B^* = \max_y\{\min\{w_i^*, B_i\}\}$$

(if the max and min norms are used) or

$$B^* = \max_y\{w_i^* B_i\} .$$

(with the product as t-norm [8]).

Finally, the Defuzzification unit generates a single output value y that is the crisp conclusion.

If this whole procedure is considered as a "black box", it can be described as a single multivariable function:

$$y = \mathcal{R}(x)$$

For a long time, very few people were interested in the concrete type of this function, as knowing the function would not help with the construction or the tuning of the model, unless transparency is given up. Recently, some investigations on the explicit functions generated by fuzzy controllers have been done [10, 11].

These results refer to real fuzzy controllers implemented in commercial products. These controllers have some important practical limitations concerning the number and concrete appearance of rules. The main difficulty is *computational complexity* that limits the possible size of the rule base.

If there are k input state variables (components of X), and it is necessary to have in every dimension of X, approximately T different linguistic terms in order to cover X densely enough, so that for every $x \in X$ there is at least one such rule that $A_i(x) \geq \alpha$ where α is a reasonably high value on $[0, 1]$, e.g. 0.5, the number of all rules covering X is

$$|\mathbf{R}| = \mathcal{O}(T^k) .$$

Consequently, the CRI has the complexity

$$C_Z == \mathcal{O}(T^{2k+1}) ,$$

while the input (rule base size) and observation complexities are

$$\mathcal{I} = \mathcal{O}((k+1)T^{k+1}) \text{ and } \mathcal{A}^* = \mathcal{O}(kT) .$$

The Mamdani- and the Larsen-algorithms have

$$C_{ML} = \mathcal{O}((k+1)T^{k+1})$$

for the number of steps, while $\mathcal{I}$ and $\mathcal{A}^*$ are the same. Comparing the two results, we get

$$\frac{C_Z}{C_{ML}} = \mathcal{O}(\frac{T^k}{k+1}) .$$

Finally, the complexity of the Takagi-Sugeno-algorithm depends on the type of $f_i(x)$, but anyway is within

$$C_{TS} = \mathcal{O}(T^k) .$$

All these results are expressed in uniform complexity, where dealing with a single rule is considered to be a single step. (For complexity calculations see [9].)

Because of these high complexities, serious limitations are introduced in the practice. The number of rules cannot be too high, so k is rather low (usually less than 6 or 10), moreover, the shape of membership functions is usually very simple, in almost all cases piecewise linear, mainly trapezoidal or even triangular, in many applications isosceles triangular. It is also very important that in such a case the support of every linguistic term in the rules is bounded.

The results referred to above deal with such special cases.

If all rules contain identical equidistant isosceles triangular terms, and the result is defuzzified by *Center of Area* method the formula obtained [10, 11] is

$$y = (i + \frac{1}{2})b + \frac{(w_1 - 1)^2 - (w_2 - 1)^2}{2w_1 - w_1^2 + w_2} \times \frac{b}{2}$$

where the ith and $(i + 1)$th rules are fired and b is the half base length of a term. If x is substituted instead of w_i the result has the form

$$y = c_1 i + c_1' + \frac{c_2 + x}{c_3 + c_4 x + c_5 x^2},$$

where $i = [\frac{x}{2b}]$, the integer part.

With the same rule base but defuzzifying by *Center of Gravity* method, where each $\min\{w_i, B_i(x)\}$ is defuzzified first separately by taking its own center, and then calculating their average each center weighted by the respective areas, we get

$$y = (i + \frac{1}{2})b + \frac{(w_1 - 1)^2 - (w_2 - 1)^2}{2w_1 - w_1^2 + 2w_2 - w_2^2} \times \frac{b}{2}$$

that has the form

$$y = c_1 i + c_1' + \frac{c_2 + c_3 x + c_4 x^2}{c_5 + c_6 x + c_7 x^2 + c_8 x^3},$$

that is rather similar in structure, but the polynomial fraction is of higher order.

The difference of these two is bounded by $0.04b$, i.e. 2% of the base length which is result of the fact that the dominant member is $(i + 1/2)b$.

If the algebraic product t-norm is used [8], the COA defuzzified version gives

$$y = (i + \frac{1}{2})b + \frac{7w_2w_1^2 - 7w_1w_2^2 + 6w_2^3 - 6w_1^3}{6(2w_1^2 + 3w_1w_2 + 2w_2^2)} \times b\,,$$

and the COG defuzzified one is

$$y = (i + \frac{1}{2})b + \frac{w_2 - w_1}{w_2 + w_1} \times \frac{b}{2}$$

It is interesting that if $w_1 + w_2 = 1$ holds (that often applies in practice), this latter reduces to

$$y = (i + \frac{1}{2})b + (1 - 2w_1) \times \frac{b}{2}$$

that is a linear function of x (as w_1 is linear).

In the Larsen-case, the difference of the two defuzzification types is bounded by $0.12b$ i.e. 6% of the base length.

All four methods with identical equidistant isosceles triangular rules (Mamdani and Larsen, COA and COG) provide approximately piecewise linear explicit transference functions, with polynomial fractions as additional nonlinear members. If the base length is bounded from below, such functions are clearly limited in their ability for approximation.

Next, the most general formula for the Mamdani-method with COG defuzzification is shown when all rules contain arbitrary (but bounded) *trapezoidal terms.*

$$y = \frac{\frac{(c_1w_1+c_2w_2^2)(c_3+c_4w_1+c_5w_1^2)}{c_6w_1} + \frac{(c_7w_2+c_8w_3^2)(c_9+c_{10}w_2+c_{11}w_2^2)}{c_{12}w_1}}{c_{13}w_1 + c_{14}w_1^2 + c_{15}w_2 + c_{16}w_2^2}$$

where c_i are constants depending on the 2×4 points defining the two trapezoidal terms involved. The whole function is constructed of polynomial fraction pieces each corresponding to the above formula (as w_i are linear functions of x between two breakpoints).

Even this rather complicated function has rather limited degrees of freedom, obviously, it is not possible to approximate arbitrary transference functions too well with such a class of fuzzy control functions, as long as the length of the pieces is *bounded from below* by the minimum length of the bases, that is again the consequence of the boundedness of the number of rules in the base.

In practical fuzzy control, the consequents never have a flat top (that is, $|core(B_i)| = 1$), and so, the possibilities are even less, as the most general formula we get in that case is of the type

$$y = \frac{p_1^{(4)}(w_1, w_2)}{p_2^{(3)}(w_1, w_2)}$$

where $p_i^{(j)}(w_1, w_2)$ denotes a polynomial of jth degree of the variables w_1 and w_2. It is easier to evaluate it if the explicit form is expressed with the help of x, as then the result is of the type

$$y = \frac{p_3^{(3)}(x)}{p_4^{(2)}(x)} .$$

This is rather similar to the former results: a linear function with a polynomial fraction as an "error" member. It is important that under the former conditions the pieces of the function determined by a single polynomial fraction are strictly monotone between two breakpoints. If only a single rule determines the result within a certain interval, the degrees of both the numerator and the denominator functions are one less, i.e. the nonlinear member is simpler in structure.

If the Sugeno-controller is considered with singleton consequents, the structure of the result is as follows:

$$y = c_1 + \frac{c_2}{c_3 + c_4 x} ,$$

a constant piece with hyperbolic "error". Here, c_3 becomes zero if $w_1 + w_2 = 1$, in this case (the facing flanks of trapezoidal antecedents are symmetrical) this model generates a step function. If a Larsen-controller is taken that has identical isosceles triangular consequents (and similar antecedents as in the previous case), the result will be identical with that of the Sugeno-controller. Using symmetrical triangles in the right hand sides of the rules does not help with constructing more general fuzzy control functions.

In the general (linear consequent) TS-controller the realized function consists of pieces in the form

$$y = c_1 + c_2 x + c_3 x^2 + \frac{c_4}{c_5 + c_6 x} ,$$

i.e. parabolic pieces with hyperbolic nonpolynomial parts are generated. If $w_1+w_2 = 1$, the hyperbolic part disappears. This type of controller realizes only functions with alternating linear and parabolic pieces (the former in the domains where only one rule is fired). The exact formulae for all above mentioned controllers can be found in [11].

After this long series of examples for explicit fuzzy control functions (covering virtually all types used in practical systems), it is interesting to refer to a theoretical result that states that the family of functions generated by a bounded number of bounded support trapezoidal rules is *nowhere dense in a compact domain* [12], cf. [13].

4 FUZZY CONTROLLERS AS UNIVERSAL APPROXIMATORS

The conclusions of the previous section are rather different from the suggestions of various authors who have shown that fuzzy controllers consisting of finite rules can approximate arbitrary continuous functions arbitrarily well on compact domains.

Kosko's theorem [14] refers to the Mamdani-controller, and states universality, however, the condition of approximating a function everywhere not worse than with accuracy ε is that the distance of two adjacent rule centres is

$$|y_i - y_{i+1}| < \frac{\varepsilon}{2p-1}$$

where p is the number of maximal overlapping antecedents (usually 2), from where the number of rules should be $|\{R_i\}| \geq \frac{|X|}{\varepsilon}$. Obviously, the number of rules is *not bounded*, and good approximation needs practically impossible many rules.

Wang's results [15, 16] give exact proof that if Gaussian (bell shaped) membership functions are used, the Larsen-method generates a family of Stone-Weierstraß functions and approximates any continuous function uniformly well in a compact domain. However, in this case the number of rules is not bounded and in addition, the *supports of terms in the rules are also not bounded* in the sense that for every rule $supp(B_i) = X$ that makes the complexity even higher as processing a single rule cannot be considered as unit time.

The same result was generalized for arbitrary non-vanishing membership function type in multiple dimensions [17]. The week points are the same: neither the support of each individual rule, nor the number of rules are bounded.

A similar statement can be formulated even for non-fuzzy expert controllers where the rules have the structure

$$R_i = \text{“If } x \in [x_{jk}, x_{jk+1}) \text{ then } y = b_i\text{”},$$

a statement that has also no validity if the number of rules is bounded [11].

Otherwise, if the shape of the term membership functions is not restricted, it is not even necessary to approximate with unbounded many rules, as using proper membership functions makes it possible to generate *exactly* arbitrary continuous function by only a few rules (minimally two, if convexity is required, the number is not more than the number of inflexion points multiplied by two plus two) [18]. As a matter of course, realizing the necessary membership function shape is equally difficult as realizing the transference function directly (without fuzzy controllers), and as usually the analytical model of the system is unknown, it is an impossible task, but the complexity of dealing with these membership functions is at least as high as the complexity of direct function generation would be.

The seemingly contradictory results in the previous two sections are not contradictory in reality: Fuzzy controllers might be considered as universal approximators, but the approximation properties are not good as soon as real controllers are considered where size and run time are strongly limited by the real time implementation and other conditions. real fuzzy controllers generate always (limited class) piecewise polynomial fractions that have rather rough approximation properties. Nevertheless, in many applications, this roughness is satisfactory, and transparency and low computational complexity seem to be still a good trade-off for the loss of this hypothetical universality...

5 RULE INTERPOLATION CONTROLLERS

The previous algorithms cannot be used if the rules do not cover X (properly), especially when there are “gaps” in the input space. Such can be obtained by automatic rule tuning [19], or simply by having not enough a priori information. High complexity might be also strongly motivating in reducing the number of rules, so that they do not cover X any more - if there is a way to conserve the essential information in the original knowledge base [20, 21, 22].

As in the case of such sparse rule bases usually no rule fires in the sense of the Mamdani-, Larsen-, and Takagi-Sugeni-controllers, other algorithms are applied that have different mathematical structures as well.

Some of the most important methods to calculate conclusions from a single consequent are *gradual reasoning, analogical reasoning*, and the *Revision Principle* [23, 24, 25, 26, 27].

A common feature of all these methods is however that they work with observations where

$$supp(A^*) \cap supp(A_i) \neq \emptyset \ ,$$

so they do not work in the gaps themselves.

However, they all interpret the semantics of If ... then rules as "fuzzy points" of a hypothetical mapping

$$\mathcal{R}: \mathcal{P}(X) \to \mathcal{P}(Y) \ ,$$

where $\mathcal{P}(S)$ denotes the fuzzy power set, the set of all fuzzy subsets, of the universe S. It is reasonable to restrict considerations to $C(X)$ and $C(Y)$, the latter ones meaning the set of all convex and normal fuzzy sets (CNF sets) of the respective universes.

If some additional properties of the universes are supposed (that are usually fulfilled with real applications), namely, the existence of full orderings and a metrics in each component dimension, the definition of *partial ordering* (po) among CNF sets and the *fuzzy distance* of two comparable sets (a family of fuzzy distances, two for each $\alpha \in [0,1]$) can be introduced. Po is defined as follows:

$$F \prec G \Leftrightarrow \forall \alpha \in [0,1]: \ \inf\{F_\alpha\} \leq \inf\{G_\alpha\} \ \& \ \sup\{F_\alpha\} \leq \sup\{G_\alpha\} \ .$$

For comparable CNF sets the lower and upper distances (d_L and d_U) for a certain $\alpha \in [0,1]$ are defined as the distances between the infima and the suprema of the respective α-cuts:

$$d_L(F,G) = \{\alpha/D(\inf\{A_\alpha\},\inf\{B_\alpha\})\} \text{ and}$$

$$d_U(F,G) = \{\alpha/D(\sup\{A_\alpha\},\sup\{B_\alpha\})\}$$

for $\alpha \in [0,1]$. For ordering and distance among CNF sets see [28].

If rules and the observation satisfy

$$A_{i1} \prec A^* \prec A_{i2} \text{ and } B_{i1} \prec B_{i2} \text{ ,}$$

it is possible to do *control by rule interpolation*, according to the following fundamental equation:

$$d(A^*,A_{i1}) : d(A^*,A_{i2}) = d(B^*,B_{i1}) : d(B^*,B_{i2})$$

where d stands for all d_L and d_U for all α-s separately. The solution of the latter equation for B^*_α:

$$\min\{B^*_\alpha\} = \frac{\frac{1}{d_L(A^*_\alpha,A_{i1\alpha})}\inf\{B_{i1\alpha}\} + \frac{1}{d_L(A^*_\alpha,A_{i2\alpha})}\inf\{B_{i2\alpha}\}}{\frac{1}{d_L(A^*_\alpha,A_{i1\alpha})} + \frac{1}{d_L(A^*_\alpha,A_{i2\alpha})}}$$

$$\max\{B^*_\alpha\} = \frac{\frac{1}{d_U(A^*_\alpha,A_{i1\alpha})}\sup\{B_{i1\alpha}\} + \frac{1}{d_U(A^*_\alpha,A_{i2\alpha})}\sup\{B_{i2\alpha}\}}{\frac{1}{d_U(A^*_\alpha,A_{i1\alpha})} + \frac{1}{d_U(A^*_\alpha,A_{i2\alpha})}}$$

the basic formulae of linear interpolation of two rules. These formulae generate "fuzzy functions", i.e. hyper-surfaces of which all α-cuts are bands limited by piecewise linear functions. After defuzzification, y is a piecewise linear function of x, if A^* is considered to be a crisp singleton value.

Interpolation of two rules can be extended to many different algorithms, e.g to the interpolation of $2n$ rules (n and n flanking the observation in the sense of $\prec$), by calculating the weighted average of several pairwise interpolated rule pairs:

$$\min\{B^*_\alpha\} = \frac{\sum_{i=1}^{2n}\frac{1}{d_L(A^*_\alpha,A_{i\alpha})}\inf\{B_{i\alpha}\}}{\sum_{i=1}^{2n}\frac{1}{d_L(A^*_\alpha,A_{i\alpha})}}$$

$$\max\{B^*_\alpha\} = \frac{\sum_{i=1}^{2n}\frac{1}{d_U(A^*_\alpha,A_{i\alpha})}\sup\{B_{i\alpha}\}}{\sum_{i=1}^{2n}\frac{1}{d_U(A^*_\alpha,A_{i\alpha})}}$$

The complexity of the simple interpolation is

$$C_I = \mathcal{O}(r + \lambda k)$$

where λ is the number of levels for which the formulae must be calculated, and r is the number of rules in the reduced base.

If the fuzzy distance of CNF sets is defined in a somewhat different way, i.e., instead of taking the infimum and supremum of α-cuts, the center and the length of the α-cuts are considered (the latter meaning the width of the fuzzy set), the solutions of the fundamental equation will take the following form:

$$centr\{B_\alpha^*\} = \frac{\frac{1}{d(A_\alpha^*,A_{i1\alpha})}centr\{B_{i1\alpha}\} + \frac{1}{d(A_\alpha^*,A_{i2\alpha})}centr\{B_{i2\alpha}\}}{\frac{1}{d(A_\alpha^*,A_{i1\alpha})} + \frac{1}{d(A_\alpha^*,A_{i2\alpha})}}$$

$$width\{B_\alpha^*\} = \frac{\frac{1}{d(A_\alpha^*,A_{i1\alpha})}\cdot\frac{width\{B_{i1\alpha}\}}{width\{A_{i1\alpha}\}} + \frac{1}{d(A_\alpha^*,A_{i2\alpha})}\cdot\frac{width\{B_{i2\alpha}\}}{width\{A_{i2\alpha}\}}}{\frac{1}{d(A_\alpha^*,A_{i1\alpha})} + \frac{1}{d(A_\alpha^*,A_{i2\alpha})}}.width\{A_\alpha^*\}$$

where

$$centr\{S_\alpha\} = \frac{\inf\{S_\alpha\} + \sup\{S_\alpha\}}{2}$$

$$width\{S_\alpha\} = \sup\{S_\alpha\} - \inf\{S_\alpha\}$$

$$d(S_{1\alpha}, S_{2\alpha}) = centr\{S_{2\alpha}\} - centr\{S_{1\alpha}\}$$

and the distance of the widths is expressed by their ratio.

This approach is extended to a general approximation algorithm where an arbitrary number of arbitrarily located rules is taken into consideration:

$$centr\{B_\alpha^*\} = \frac{\sum_{i=1}^{r}\frac{1}{d(A_\alpha^*,A_{i1\alpha})}[centr\{B_{i\alpha}\} + d(A_\alpha^*, A_{i\alpha}).\frac{width\{B_{i\alpha}\}}{width\{A_{i\alpha}\}}]}{\sum_{i=1}^{r}\frac{1}{d(A_\alpha^*,A_{i\alpha})}}$$

$$width\{B_\alpha^*\} = \frac{\sum_{i=1}^{r}\frac{1}{d(A_\alpha^*,A_{i1\alpha})}\cdot\frac{width\{B_{i\alpha}\}}{width\{A_{i\alpha}\}}}{\sum_{i=1}^{r}\frac{1}{d(A_\alpha^*,A_{i\alpha})}}.width\{A_\alpha^*\}$$

This is a single polynomial fraction for the whole of X fitting all rule points (the ends of the cuts), and the degree of the numerator is identical or one higher as the degree of the denominator.

Further details on rule interpolation algorithms (including not exactly fitting approximations, as well) can be found in [29, 30, 32, 31, 33, 34, 35]. Lagrangian interpolation is discussed especially in [21, 20, 22].

Interpolation controllers also have the universal approximator property (cf. [36]), however, with the same strong limits, i.e. in practice they can be used only as rough approximators.

In the complexity of the interpolation controller a crucial point is the value of λ. If it is high, it increases resulting complexity even though it cannot grow that fast as the other component does, as $r = (T/s)^k$, where s is the reduction factor compared to the number of rules in a dense rule base. λ depends on how accurately the shape of the conclusion has to be reconstructed (in order to get good defuzzified results).

In the practice controllers use piecewise linear rules, the question is whether the interpolated conclusion is piecewise linear, as well. If so, $|lambda| = 2$ is satisfactory (calculating only for $\alpha = 0, 1$), and this will bring in no considerable increase of the runtime.

If both interpolated rules are trapezoidal, the equation of the (left) flank conclusion is as follows:

$$y^*_{\alpha L} = \frac{C_1\alpha^2 + C_2\alpha + C_3}{C_4\alpha + C_5}$$

that is obviously *not linear*. The constants C_i are determined by the parameters of the trapezoidal membership functions in the rules. The flanks of B^* are piecewise quadratic if and only if the two antecedents A_1 and A_2 have symmetrical (linear) slopes. The flanks of B^* are linear if either both the antecedents A_i and the consequents B_i of the rules are equivalent pairwise or if the antecedents and the observation A^* are all equivalent. This latter is not possible if the observation is a singleton x.

These results seem to be rather disappointing at first sight as in the practically important cases the result has nonlinear flanks that can increase $|\lambda|$. However, numerical examples and analytical considerations show that nonlinearity of the flanks is in reality very small. The explanation for this is that the expression of the flanks can be divided into two parts: a linear and a hyperbolic part, where the hyperbolic part represents a segment of a hyperbola that is taken from a region that is far from the focal area, i.e. it behaves "rather linearly". Some examples and exact calculations on this matter can be found in [37, 38].

It is also interesting to analyze the shape of the conclusion obtained by interpolation. Often it is an "abnormal" membership function (it has a loop, or it has to increasing or two decreasing flanks), and in order to obtain the corresponding regular membership function, it is necessary to truncate the original construction. As a consequence, the result will be subnormal. Convexity is always true because of the choice of the algorithms. details on normality conditions can be found in [37, 39, 40].

6 CONTROL WITH STRUCTURED RULE BASES

Computational complexity is further on one of the key issues of fuzzy control. By using sparse rule bases and interpolation controllers, the exponent of the exponential complexity cannot be decreased, only the base. A method to reduce the exponent by applying the "Divide and conquer" philosophy, by dividing the state space into subdomains and constructing hierarchical rule bases was presented in the unmanned helicopter control application by Sugeno [41, 42].

The engineering background of this method is that the model might be dominated *locally* by only a *subset of variables* (this subset changing with the change of the state space domain in question). The total state space is subdivided into such domains, and for every domain another "local model" (with possibly very restricted validity) is constructed. This method is effective if every local model contains a (possibly considerably) lower number of variables and consequently also a much lower number of rules.

Every local model is determined by a sub rule base. On the higher level(s) called meta level(s) first it is decided which sub rule base is adequate for the neighborhood of the observation. In order to do so, a set of meta rules is given of which the output is the domain variable. The rules decide on some input variables separating the different *sub models*.

In the original form of this method, the domains form a classical (crisp) *partition* of the state space, so for every situation exactly one domain is adequate, consequently always exactly one sub rule base is used. The same principle can be extended to general covers of the state space. Even, the cover might be fuzzy, so that its elements (and the validity of the corresponding sub rule base) have decreasing membership values as the observation is getting farther from the typical validity area (core) of that particular sub rule base. In the general case, usually several sub rule bases must be combined with different degree of matching, in a similar way as single consequents are combined in the Mamdani-model.

A simple example for a structured rule base with a single meta level and very few rules is the following one:

Metalevel (R_0):

"If x_1 is A_{11} and x_2 is A_{21} then take domain D_1"

"If x_1 is A_{12} and x_2 is A_{22} then take domain D_2"

Sub rule base R_1 for D_1:

"If x_3 is A_{31} and x_4 is A_{41} then Y is B_1"

"If x_3 is A_{32} and x_4 is A_{42} then Y is B_2"

Sub rule base R_2 for D_2:

"If x_5 is A_{51} then Y is B_3"

"If x_5 is A_{52} then Y is B_4"

If there are k input variables that can be divided into n groups, the first group containing the first (k_1) meta level variables, the second one (k_2) the second meta level, etc., and the last one (k_n) the real control level inputs, in the worst case the total number of rules in the base is

$$r = \mathcal{O}(T^{k_1} + T^{k_1}T^{k_2} + \ldots + T^{k_1}T^{k_2}\ldots T^{k_n}) = \mathcal{O}(T^k)$$

however, even so, for any concrete observation, only

$$r_{A^*} = \mathcal{O}(T^{k_1} + T^{k_2} + \ldots + T^{k_n}) = \mathcal{O}(nT^{\max_{i=1}^n k_i})$$

rules must be processed which is an effective decrease in the run time. Moreover, if the individual sub rule bases contain only a reduced number of variables, the role of $\max_{i=1}^n k_i$ is taken over by the largest number of variables occurring in the *same sub rule base.*

The idea of interpolation and approximation within the meta levels, applied on sub rule bases is discussed in some detail in [43, 44, 45]. This topic however has more unsolved questions than solved ones and needs serious mathematical considerations. The general solution of how to divide large rule bases into sub levels could bring a new domain of applications, where the number of variables is much higher than in the presently existing ones.

7 CONCLUSIONS

Fuzzy controllers are based on logical heuristics. The use of linguistic variables and terms for values is an attempt to mimic human ways of solving control problems by rule type experience. Behind these intuitive methods there are however clear mathematical formulae, and by analyzing them, some further conclusions for the applicability of fuzzy control can be obtained. The analysis of real controllers helps especially much with realizing the limits of such, that are in crisp contradiction with the rather optimistic conclusions taken after some results on idealistic fuzzy controllers that have nice analytic properties. the real difficulty to overcome is computational complexity. Using interpolative controllers helps somewhat but this also does not bring final solutions. The next step is hierarchical control where a real reduction of the exponent in the exponential expression of complexity is possible if the system is decomposable in this sense. A farther possibility is promised by using fuzzy covers that do not require strict decomposability just varying local importance of the sub models (as a matter of course, with some of them vanishing locally, otherwise the method does not help). Here the problem of combining interpolation and structured control offers a wide range of interesting unsolved problems.

However, this paper has been far from covering all aspects of the mathematical analysis of fuzzy controllers and related topics. There are much more questions there than answers, among them, an exact and clear description of fuzzy models and fuzzy mappings $\mathcal{R}$ is missing. I hope, these pages will give some stimulus to start further research in this direction.

Anyway, there are many unsolved problems in fuzzy mathematics. More than enough to fill in the next fifty years with new challenges and interesting results that I sincerely wish to my celebrating Friend.

So, once again, happy birthday, and many happy new results in the fuzzy field, Etienne!

Acknowledgements

This paper was partly written during a visiting appointment at the Department of Computer and Management Science, The University of Trento, Italy.

REFERENCES

[1] Zadeh, L. A.: Outline of a new approach to the analysis of complex systems and decision processes. *IEEE Tr. SMC*, 28-44, 1973.

[2] Mamdani, E. H. and Assilian, S.: An experiment in linguistic synthesis with a fuzzy logic controller. *Int. J. of Man-Machine Studies*, 1-13, 1975

[3] Takagi, T. and Sugeno, M.: Fuzzy identification of systems and its applications to modelling and control. *IEEE Tr SMC 15*, 116-132, 1985.

[4] Driankov, D., Hellendoorn, H, and Reinfrank, M.: *An Introduction to Fuzzy Control*, Springer, Berlin, etc., 1993.

[5] Pedrycz, W.: *Fuzzy Control and Fuzzy Systems*, 2nd edition, J. Wiley, Chichester, 1992.

[6] Yager, R., and Filev, D.: *Essentials of Fuzzy Modeling and Control*, J. Wiley, Chichester, 1994.

[7] Wang, L. X.: *Adaptive Fuzzy Systems and Control*, Prentice Hall, Englewood Cliffs, N.J., 1994.

[8] Larsen, P. M.: Industrial application of fuzzy logic control. *Int. J. of Man Machine Studies 12*, 3-10, 1980.

[9] Kóczy, L. T.: Computational complexity of various fuzzy inference algorithms. *Annales Univ. Sci. Budapest, Sect. Comp. 12*, 151-158, 1991.

[10] El Hajjaji, A. and Rachid, A.: Explicit formulas for fuzzy controller. *FSS 62*, 135-141, 1994.

[11] Kóczy, L. T., and Sugeno, M.: Explicit formulas for fuzzy control systems and the approximator property, Technical Report # TR 93-94/408, LIFE Chair of Fuzzy Theory, Tokyo Institute of Technology, 1994.

[12] Moser, B.: On the nowhere denseness of fuzzy control functions, Technical Report, Dept. of Mathematics, J. Kepler University, Linz, 1994.

[13] Klement, E. P., Kóczy, L. T., and Moser, B.: Are fuzzy systems universal approximators?, *Int J. of General Systems*, to appear.

[14] Kosko, B.: Fuzzy systems as universal approximators, *FUZZ-IEEE '92*, San Diego, 1153-1162, 1992.

[15] Wang, X. L., and Mendel, J.: Generating fuzzy rules from numerical data with applications, Technical Report TR USC-SIPI #169, Signal and Image Processing Institute, Univ. of Southern Calif., 1991.

[16] Wang, X. L.: Fuzzy systems are universal approximators, *FUZZ-IEEE '92*, San Diego, 1163-1169, 1992.

[17] Nguyen, H. T., and Kreinovich, V.: On approximations of controls by fuzzy systems, Technical Report # TR 92-93/302, LIFE Chair of Fuzzy Theory, Tokyo Institute of Technology, 1992.

[18] Bauer, P., Klement, E. P., Moser, B., and Leikermoser, A.: Modeling of fuzzy functions by fuzzy controllers, in *Theoretical Aspects of Fuzzy Control* (H. T. Nguyen, M. Sugeno, R. Tong, and R. Yager, Eds.), J. Wiley, Chichester, to appear in 1995.

[19] Burkhardt, D. G., and Bonissone, P. P.: Automated fuzzy knowledge base generation and tuning. *FUZZ-IEEE '92*, San Diego, 179-196, 1992.

[20] Kóczy, L. T., and Hirota, K.: Interpolation and size reduction in fuzzy rule bases. Technical Report # TR93-94/401, LIFE Chair of Fuzzy Theory, Tokyo Institute of Technology, 1993.

[21] Kóczy, L. T.: Compression of fuzzy rule bases by interpolation. *First Asian Fuzzy Systems Symposium*, Singapore, [500-507], 1993.

[22] Kóczy, L. T., and Hirota, K.: Size reduction by interpolation in fuzzy rule bases, *IEEE Tr. SMC*, to appear.

[23] Dubois, D., and Prade, H.: Gradual rules in approximate reasoning. *Information Sciences 61*, 103-122, 1992.

[24] Dubois, D., Grabisch, M., and Prade, H.: Gradual rules and the approximation of functions. *2nd Int. Conf. Fuzzy Logic and Neural Networks, IIZUKA '92*, Iizuka, 629-632, 1993.

[25] Türkşen, I. B., and Zhong, Z.: An approximate analogical reasoning approach based on similarity measures. *IEEE Trans. SMC*, 1049-1056, 1988.

[26] Mukaidono, M., Ding, L., and Shen, Z.: Approximate reasoning based on Revision Principle. *NAFIPS '90*, Toronto, 94-97, 1990.

[27] Ding, L., Shen, Z. and Mukaidono, M.: Principle for approximate reasoning - based on Revision Principle. *2nd Int. Conf. Fuzzy Logic and Neural Networks, IIZUKA '92*, Iizuka, 305-308, 1992.

[28] Kóczy, L. T., and Hirota, K.: Ordering, distance and closeness of fuzzy sets. *Fuzzy Sets and Systems 59*, 281-293, 1993.

[29] Kóczy, L. T., Hirota, K., and Juhász, Á.: Interpolation of 2 and $2k$ rules in fuzzy reasoning. *IFES '91, Fuzzy Engineering toward Human Friendly Systems I*, Yokohama, 206-217, 1991.

[30] Kóczy, L. T., and Hirota, K.: Approximate reasoning by linear rule interpolation and general approximation. *Int. J. of Approximate Reasoning 9*, 197-225, 1993.

[31] Kóczy, L. T., and Hirota, K.: Interpolative reasoning with insufficient evidence in sparse fuzzy rule bases. *Information Sciences 71*, 169-201, 1993.

[32] Vass, Gy., Kalmár, L., and Kóczy, L. T.: Extension of the fuzzy rule interpolation method. *Int. Conf. on Fuzzy Sets Theory and Applications*, Liptószentmiklós (Czechoslovakia), 1992.

[33] Kóczy, L. T., and Hirota, K.: Reasoning by analogy with fuzzy rules. FUZZ-IEEE '92, San Diego, 263-270, 1992.

[34] Kóczy, L. T.: Inference in fuzzy rule bases with conflicting evidence. *NAFIPS '92*, Puerto Vallarta, 608-617, 1992.

[35] Kóczy, L. T.: Techniques of inference in insufficient and inconsistent fuzzy rule bases. *Linz-92, Non-Classical Logics and Their Applications*, Linz, 46-50, 1992.

[36] Kacprzyk, J.: Fuzzy dynamic programming - battling against the curse of dimensionality via interpolative reasoning, *Third Int. Conf. on Fuzzy Logic, Neural Nets and Soft Computing*, Iizuka, Japan, 245-246, 1994

[37] Kóczy, L. T., and Kovács, Sz.: On the preservation of convexity and piecewise linearity in linear fuzzy rule interpolation. Technical Report, # TR93-94/402, LIFE Chair of Fuzzy Theory, Tokyo Institute of Technology, 1993.

[38] Kóczy, L. T.: Algorithmic aspects of fuzzy control, *Int. J. of Approximate Reasoning*, to appear.

[39] Shi, Y., and Mizumoto, M.: On Koczy's interpolative reasoning method in sparse fuzzy rule bases, *10th Fuzzy Systems Symposium*, Osaka, 221-224, 1994.

[40] Shi, Y., and Mizumoto, M.: Some considerations on Koczy's fuzzy interpolative reasoning method, To appear in *FUZZ-IEEE/IFES '95*, Yokohama, 1995.

[41] Sugeno, M.: Advanced fuzzy control of unmanned helicopters, Plenary Lecture, *FUZZ-IEEE '93*, San Francisco, 1993.

[42] Sugeno, M., Griffin, M. F., and Bastian, A.: Fuzzy hierarchical control of an unmanned helicopter, *Fifth IFSA World Cong.*, Seoul, 179-182, 1993.

[43] Kóczy, L. T., and Hirota, K.: Interpolation in structured fuzzy rule bases. *FUZZ-IEEE '93,* San Francisco, 402-405, 1993.

[44] Kóczy, L. T., and Hirota, K.: Modular rule bases in fuzzy control. *EUFIT '93,* Aachen, 606-610, 1993.

[45] Kóczy, L. T., and Hirota, K.: Approximative inference in hierarchical structured rule bases. *Fifth IFSA World Congress,* Seoul, 1262-1265, 1993.

11

KEEP FUZZY CONTROLLERS FUZZY

Ann Nowé

Department Info
Vrije Universiteit Brussel
Pleinlaan 2
B-1050 Brussel, Belgium

ABSTRACT

In this paper we argue that fuzzy controllers should be looked at as fuzzy valued mappings rather than crisp valued mappings. As a result using an explicit interpolation method as an inference mechanism becomes more natural than applying fuzzy reasoning. We propose a fuzzy piecewise Lagrange interpolation approach which is a generalisation of the conventional piecewise Lagrange interpolation. We also prove that such controllers are universal approximators.

1 INTRODUCTION

Today, the most sophisticated autonomous controllers known, are human controllers. Therefore, a natural approach in the attempt of building controllers is to try to emulate human decision making as directly as possible. This is in fact the basic idea of fuzzy control. The human approach, unlike the conventional analytical approaches, is not based on the solution of differential equations. For example, when people are driving a car, they do not solve a single differential equation. Instead they apply rules of thumb which are often invaded by imprecision and uncertainty. Nevertheless, people are quite successful in this kind of tasks.

Fuzzy controllers model expert knowledge in a natural way by blending rule based control and fuzzy set theory. The rules encountered in a fuzzy controller have the following canonical form "IF situation is A THEN action is B IS τ" where A and B are fuzzy predicates modelled as fuzzy sets and

τ is an optional truth qualification which is also modelled as a fuzzy set. The inference mechanism is based on fuzzy set theory.

In the relatively young domain of fuzzy control three phases can be identified [15], i.e. the academic phase (1965-1990), the transformation phase (1978-1988) and the fuzzy boom phase (1988-). The academic phase started with the seminal publication [11] in which the basic concepts of the fuzzy set theory were introduced. In 1973, L.A. Zadeh published a paper that pointed out the potential use of fuzzy set theory in the field of analysis of complex systems and decision processes. Inspired by this paper E.H. Mamdani further elaborated on the subject and applied these ideas to control problems [6]. He called the new approach to automatic control: fuzzy control. The first published industrial application was in the area of cement kiln control and dates from 1980. This domain is very well suited for the application of fuzzy control since the process' state can only be measured indirectly, and therefore conventional control techniques as such cannot be applied. In the eighties fuzzy control became very popular in Japan and was applied to a broad range of consumer goods such as camcorders, washing machines, rice cookers, electric shavers and vacuum cleaners. The first fuzzy analogue computer and the first fuzzy chip became available at that time. These years are known as the Japanese fuzzy boom. Since the nineties, the other continents have also fallen under the spell of the fuzzy boom. This has resulted in the foundation of several international conferences and a number of international journals on fuzzy systems. Today, fuzzy control has been adapted into a broad range of applications, including train control, vehicle automatic transmission, remote control of an unmanned helicopter, elevator control, image analysis, etc. Currently, a second academic phase is about to be established. Modern control techniques originating from different areas such as fuzzy set theory, artificial neural nets, evolutionary algorithms, rule based systems and non-linear control theory are coalescing. It is expected that the integration of all this expertise will result in a synergetic effect and revive the domain of control theory.

Despite its original motivation, fuzzy control is frequently used for tasks which can reasonably well be controlled by conventional linear controllers. These control tasks are typically setpoint oriented. Fuzzy controllers dealing with this type of control problems are referred to by L.A. Zadeh as fuzzy controllers in the narrow sense. In literature the term fuzzy control has almost exclusively been used in this context. This type of controllers usually deduce the control action in one stage of fuzzy inference and no fuzzy chaining is involved. Fuzzy controllers, in the narrow sense, typically have an interpolative behaviour. Hence, the idea that the inference mechanism is based on fuzzy logic has been questioned. A more recent viewpoint is to consider these controllers as fuzzy graphs [14]. Their inference mechanism can then be explained as a direct application of fuzzy relation theory. No fuzzy logic is involved.

Opposed to fuzzy control in the narrow sense, there is fuzzy control in the wide sense [14]. The latter refers to controllers of a higher level. These controllers, also called approximately reasoning systems, exploit fuzzy logic in its whole as an inference mechanism. The structure of these controllers is close to the architecture usually encountered in rule based systems. The rules are structured in one way or the other and meta fuzzy rules might be used to guide the search process. Such systems usually advise, supervise and co-ordinate controllers on a lower level.

In this paper we will focus on fuzzy controllers in the narrow sense and in particular on Mamdani fuzzy controllers.

2 FUZZY CONTROLLER LAWS ARE FUZZY VALUED MAPPINGS RATHER THAN CRISP VALUED MAPPINGS

As mentioned above the basic idea of fuzzy control is to model human operating knowledge in a natural way. However due to its success, this control technique has also been applied to systems, such as linear systems, which can also be controlled by linear controllers. In se this it is no offense to apply fuzzy control to such systems, provided the synthesis method respects the nature of fuzzy control.

However, more and more people have started to consider a fuzzy controller as a bare approximation space. Very often the parameters describing the fuzzy rules are tuned to get the optimal representative in the space of fuzzy controllers. Hereby fuzzy controllers are considered as crisp mappings from states to actions, without taking into account the meaning of the rules. It is here that fuzzy controllers and artificial neural networks meet each other. By the crisp mapping generated by a fuzzy controller we mean the mapping one gets from the crisp input reading to the crisp control action which is obtained after the defuzzification phase, see Figure 1. As will be

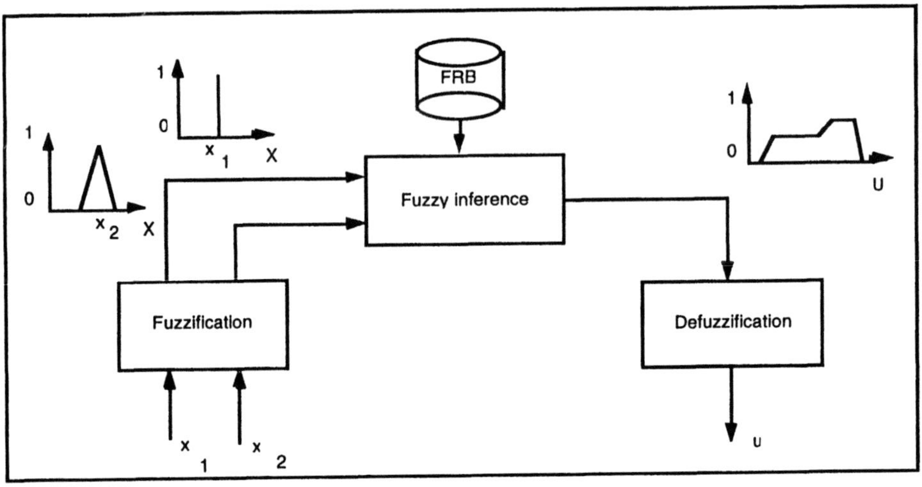

Figure 1 The inference mechanism of fuzzy controllers. The inference mechanism of fuzzy controllers consists of three phases. In the first phase, the fuzzification phase, the crisp input readings are translated into fuzzy sets. In the inference phase the fuzzy input data is given to the fuzzy rule base and an overall fuzzy conclusion is produced. This phase can be considered to be based on the generalised modus ponens or as an application of the fuzzy graph theory. Finally, a crisp control action is obtained in the defuzzification phase by applying one of the many defuzzification methods. For more details on the inference mechanism of fuzzy controllers, we refer to for instance [8].

explained below, from a semantical point of view, the fuzzy set one gets before defuzzification turns out to be very important.

From a pure mathematical perspective there is no real problem to use a fuzzy controller to express a crisp mapping. It can be shown that fuzzy controllers can approximate any continuous function on a compactum to any degree of accuracy, see e.g. [1] and [10]. The approximation space of fuzzy controllers, in the narrow sense, is similar to the space of piecewise linear functions. Indeed, in [7] it is shown that the crisp mapping from states to actions is usually continuous but not continuous differential. In the case of Sugeno fuzzy controllers we even have an perfect correspondence to piecewise linear interpolation, provided the multiplication is used as T-norm. What Mamdani fuzzy controllers is concerned, it is well known that the width of the fuzzy conclusions only plays a second class role. Therefore if the objective is to express a crisp mapping by means of a fuzzy controller no to much effort should be put in to fine tuning the width

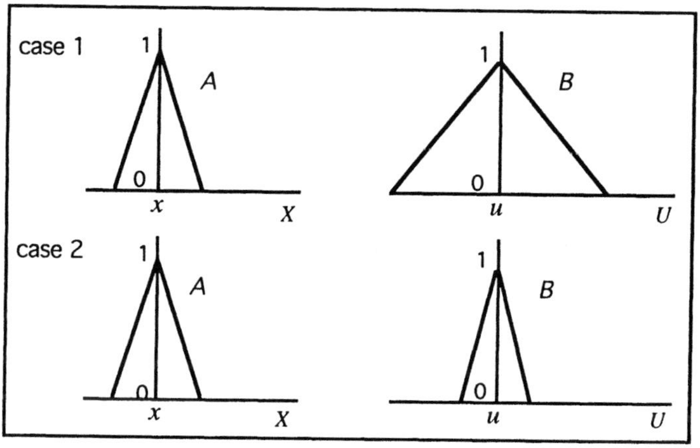

Figure 2 The semantic of fuzzy rules. The explanation is given in the text.

of the conclusions rather one could consider to use a more dense covering of the input domain, i.e. using more rules.

But there is more to be said about fuzzy rules. Let's consider the two rules represented in Figure 2. Both rule are of the form IF A THEN B, with A and B fuzzy sets. Now consider the case of a crisp input reading x as shown in the figure. From a crisp point of view both rules are the same in the sense that these rules recommend the same action u as the action that should be applied. However, from a semantical or fuzzy point of view the rules are quite different. If we look at the early papers by L.A. Zadeh we can read that a fuzzy predicate induces a fuzzy restriction on the possible values a variable can take. Therefore, with respect to control, the conclusion of a fuzzy rule expresses a fuzzy restriction on the control actions available to the region described by the corresponding fuzzy antecedent. If we now return to Figure 2, we see that both rule are quite different. In the first case the restriction is much less tight than in the second case. As shown in [8] a fuzzy conclusion, if expressing a fuzzy restriction on the available control actions, contains important information on the system to be controlled. This information is very useful when so-called viability problems are concerned.

Viability problems are control problems expressed in terms of state constraints which must be obeyed at each instant under death penalty. These state constraints, named viability constraints, might for example be homeostatic constraints in biological regulation, scarcity constraints in economics, state constraints in control, power relations in game theory, ecological constraints in genetics or sociability constraints in sociology [3]. The objective here is to find solutions which are viable, i.e. solutions which obey, at each instant, the viability constraints. As shown in [8] the larger the spectrum of viable control actions of a state, the better the system handles unexpected disturbances coming from the environment, and the more robust this systems'state is. Therefore, if we want to keep a process within a constraint domain in a robust manner, it is important to locate the safe states within the domain and to find out how we can drive the system into these safe states. A reinforcement learning type of algorithm which does the job is explained in detail in [8].

3 FUZZY CONTROLLERS AS FUZZY GRAPHS

In this section we elucidate on a very recent viewpoint on fuzzy controllers by L.A. Zadeh [14]. He states that Mamdani fuzzy controllers should not be considered as being based on fuzzy logic but rather should be looked at as fuzzy graphs. A fuzzy graph [13] may be represented as:

$$G = (\text{"}a_1\text{"} \times \text{"}b_1\text{"}), (\text{"}a_2\text{"} \times \text{"}b_2\text{"}), \ldots, (\text{"}a_n\text{"} \times \text{"}b_n\text{"})$$

i.e. the fuzzy graph G is defined by a union of couples (a_i, b_i) where a_i and b_i are points in the universes of discourse X, U respectively, and "a_i" and "b_i", $1 \leq i \leq n$, represent fuzzy sets labelled as close to a_i and close to b_i. The fuzzy sets a_i and b_i might for example be represented by triangular fuzzy sets T_{a_i-d,a_i,a_i+d} and T_{b_i-d,b_i,b_i+d} respectively, with d any positive constant. Figure 3 shows a fuzzy graph consisting of 3 fuzzy points.

A fuzzy graph can be interpreted as a fuzzy relation R, as follows,

$$\begin{aligned} R : X \times Y &\rightarrow [0,1] \\ (x,u) &\rightarrow \mu_{\bigcup_{i=1}^{n} A_i \times B_i}(x,u) \end{aligned}$$

Therefore the compositional rule of inference can be applied to obtain the image of an input under the fuzzy graph. In figure 3 the image is shown for the crisp input $x = 2$ when the Cartesian product is implemented as the minimum operator and the union as the maximum operator. If we would

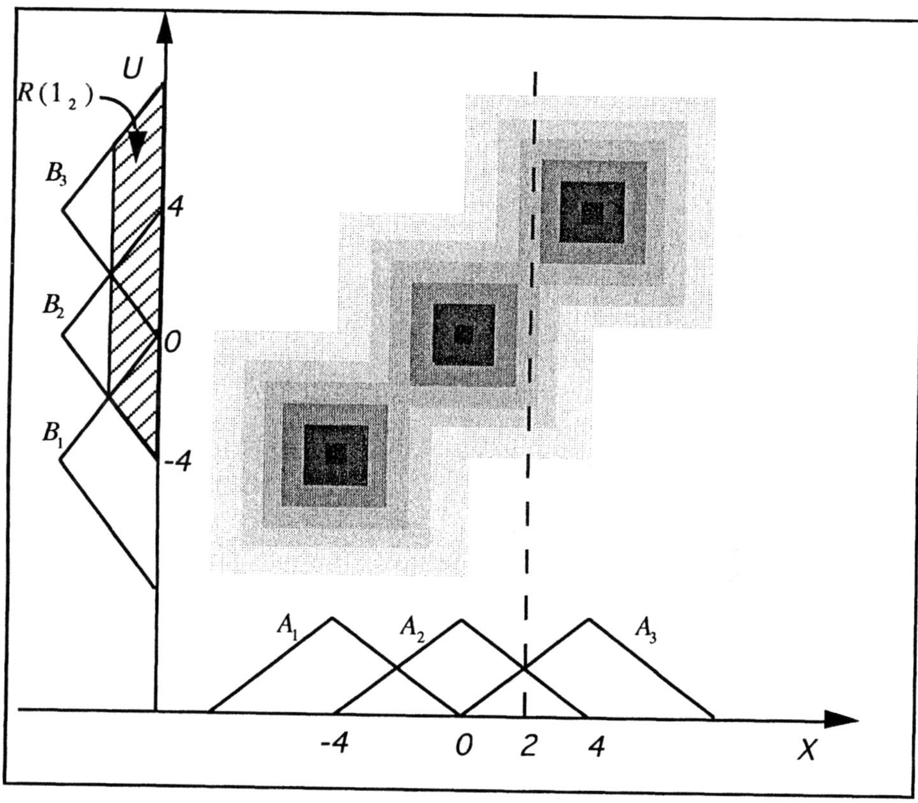

Figure 3 A fuzzy graph represented as a union of 3 fuzzy points.

now consider a fuzzy controller containing the rules $A_i \rightarrow B_i, 1 \leq i \leq 3$, then the same output set is obtained when the usual inference mechanism is applied, under the assumption that the implication operator is the same as the intersection operator which is used to implement the Cartesian product and the rules are combined prior to inference. Therefore one can think about the set of rules $A_i \rightarrow B_i$ as a fuzzy graph represented by a union of fuzzy points (A_i, B_i). By doing so we do not longer look at the intersection operator as being an implication operator, but rather as a coupling operator, i.e. $A_i \rightarrow B_i$ should be read as A_i is coupled with B_i. It should be noted that with respect to control problems the most commonly used implication operators are the intersection operators minimum and product.

Considering a fuzzy controller which uses an intersection operator as implication rather than a genuine implication operator as a fuzzy graph, entirely agrees with the remark made by L.A. Zadeh on his definition of the original fuzzy implication. In [13] the fuzzy rules of the form IF A THEN B are treated as a special case of IF A THEN B ELSE C. The former expression results from the latter by assigning $C = U$. It is remarked that another option could have been to set C equal to the empty fuzzy set $\emptyset$ instead of assigning it to U. If so, the expression IF A THEN B ELSE C, which is defined by: $(A \times B) \cup (\neg A \times C)$ reduces to the Cartesian product $A \times B$, which should be interpreted as A is coupled with B, rather than A entails B.

Related to this viewpoint is an alternative inference mechanism discussed in [8]. The method is based on fuzzy interpolation and is outlined below. As might have been recognized by the reader, the control law underlying this fuzzy graph R is R' : "u is more or less x". This fuzzy relation is presented in figure 4. If we now compute the image of the fuzzy singleton 1_2 by applying the compositional rule of inference on R', with the minimum as implementation of the Cartesian product, the result is quite different from what we obtain using the relation R.

Another way to use the information of the relation R is as follows: The sample data of the relation R may be considered to be most opportune for inputs x, which most strongly belong to one of the A_i, i.e. for x = -4, 0, 4. In case the fuzzy antecedents are chosen to be equivalent to a piecewise linear Lagrange interpolation basis, which is the case for the example under consideration, fuzzy piecewise linear Lagrange interpolation can be used to approximate the image of input x under the relation R', which is assumed to be only known at a number of sample points, here this number is 3.

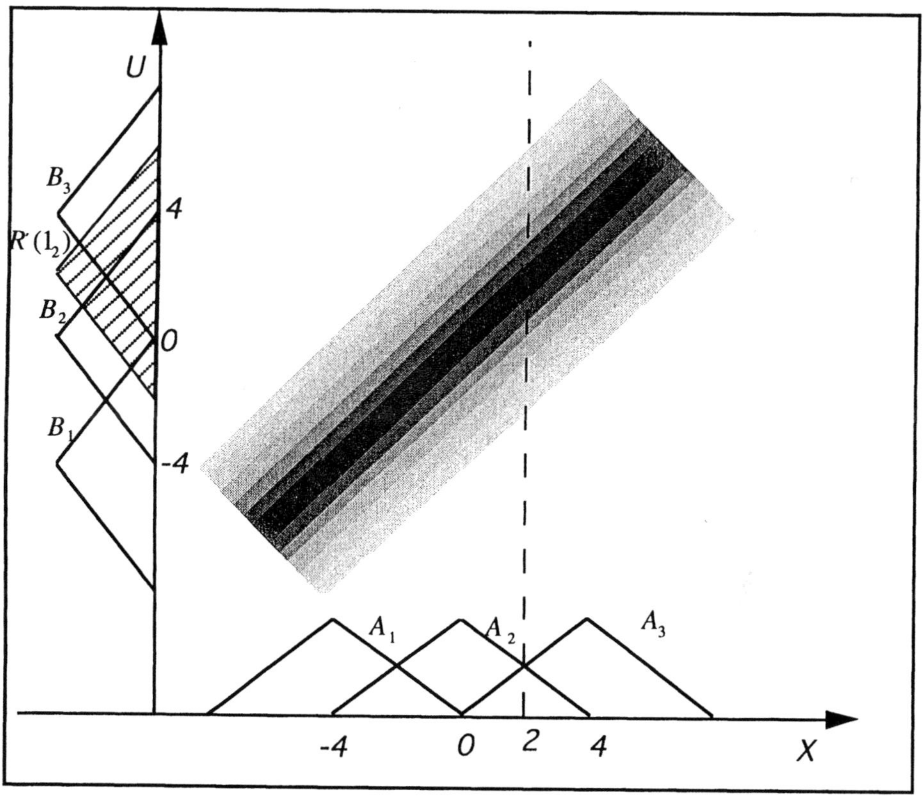

Figure 4 Fuzzy interpolation as an alternative inference method. The explanation is given in the text.

Fuzzy Lagrange, which is the subject of the following section, is a method to construct a continuous fuzzy valued function which coincides on n given points with n given fuzzy values. If fuzzy piecewise Lagrange interpolation is applied to the data points (-4, B_1), (0, B_2) and (4, B_3), the image of the input $x = 2$ is exactly the result obtained by computing the image of the fuzzy singleton 1_2 under the relation R'. This is because the relation R' has a true linear behaviour. In general, we can only conclude that if the fuzzy partitioning is taken infinitely dense, the fuzzy interpolation converges to the underlying fuzzy relation (or fuzzy function) which underlies the sample data, see below. It should be noted that this is generally not the case if the compositional rule of inference is applied to compute the image of a fuzzy singleton under a fuzzy graph [8].

4 AN INFERENCE MECHANISM BASED ON FUZZY PIECEWISE INTERPOLATION

In [2] fuzzy controllers are looked at as fuzzy input-output mappings. The fuzzy output of the controller is computed based on a weighted sum with the weights obtained by an equality measure. It is shown that any uniformly continuous function on a compactum can be approximated by a fuzzy controller using the inference method as mentioned just above. The resulting fuzzy controller is however not continuous with the respect to the particular metrics used. The inference method we propose is based on linear Lagrange interpolation

Fuzzy Lagrange interpolation [5] is an answer to the following problem. "Suppose we are given $m + 1$ points $x_0, x_1, \ldots, x_m$ in $\Re$, and for each of these points a 'fuzzy value' Y_i in $\Re$, rather than a crisp one. Construct some function on $\Re$ with range also a collection of 'fuzzy values'; which coincides, on the given $m + 1$ points, with the given 'fuzzy values'; and which fulfils some natural 'smoothness' condition?"

Fuzzy values are in [5] defined as fuzzy sets on $\Re$ which are normalised, convex and compact. Below we give the definitions of convexity, normality and compactness.

Definition 1 *A fuzzy set A is said to be normalised if and only if*

$$\exists x \in X, \mu_A(x) = 1.$$

Definition 2 *A fuzzy set A is called convex if and only if its α-cuts A_α are convex. Or more formally, a fuzzy set A on X is convex if and only if*

$$\forall x_1 \in X, \forall x_2 \in X, \forall \lambda \in [0,1]$$
$$\mu_A(\lambda x_1 + (1-\lambda)x_2) \geq min(\mu_A(x_1), \mu_A(x_2)).$$

Definition 3 ([5]) *A fuzzy set A is compact with respect to a generated fuzzy topology on $\Re$, if and only if $\mu_A^{-1}[\alpha, 1]$ is compact $\forall \alpha \in]0,1]$.*

The set of all compact, normalised and convex fuzzy sets is denoted by $\Gamma(\Re)$.

The fuzzy Lagrange interpolant P, which meets the above requirements is constructed as follows:

$$P : \Re \rightarrow \Im(\Re)$$
$$x \rightarrow P(x)$$

with

$$P(x) : \Re \rightarrow [0,1]$$
$$y \rightarrow \sup\{\alpha \in]0,1] \mid \exists t \in \prod_{i=0}^{m} \mu_{y_i}^{-1}[\alpha, 1] \ : \ p^t(x) = y\}$$

where $\Im(\Re)$ denotes all fuzzy sets in $\Re$ and p^t the unique polynomial of degree smaller or equal to m which passes through the points $(x_i, b_i)_{i=0}^m$, i.e. $P(x_i) = b_i$, $\forall i = 0, \ldots, m$.

In [5] it is shown that the fuzzy valued function P, fulfils the following properties:

1) $\forall i = 0, \ldots, m \ : \ P(x_i) = Y_i$.

2) P is continuous with respect to the restriction of the hyperspace structure on $\Im(\Re)$ [5] as fuzzy topological structure on $\Gamma(\Re)$.

3) If $\forall i = 0, \ldots, m \ : \ Y_i = 1_{b_i}$ then $P(x) = 1_{p^b(x)}, \forall x \in \Re$, with $b = (b_1, \ldots, b_m)$. This means that the fuzzy interpolant reduces to the crisp interpolant if the fuzzy points reduce to crisp points.

The fuzzy Lagrange interpolant P can also be written as:

$$P : \mathfrak{R} \to \mathfrak{I}(\mathfrak{R})$$
$$x \to \frac{\prod_{j=0}^{m}(x - x_j)}{\prod_{j=0}^{m}(x_0 - x_j)} Y_0 + \cdots + \frac{\prod_{j=0, j\neq i}^{m}(x - x_j)}{\prod_{j=0, j\neq i}^{m}(x_0 - x_j)} Y_i + \cdots + \frac{\prod_{j=0}^{m-1}(x - x)}{\prod_{j=0}^{m-1}(x_0 - x_j)} Y_m$$

where the sum and scalar product are defined by the extension principle [12].

We now turn to a well known problem of global interpolation. This problem is called Runge's phenomena and we will illustrate it by means of the following example. Since fuzzy interpolation is a generalisation of conventional interpolation the phenomena also applies to the former. Suppose the values of the function

$$f : \mathfrak{R} \to \mathfrak{R}$$
$$x \to \frac{1}{1+25x^2}$$

are known at eleven equidistant points of the interval $[-1, 1]$. $x_i = -1 + \frac{2i}{m}, i = 0, 1, \ldots, m$ with $m = 10$.

The plain curve in Figure 5 shows the function f. The Lagrange interpolant on the 11 equidistant points is shown by the dashed curve. The agreement of the interpolant with f is bad, especially near the ends of the interval. This behaviour is typical for global equidistant interpolation on a large number of points. The agreement of the interpolant and the curve can be approved by considering interpolation in the so-called Tchebycheff abscissae. These points are defined as:

$$x_i = \cos \frac{2i+1}{m+1} . \frac{\pi}{2}, i = 0, 1, \ldots, m$$

In Figure 5 the interpolant on the Tchebycheff abscissae is shown by the dotted curve. The agreement is now much better, however still not good. This is because the function f is not at all suited for approximation by one polynomial over the entire interval. Actually there are a lot of functions for which this applies, even for very regular functions as the function f. It can be shown that the approximation is not necessarily made better by increasing the number of points on which the interpolant is taken. This problem is called Runge's phenomena and can theoretically be explained [9].

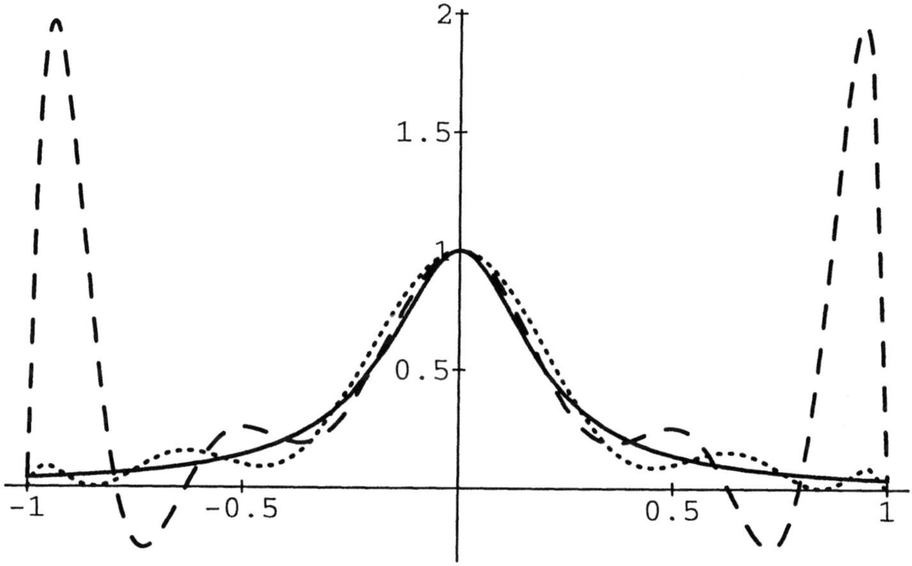

Figure 5 An illustration of Runge's phenomena. The explanation is given in the text.

On the other hand a lot of functions can suitably be approximated piecewise using many different simple functions in various parts of the interval. An example of piecewise approximations is the piecewise linear Lagrange interpolation. This is an approximation of a curve by secants which is obtained by taking the linear Lagrange interpolant on subintervals of the given interval. The piecewise linear Lagrange interpolant is defined as:

$$\begin{aligned} P &: [a,b] \to \Re \\ x &\to \frac{(x_{i+1}-x)}{(x_{i+1}-x_i)} y_i + \frac{(x-x_i)}{(x_{i+1}-x_i)} y_{i+1}, \forall x \in [x_i, x_{i+1}], 0 \le i \le m-1 \end{aligned}$$

where $x_0 = a$ and $x_m = b$. The m subintervals $[x_i, x_{i+1}]$ are called elements and the points $x_i, 0 \le i \le m$, are called knots or nodes. The technique of approximating a function by piecewise polynomial interpolation is called spline interpolation. During the last decades, this technique has got a solid mathematical basis. Piecewise linear Lagrange interpolation is the simplest form of spline interpolation. The function is approximated on each element by a polynomial of degree 1.

The piecewise linear Lagrange interpolant $p(x)$ can also be written as:

$$P : [a, b] \to \Re$$
$$x \to \sum_{i=0}^{m} y_j \varphi_j(x)$$

where $\varphi_j(x), 1 \leq i \leq m - 1$ are defined as:

$$\varphi_j : [a, b] \to \Re$$
$$x \to \begin{cases} 0 & \forall x \ni [x_{i-1}, x_{i+1}] \\ \frac{(x - x_{i-1})}{(x_i - x_{i-1})} & \forall x \in [x_{i-1}, x_i] \\ \frac{(x_{i+1} - x)}{(x_i - x_{i-1})} & \forall x \in [x_i, x_{i+1}] \end{cases}$$

and

$$\varphi_0 : [a, b] \to \Re$$
$$x \to \begin{cases} \frac{(x_1 - x)}{(x_1 - x_0)} & \forall x \in [x_0, x_1] \\ 0 & \forall x \ni [x_0, x_1] \end{cases}$$

and

$$\varphi_m : [a, b] \to \Re$$
$$x \to \begin{cases} \frac{(x - x_{m-1})}{(x_m - x_{m-1})} & \forall x \in [x_{m-1}, x_m] \\ 0 & \forall x \ni [x_{m-1}, x_m] \end{cases}$$

The functions $\varphi_i, 0 \leq i \leq m$, are so-called basis functions. φ_i and φ_{i+1} are triangular shaped functions as shown in Figure 6. Therefore, these basis functions can be considered as membership functions of fuzzy sets, more precisely of triangular fuzzy sets.

It is well known that by using piecewise linear Lagrange interpolation one can approximate arbitrarily well any continuous function on a compactum, provided that sufficient subintervals are considered. Important hereby is to recognize the constructive nature of the result. This opposed to the non-constructive Weierstrass approximation theorem that states that every continuous function can be approximated to any degree of accuracy by a single polynomial.

We will now generalize the concept of piecewise linear Lagrange interpolation to fuzzy piecewise linear Lagrange interpolation based on the results reported in [5]. We will show that a continuous fuzzy valued function F can be approximated to any degree of accuracy by a fuzzy piecewise linear Lagrange interpolant. The fuzzy piecewise linear Lagrange interpolant is

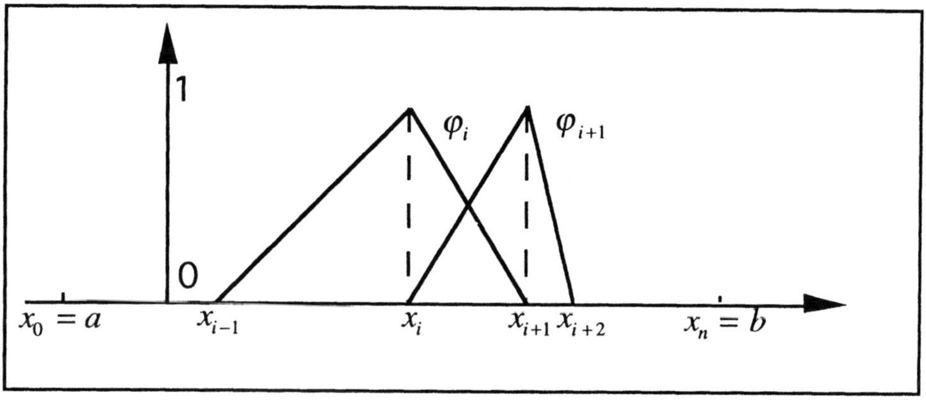

Figure 6 Basis function for piecewise linear Lagrange interpolation. The explanation is given in the text.

defined as:

$$P : [a, b] \to \Gamma(\Re)$$
$$x \to \sum_{i=0}^{m} Y_j \varphi_j(x)$$

with $Y_i = F(x_i), 0 \leq i \leq m$, the fuzzy values of the function F in the m knots.

Theorem 1 *For each $x \in [a, b]$ we have $F(x) \in \Gamma(\Re)$, i.e. the interpolant is well defined. Moreover we have:*

1) $\forall i = 0, \ldots, m \; : \; P(x_i)) = Y_i$.

2) *P is continuous with respect to the hyperspace structure on $\Im(\Re)$ as fuzzy topological structure on $\Gamma(\Re)$.*

3) *If $\forall i = 0, \ldots, m \; : \; Y_i = 1_{y_i}$ then the fuzzy piecewise interpolant reduces to the crisp piecewise linear Lagrange interpolant.*

proof

We first prove 1)

$$\forall t \in \Re, \mu_{P(x_i)}(t)$$

$$= \sup_{(q_0,\ldots,q_m)\in\Re^m} \left\{ min \left\{ \mu_{Y_j}(q_j) \right\}_{0\leq j\leq m} \mid t = \sum_{j=0}^{m} \varphi_j(x_i) q_j \right\}$$

$$
\begin{aligned}
&= \sup_{(q_0,\ldots,q_m)\in\Re^m} \left\{ min \left\{ \mu_{Y_j}(q_j) \right\}_{0\le j\le m} \mid t = q_j \right\} \\
&= \mu_{Y_j}(t)
\end{aligned}
$$

The first step applies since only the ith basis function has a value different from 0, namely 1, at x_i. The second step is valid because the fuzzy values Y_i are normalised.

Since the fuzzy piecewise linear Lagrange interpolant is a global Lagrange interpolant on two points when considered per element, it follows from [5] and 1) that P is well defined.

The continuity of P per element follows from the continuity of global Lagrange interpolation. The global continuity in $[a, b]$ follows from 1).

Proposition 3) can be proven as:

$$
\begin{aligned}
&\forall t \in \Re, \mu_{P(x_i)}(t) \\
&= \sup_{(q_0,\ldots,q_m)\in\Re^m} \left\{ min \left\{ \mu_{Y_j}(q_j) \right\}_{0\le j\le m} \mid t = \sum_{j=0}^{m} \varphi_j(x_i) q_j \right\} \\
&= max \left\{ min \left\{ \mu_{Y_j}(y_j) \right\}_{0\le j\le m} \mid t = \sum_{j=0}^{m} \varphi_j(x_i) y_j, \right. \\
&\qquad \left. \sup_{(q_0,\ldots,q_m)\in\Re^m/(y_0,\ldots,y_m)} \left\{ min \left\{ \mu_{Y_j}(q_j) \right\}_{0\le j\le m} \mid t = \sum_{j=0}^{m} \varphi_j(x_i) q_j \right\} \right\} \\
&= \begin{cases} 0 & \text{if } t \ne \sum_{j=0}^{m} \varphi_j(x_i) y_j \\ 1 & \text{if } t = \sum_{j=0}^{m} \varphi_j(x_i) y_j \end{cases}
\end{aligned}
$$

i.e., $P(x) = 1_{p(x)}$ with $p(x)$ the crisp piecewise linear Lagrange interpolant on the data points, $(x_i, y_i), 0 \le i \le m$. Q.E.D.

Now we will prove that if $F : [a, b] \to \Im(\Re)$ is continuous with respect to the hyperspace structure on $\Im(\Re)$ as fuzzy topological structure on $\Gamma(\Re)$, then the function F can be approximated arbitrarily well by a fuzzy piecewise linear Lagrange interpolant if the number of subintervals is taken sufficiently large. Before giving the proof we first introduce the following definition.

Definition 4 *[4] Let Y be any element of $\mathfrak{I}(\mathfrak{R})$ then the set $\mathcal{G}(Y) = \{(t,\alpha) \mid \alpha \leq \mu_Y(t)\}$, is called the endograph of Y.*

In [5] it has been shown that since $\mathfrak{R}$ is equipped with its usual generated fuzzy topology and compact fuzzy sets are closed, the continuity of a mapping $F : \mathfrak{R} \to \Gamma(\mathfrak{R})$ follows from the continuity of the mapping

$$\begin{aligned} \mathcal{G} \circ F &: \mathfrak{R} \to \tau(\mathfrak{R} \times I) \\ x &\to \mathcal{G}(F(x)) \end{aligned}$$

with $\tau(\mathfrak{R}\times I)$ the set of closed subsets of $\mathfrak{R}\times I$ endowed with the Hausdorff-Bourbaki hyperspace structure.

The Hausdorff-Bourbaki uniformity [5] on $\tau(\mathfrak{R}\times I)$ is generated by the basis

$$\{B_\epsilon \otimes B_\epsilon \mid \epsilon > 0\}$$

where

$$B_\epsilon = \{(x,y) \mid |x-y| < \epsilon\}$$

and

$$A \otimes B = \{((x,t)(x',t')) \mid (x,x') \in A, (t,t') \in B\}$$

We will now prove that any continuous function $F : [a,b] \to \Gamma(\mathfrak{R})$ can be approximated arbitrarily well by a fuzzy piecewise linear Lagrange interpolant.

Theorem 2 *For any continuous function $F : [a,b] \to \Gamma(\mathfrak{R})$ and any $\epsilon > 0$ there exists a fuzzy piecewise linear Lagrange interpolant $P(x)$ such that $\forall x \in [a,b], (\mathcal{G}(P(x)), \mathcal{G}(F(x))) \in B_\epsilon \otimes B_\epsilon$.*

proof

Because $F(x)$ is continuous we know that for any $\epsilon > 0$ there exist m subintervals $[x_i, x_{i+1}]$ of $[a,b]$ so that

$$\forall x, x' \in [x_i, x_{i+1}], 0 \leq i \leq m-1, (\mathcal{G}(F(x)), \mathcal{G}(F(x'))) \in B_{\epsilon/2} \otimes B_{\epsilon/2}$$

We will now proof that

$$\forall x \in [x_i, x_{i+1}], i, 0 \leq i \leq m-1, (\mathcal{G}(P(x)), \mathcal{G}(F(x))) \in B_{\epsilon/2} \otimes B_{\epsilon/2}$$

For this purpose we first proof that

$$\forall x \in [x_i, x_{i+1}], 0 \leq i \leq m-1, (\mathcal{G}(P(x)), \mathcal{G}(F(x_i))) \in B_{\epsilon/2} \otimes B_{\epsilon/2}$$

this means we have to prove that

$$\forall x \in [x_i, x_{i+1}], 0 \le i \le m-1, \forall (t,\alpha) \in \mathcal{G}(P(x)), \exists (t',\alpha') \in \mathcal{G}(F(x_i))$$

so that

$$|t - t'| \le \epsilon/2 \text{ and } |\alpha - \alpha'| \le \epsilon/2,$$

i.e.

$$\mathcal{G}(P(x)) \subseteq B_{\epsilon/2} \otimes B_{\epsilon/2}\mathcal{G}(F(x_i)) \tag{11.1}$$

and

$$\forall x \in [x_i, x_{i+1}], 0 \le i \le m-1, \forall (t,\alpha) \in \mathcal{G}(F(x_i)), \exists (t',\alpha') \in \mathcal{G}(P(x))$$

so that

$$|t - t'| \le \epsilon/2 \text{ and } |\alpha - \alpha'| \le \epsilon/2,$$

i.e.

$$\mathcal{G}(F(x_i)) \subseteq B_{\epsilon/2} \otimes B_{\epsilon/2}(\mathcal{G}(P(x))) \tag{11.2}$$

We first prove 11.1. Take any $x \in [x_i, x_{i+1}]$, $0 \le i \le m-1$ and any $(t,\alpha) \in \mathcal{G}(P(x))$. By definition we have that

$$\mu_{P(x)}(t) = \sup \{\alpha \in]0,1] \mid \exists (y_i, y_{i+1}) \in Y_{i_\alpha} \times Y_{i+1_\alpha} : p^{y_i,y_{i+1}}(x) = t\}$$

with

$$\begin{array}{l} p^{y_i,y_{i+1}} : \mathfrak{R} \to \mathfrak{R} \\ x \to \frac{(x_{i+1}-x)}{(x_{i+1}-x_i)} y_i + \frac{(x-x_i)}{(x_{i+1}-x_i)} y_{i+1} \end{array}$$

Furthermore we know form [5] that $\forall x \in [x_i, x_{i+1}], 0 \le i \le m-1$ and $\forall t \in \mathfrak{R}$,

$$\mu_{P(x)}(t) = \sup \{\alpha \in]0,1] \mid \exists (y_i, y_{i+1}) \in Y_{i_\alpha} \times Y_{i+1_\alpha} : p^{y_i,y_{i+1}}(x) = t\},$$

this means that the supremum is attained. This implies that

$$\exists (t_i,\alpha) \in \mathcal{G}(F(x_i)), \exists (t_{i+1},\alpha) \in \mathcal{G}(F(x_{i+1})),$$

so that

$$t = \frac{(x_{i+1}-x)}{(x_{i+1}-x_i)} t_i + \frac{(x-x_i)}{(x_{i+1}-x_i)} t_{i+1}$$

holds.

Since

$$(\mathcal{G}(F(x_i)), \mathcal{G}(F(x_{i+1}))) \in B_{\epsilon/2} \otimes B_{\epsilon/2}$$

we know that

$$\exists (t', \alpha') \in \mathcal{G}(Y_i)$$

so that

$$|t_{i+1} - t'| \leq \epsilon/2 \text{ and } |\alpha - \alpha'| \leq \epsilon/2$$

First suppose $t_i \leq t_{i+1}$, then if $t' \leq t$ it follows from $t' \leq t \leq t_{i+1}$ that

$$|t - t'| \leq \epsilon/2 \text{ and } |\alpha - \alpha'| \leq \epsilon/2$$

therefore

$$\mathcal{G}(P(x)) \subseteq B_{\epsilon/2} \otimes B_{\epsilon/2}(\mathcal{G}(F(x_i)))$$

else if $t' > t$ then since

$$(t_i, \alpha) \in \mathcal{G}(F(x_i)) \text{ and } (t', \alpha') \in \mathcal{G}(F(x_i)),$$

we have that

$$(t_i, \min(\alpha, \alpha') \in \mathcal{G}(F(x_i))$$

because fuzzy values are convex.

Moreover

$$|t - t| \leq \epsilon/2 \text{ and } |\alpha - \min(\alpha, \alpha')| \leq \epsilon/2$$

therefore we also have that

$$\mathcal{G}(P(x)) \subseteq B_{\epsilon/2} \otimes B_{\epsilon/2} (\mathcal{G}(F(x_i)))$$

For $t_i > t_{i+1}$ we can proceed analogically, such that we can conclude that 11.1 holds.

We now prove the other inclusion 11.2. Take any $(t, \alpha) \in \mathcal{G}(F(x_i)), 0 \leq i \leq m - 1$. We know that

$$\exists(t_{i+1}, \alpha_{i+1}) \in \mathcal{G}(F(x_{i+1})) \text{ with } |t - t_{i+1}| \leq \epsilon/2 \text{ and } |\alpha - \alpha_{i+1}| \leq \epsilon/2$$

take

$$t' = \frac{(x_{i+1}-x)}{(x_{i+1}-x_i)} t + \frac{(x-x_i)}{(x_{i+1}-x_i)} t_{i+1} \text{ and } \alpha' = \min(\alpha, \alpha_{i+1})$$

then we have that

$$|t' - t_{i+1}| \leq \epsilon/2, |\alpha - \alpha_{i+1}| \leq \epsilon/2 \text{ and } (t', \alpha') \in \mathcal{G}(P(x))$$

i.e.

$$(t, \alpha) \subseteq B_{\epsilon/2} \otimes B_{\epsilon/2}(\mathcal{G}(P(x)))$$

which proves 11.2.

We now prove that

$$\forall x \in [x_i, x_{i+1}], 0 \leq i \leq m-1, (\mathcal{G}(F(x)), \mathcal{G}(P(x))) \in B_{\epsilon/2} \otimes B_{\epsilon/2}$$

By construction of the intervals, we have that

$$\forall x \in [x_i, x_{i+1}], 0 \leq i \leq m-1, (\mathcal{G}(F(x)), \mathcal{G}(F(x_i))) \in B_{\epsilon/2} \otimes B_{\epsilon/2} \quad (11.3)$$

and above we have proven that

$$\forall x \in [x_i, x_{i+1}], 0 \leq i \leq m-1, (\mathcal{G}(F(x_i)), \mathcal{G}(P(x))) \in B_{\epsilon/2} \otimes B_{\epsilon/2} \quad (11.4)$$

Take any $x \in [x_i, x_{i+1}], 0 \leq i \leq m-1$, and $(t, \alpha) \in \mathcal{G}(P(x))$, then since 11.3

$$\exists(t_i, \alpha_i) \in \mathcal{G}(F(x_i)) \text{ so that } |t - t_i| \leq \epsilon/2 \text{ and } |\alpha - \alpha_i| \leq \epsilon/2$$

and because of 11.4

$$\exists(t', \alpha') \in \mathcal{G}(F(x)) \text{ so that } |t_i - t'| \leq \epsilon/2 \text{ and } |\alpha_i - \alpha'| \leq \epsilon/2$$

Therefore we have that

$$\exists(t', \alpha') \in \mathcal{G}(F(x)) \text{ so that } |t - t'| \leq \epsilon \text{ and } |\alpha - \alpha'| \leq \epsilon$$

which proves that

$$\forall x \in [x_i, x_{i+1}], 0 \leq i \leq m-1, \mathcal{G}(P(x)) \subset B_\epsilon \otimes B_\epsilon(\mathcal{G}(F(x)))$$

on the analogy of the above we can prove that

$$\forall x \in [x_i, x_{i+1}], 0 \le i \le m-1, \mathcal{G}(F(x)) \subset B_\epsilon \otimes B_\epsilon(\mathcal{G}(P(x)))$$

Thus we can conclude that for any continuous function $F(x)$ and any $\epsilon > 0$ we can construct a fuzzy piecewise linear Lagrange interpolant such that

$$\forall x \in [x_i, x_{i+1}], 0 \le i \le m-1, (\mathcal{G}(P(x)), \mathcal{G}(F(x))) \in B_\epsilon \otimes B_\epsilon \text{ Q.E.D.}$$

The idea of piecewise linear Lagrange interpolation can easily be generalised to the multi-dimensional case. Below we consider the case for fuzzy piecewise bi-linear Lagrange interpolation on a rectangular interpolation net. The generalisation to the multi-dimensional case is straightforward.

Let $C = [a,b] \times [c,d]$ and π_x and π_y a partitioning of $[a,b]$ and $[c,d]$, respectively so that $\pi_x : a = x_0 < x_1 < \ldots < x_n = b$ and $\pi_y : c = y_0 < y_1 < \ldots < y_n = d$. Based on these two partitionings, a partitioning of C can be obtained noted as $\pi = \pi_x \pi_y$. The partitioning π contains $(n+1).(m+1)$ knots, i.e. $\pi : \{(x_i, y_i) \mid 0 \le i \le n, 0 \le j \le m\}$

If $F(x,y)$ is a fuzzy valued function defined on C, or at least on all knots of π, then the bi-linear fuzzy Lagrange interpolant $P_{\pi,f}$ with respect to the partitioning π and the function F is given by:

$$\begin{aligned} P : C &\to \Gamma(\mathfrak{R}) \\ (x,y) &\to \sum_{i=0}^{n} \sum_{j=0}^{m} Y_{ij}.\varphi_{ij}(x,y) \end{aligned}$$

where the basis functions $\varphi_{ij}(x,y), 0 \le i \le n, 0 \le j \le m$ are defined as: $\varphi_{ij}(x,y) = \varphi_i(x).\varphi_j(y)$ with $\varphi_i(x)$ and $\varphi_j(y)$ the one-dimensional basis functions as defined above and Y_{ij} the fuzzy values of F on the knots $(x_i, y_i), 0 \le i \le n, 0 \le j \le m$.

In an analogous way as above, it can be proven that any continuous function can be approximated to any degree of accuracy on a compactum by a piecewise n-linear fuzzy Lagrange interpolant.

5 CONCLUSION

In this paper we have argued that fuzzy controllers should be looked at as fuzzy valued mappings rather than crisp valued mappings. By doing so, applying an explicit interpolation method becomes very natural. If the antecedents of the fuzzy rules are chosen to be dual to a piecewise Lagrange interpolation net, and this is indeed very often the case, fuzzy piecewise Lagrange interpolation can be applied. It has been shown that this alternative inference method yields universal approximators which are continuous.

Acknowledgements

I would like to take the opportunity to thank Prof. Etienne Kerre. Not only did he introduced me with great enthusiasm into the field of fuzzy set theory, he also taught me to continuously question everything in my work as a scientific researcher.

REFERENCES

[1] J.J.. Buckley. Sugeno type controllers are universal controllers. *Fuzzy Sets and Systems*, 53:299–303, 1993.

[2] J.J.. Buckley and Y. Hayashi. Fuzzy input-output controllers are universal approximators. *Fuzzy Sets and Systems*, 58:273–278, 1993.

[3] Aubin J.P. *Viability Theory*. Systems and Control: Foundations and Applications. Brikhäuser, 1991.

[4] P.E. Kloeden. Compact supported endographs and fuzzy sets. *Fuzzy Sets and Systems*, 4:621–633, 1980.

[5] R. Lowen. A fuzzy lagrange interpolation theorem. *Fuzzy Sets and Systems*, 34:33-38, 1990.

[6] E.H. Mamdani. Application of fuzzy algorithm for control of a simple dynamic plant. *Proceedings of the IEEE*, 121:1585–1588, 1974.

[7] A. Nowé. Some mathematical aspects of fuzzy controllers. Technical Report 1, Department INFO, Vrije Universiteit Brussel, 1991.

[8] A. Nowé. *Synthesis of śafefuzzy controllers based on reinforcement learning.* PhD thesis, Vrije Universiteit Brussel, 1994.

[9] A. Ralston and P. Rabinowitz. *A first course in numerical analysis.* Mathematical Series. McGraw-Hill International Editions, 1986.

[10] L. Wang. Fuzzy systems are universal approximators. In *the second IEEE International conference on Fuzzy Systems*, pages 1163–1169, San Diego, 1992.

[11] L.A. Zadeh. Fuzzy sets. *Information Control*, 8:338–353, 1965.

[12] L.A. Zadeh. The concept of a linguistic variable and its application to approximate reasoning-I. *Information Sciences*, 8:199–249, 1975.

[13] L.A. Zadeh. The concept of a linguistic variable and its application to approximate reasoning-III. *Information Sciences*, 9:43–80, 1975.

[14] L.A. Zadeh. The role of fuzzy logic and soft computing in the conception and design of intelligent systems. In *Invited talk at the 8th Austrian Artificial Intelligence Conference, FLAI'93*, Linz, 1993.

[15] H.J. Zimmermann. Recent development of fuzzy technology in europe, japan and the u.s.a. In *Opening Keynote Speech at the Third International Conference on Industrial Fuzzy Control and Intelligent Systems*, Houston, 1993.

12

COMPARATIVE STUDY OF FUZZY CONTROL, NEURAL NETWORK CONTROL AND NEURO-FUZZY CONTROL

Jelena Godjevac

Microcomputing Laboratory
Swiss Federal Institute of Technology
IN-F Ecublens
1015 Lausanne, Switzerland

ABSTRACT

The goal of this work is to compare fuzzy, neural network and neuro-fuzzy approaches to the control of mobile robots. The first part of this paper is devoted to the formal framework of fuzzy controllers. Results of an example of their use for a mobile robot are discussed. As an experimental platform, the Khepera mobile robot is used. The same example is studied using artificial neural networks. For that purpose, fundamentals of artificial neural networks are outlined. Similarities and differences between fuzzy systems and neural networks are discussed as well as the respective advantages and drawbacks, and reasons for merging these two approaches are developed. Three models of fuzzy neurons, the learning methods and an architecture of neuro-fuzzy controller are presented. A learning procedure for the controller is described. To conclude, the application of a neuro-fuzzy controller on Khepera is discussed.

1 INTRODUCTION

The word *robot* (robota) has Slavic origins and means work. Robots are built to replace human beings in performing tasks that humans cannot or prefer not to do. They can clean dangerous areas in nuclear plants, transport goods, clean floors, search for mines or harvest crops in fields. Present technology is far too deficient to make robots accomplish these complicated tasks. This paper surveys very simple controllers based on *fuzzy logic* and/or *neural networks* for the *control of mobile robots*. The purpose is to investigate those solutions most likely to be found in nature which, notwithstanding an astonishing simplicity can produce complex behaviours. For instance, in spite of a relatively simple nervous system, an ant may show many different behaviours. In the same way as animals in their natural environment, real mobile robots have to achieve tasks in unpredictable conditions:

Fuzzy logic was one of the major developments of *Fuzzy Set Theory* and was primarily designed to represent and reason with knowledge that cannot be expressed by quantitative measures. The main idea of algorithms based on fuzzy logic -loosely called *fuzzy systems* or *fuzzy controllers-* is to imitate the human reasoning process to control ill-defined or hard-to-model plants. Fuzzy inference systems model the qualitative aspects of human knowledge through *linguistic if-then rules* [Zad65]. They have to capture the imprecision of the reasoning process without using exact quantitative analysis. We do not pretend the fuzzy logic approach to be a methodology borrowed from biology or having biological foundations. It is a rigorous mathematical branch offering solutions for control. An example of *fuzzy control* is shown in Section 3.1. We report experiments with a small mobile robot, Khepera, developed in our laboratory.

Neural networks were developed as an attempt to realize simplified mathematical models of brain-like systems. The key advantage is their ability to learn from examples instead of requiring an algorithmic development from the designer. One of the main motivations of the interest in neural networks is to build machines with the intelligence of biological organisms. In Section 4.3, we describe a neural network approach applied to the Khepera robot.

The main drawback of fuzzy controllers is the lack of a systematic methodology for their design. Usually, tuning parameters is a time consuming task. Neural network learning techniques can automate this process, significantly reduce development time, and result in a better performance. The merge of neural networks and fuzzy logic led to the creation of *neuro-fuzzy controllers* which are one of the most popular research fields today. Lotfi Zadeh proposed the name *soft computing* for these techniques [Zad94]. The principal constituents of soft computing are: fuzzy logic, neural networks and probabilistic reasoning. An example of a neuro-fuzzy controller is presented in Section 6 as well as its implementation on Khepera.

2 FUZZY CONTROL

Humans, when making decisions tend to work with vague or imprecise concepts which can often be expressed linguistically. One of the ways of modelling this decision making process has been proposed by Zadeh [Zad65] and is based on the *Theory of Approximate Reasoning* which enables certain classes of linguistic statements to be treated mathematically. First investigations by Prof. Zadeh [Zad73] concerned how to use mathematical tools to represent a human language and human knowledge. He was the first to introduce the fuzzy set theory in the field of control. He proposed that all problems in which the data, the objectives and the constraints are too complex, or too ill-defined to admit a precise mathematical analysis have to be treated by approximate (fuzzy) solutions. Fuzzy control has received a lot of attention since it was applied for the first time by Mamdani and Assilian [Mam75].

The main argument in favour of fuzzy control is that conventional mathematical tools are not well suited for dealing with ill-defined and uncertain systems that are typically difficult to model. Some authors argue that fuzzy controllers are suitable where a precise mathematical model of the process being controlled is not available [Kic78, Li88]. However, it is not possible to build a controller which does not utilize environmental information.

2.1 Linguistic variables and membership functions

Linguistic variables represent process states and control variables. We can use *fuzzy sets* to represent them. Every fuzzy set can be represented by its *membership function*. If the referential set is a finite set, membership values are discrete values defined in the range [0,1]. If the referential set is an infinite set, we can represent these values as a continuous membership function. In general, the shape of a membership function depends on the application and can be monotonous, triangular, trapezoidal or bell-shaped as shown in Figure 1.

One of the first steps in every fuzzy application is to define the *universe of discourse* (dynamic range) for every linguistic variable. Every fuzzy set on a universe of discourse represents one *linguistic value* or *label*. Figure 2 illustrates one example of the universe of the discourse for the linguistic variable speed. Linguistic values, which define this variable are: NEGATIVE BIG, NEGATIVE MEDIUM, etc.

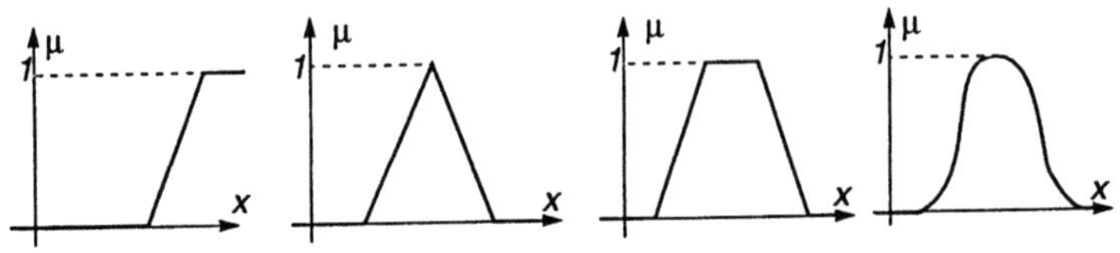

Figure 1 Different shapes of membership functions: monotonous, triangular, trapezoidal and bell-shaped

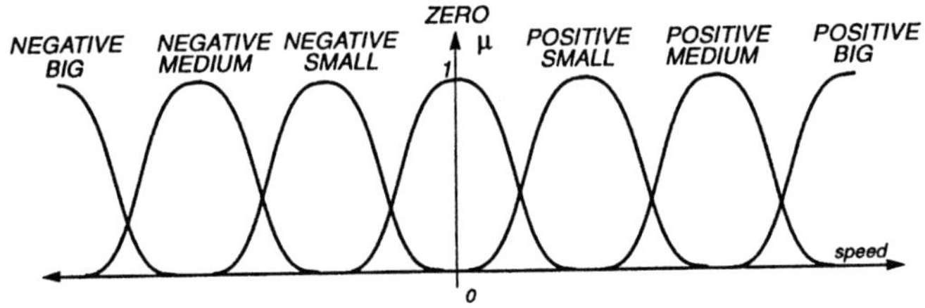

Figure 2 Universe of discourse for linguistic variable: speed

2.2 Notion of linguistic rule

As mentioned above, the principal idea of fuzzy logic systems is to express human knowledge in the form of linguistic if-then rules. Every rule has two parts:

- antecedent part (premise), expressed by ***If***...
- consequent part, expressed by ***then***...

The antecedent part is the description of the state of the system which should turn on the rule, and the consequent is the action that the operator who controls the system must take. There are several forms of if-then rules. The general one is:

> ***If*** a set of conditions is satisfied ***then*** a set of consequences can be inferred.

Zadeh was the first to introduce a notion of fuzzy rule [Zad73] of the form:

> R_1: ***If*** x is A, ***then*** y is B.

A and B are linguistic values or labels characterized by appropriate membership functions of fuzzy sets defined on the universe of discourse of the linguistic variables x and y respectively. Every control rule is implemented by a *fuzzy implication (fuzzy relation)* and is defined as follows:

$$\mu_{R_1} = \mu_{A \to B}(x, y) = \mu_A(x) \to \mu_B(y) \tag{12.1}$$

where $\mu_{R_1} = \mu_{A \to B}(x, y)$ is the *truth value* for the statement R_1 and μ_A and μ_B are the *levels of membership* (*membership values*) of the variables x and y in fuzzy sets A and B respectively.

There are many ways in which the fuzzy implication may be defined. Zadeh suggested the *compositional rule of inference* [Zad73]. Nearly 40 distinct fuzzy implication functions have been described in the literature [Lee90, Zim90].

Takagi and Sugeno [Tak83] proposed a form which has fuzzy sets only in the premise part of the rule, and the consequent part is described by a non-fuzzy equation of the input variable:

R_2: ***If*** x is A, ***then*** y is $k_1 x^2 + k_2 x + k_3$.

Like in Zadeh's rule, A is the linguistic label. k_1, k_2, k_3 are predefined constants.

The generalization of a control rule which has two conditions in the antecedent part has the following form:

R_3: ***If*** x is A ***and*** y is B, ***then*** z is C.

Following the previous definition (eq. 12.1) we can conclude that the truth value of this statement is:

$$\mu_{R_3} = \mu_{A\,and\,B \to C}(x, y, z) = [\mu_A(x) \ \ \textbf{\textit{and}} \ \ \mu_B(y)] \to \mu_C(z) \tag{12.2}$$

where A, B and C are fuzzy sets defined on the universes of discourse for x, y and z. μ_A, μ_B and μ_C are the levels of membership of the variables x, y and z in fuzzy sets A, B, C respectively.

For the fuzzy intersection (***and*** operator) Zadeh [Zad65] suggested the ***min*** operator. Generally, intersection operators are called *T-norms* and union operators are called *T-conorms*.

2.3 Reasoning based on fuzzy logic vs. reasoning based on Boolean logic

In order to present the idea of fuzzy control to the reader, we will give a very simple toy problem from the field of mobile robotics.

Suppose that the goal of our experiment is to control a mobile robot shown in Figure 3 using a trivial algorithm. The robot is equipped with a single

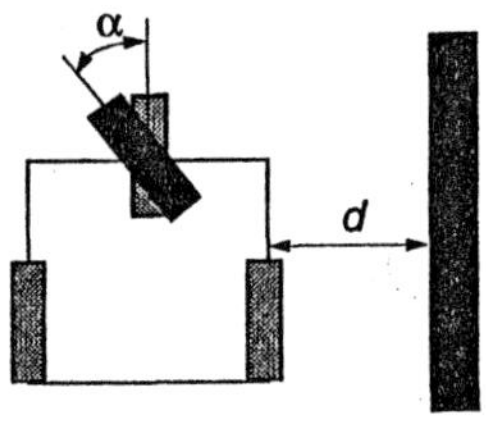

Figure 3 Mobile robot

sensor which gives information about the distance between the robot and the obstacle. The first step in the design of a controller is to define its inputs and outputs i.e. linguistic variables.. In this example, we consider a one input/one output controller: the input is a distance between the robot and the obstacle (d) and the output is the steering angle of the robot's wheel (α).

If we apply the Boolean logic approach, we can establish the following linguistic rules:

- ***If*** the distance between the robot and the obstacle is less than 10 cm, ***then*** steer for +10°.
- ***If*** the distance between the robot and the obstacle is more than 10 cm, ***then*** steer for -10°.
- ***If*** the distance between the robot and the obstacle is 10 cm, ***then*** go straightforward.

These Boolean rules are intended to result in a wall following behaviour. The universes of discourse for these two linguistic variables are shown in Figure 4. Since we are in the binary domain, membership functions are crisp. This algorithm is quite primitive and the robot will unavoidably oscillate.

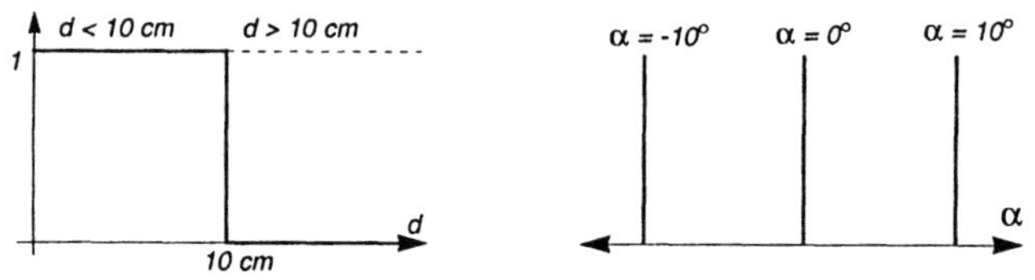

Figure 4 Universe of discourse for the distance and the steering angle

For the fuzzy controller, we define the following linguistic rules:

- ***If*** the distance between the robot and the obstacle is less than 10 cm, ***then*** turn to the left (α is negative).
- ***If*** the distance between the robot and the obstacle is more than 10 cm, ***then*** turn to the right (α is positive).
- ***If*** the distance between the robot and the obstacle is nearly 10 cm, ***then*** keep the direction (α is around 0).

In fact, with these rules, we defined three linguistic values for the linguistic variable distance: More than 10 cm, Nearly 10 cm (optimal distance) and Less than 10 cm. For the steering angle, we have chosen three linguistic values: Negative for negative angles, Zero for going straightforward and Positive for positive angles.

With this controller, we expect to have a more "intelligent" behaviour. The universes of discourse for the distance and the steering angle are shown in Figure 5. The chosen membership functions have triangular and mono-

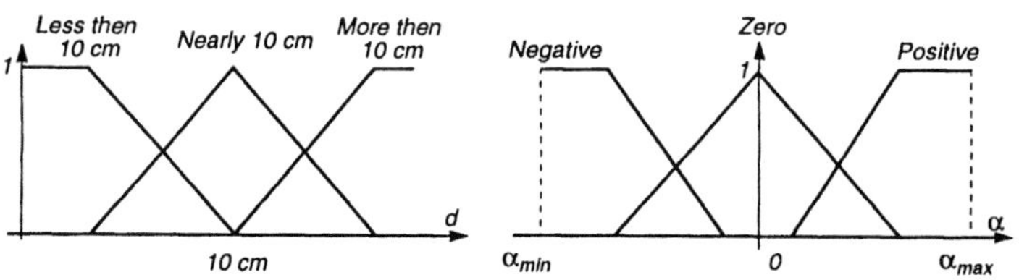

Figure 5 Universes of discourse for the distance and the steering angle

tonous shapes. We can calculate the control value by applying the fuzzy implication we mentioned in Section 2.2.

Suppose the sensor gave 5 cm for the distance. This distance belongs to the fuzzy set Less than 10 cm with a degree of truth 0.7, or 70% (Figure 6), and

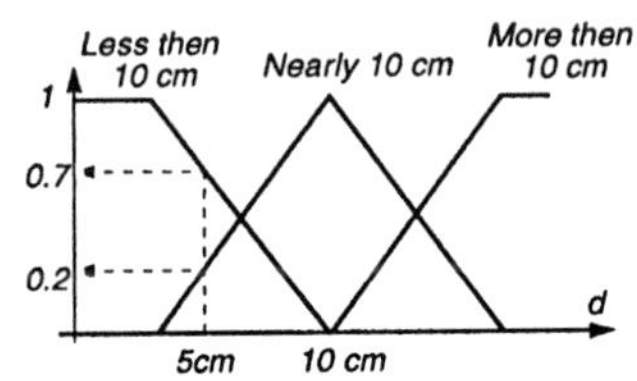

Figure 6 Fuzzification

we can say that the truth value of the statement It is true that the distance is less than 10 cm is 0.7. At the same time, it is true but only 0.2, or 20%, that this distance is Nearly 10 cm.

We apply the rules. The antecedent part of the first rule is satisfied but not 100%. It is true only 70%. This value is the firing strength of this rule. It means that the steering angle has to be Negative 70%. For the second rule, we will proceed in an analogous way. The distance is Nearly 10 cm, but only 20%. The robot has to keep the direction with this degree of truth. The premise part of the third rule is not satisfied so the rule is not used.

We can conclude that the robot has to steer 70% to the left and 20% to go straightforward. It means that we have to "modulate" the appropriate fuzzy sets with these two values. The methods mostly used are: *linear modulation* (Figure 7a) and *modulation by clipping* (Figure 7b).

There are several methods to find a crisp value of the command. Figures 7a and 7b illustrate two of them: *centre of gravity* and *mean of maxima*.

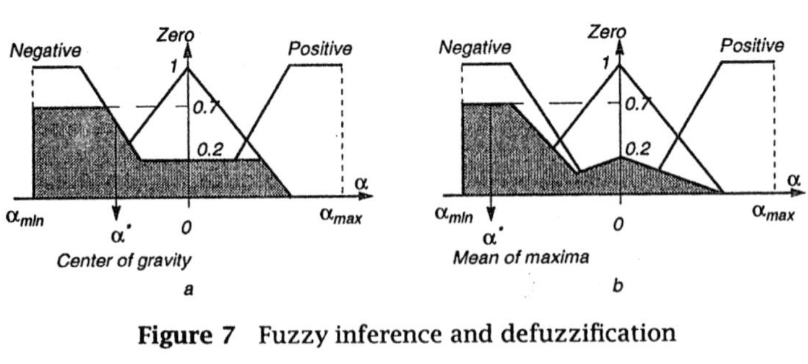

Figure 7 Fuzzy inference and defuzzification

Although trivial, this example is introduced to give a general idea of fuzzy reasoning and the design of a fuzzy controller.

2.4 Procedure for fuzzy reasoning

From the previous example, we summarize the steps in fuzzy reasoning:

1. **Fuzzification**: to each measure of an input variable is attributed a membership value for all the fuzzy sets defined over the universe of discourse of that variable.
2. **Application of the T-norm** (usually this operator is ***min*** or product) on the membership values of the premise part of the rules to get *firing strength* or the *weight* for each rule ($\mu_A(x)$ in eq. (12.1) or $[\mu_A(x)$ ***and*** $\mu_B(y)]$ in eq. (12.2)). [1]
3. **Generation of the consequent value** ($\mu_B(y)$ in eq. (12.1) or $\mu_C(z)$ in eq. (12.2)) of each rule. It can be crisp or fuzzy.
4. **Defuzzification**: generate the crisp output values.

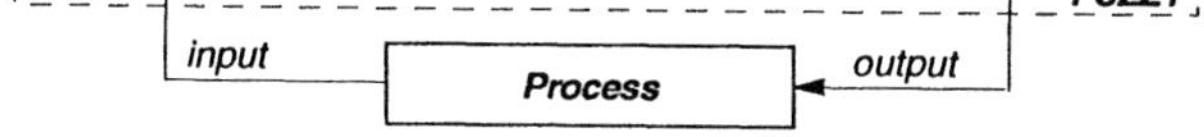

Figure 8 General structure of fuzzy inference system

This means that every fuzzy controller is composed of four principal blocks (Figure 8):

1. **Knowledge base** (parameters of membership functions and definitions of rules)
2. **Fuzzification interface** (transformation of crisp inputs into membership values)

[1] In the previous example we had only one input and we didn't need to apply the T-norm.

3. **Decision making unit** (inference operations on the rules)
4. **Defuzzification interface** (transformation of the fuzzy result of the inference into a crisp output)

2.5 Advantages and disadvantages of fuzzy control

- No need to have a mathematical model of the system.
- It is possible to implement expert human knowledge and experience using comprehensible linguistic rules.
- It is possible to control non-linear plants.
- Thanks to dedicated processors, it is possible to control fast processes.

One of the results of the fuzzy "boom" in 90's was the general opinion that the fuzzy approach can solve all control problems and that classical control approaches will be soon abandoned. However, it is definitely not possible to solve all control problems using fuzzy logic because there are some very important drawbacks:

- There is no standard and systematic method for the transformation of the human knowledge or experience into the rule base of a fuzzy inference system, no general procedure for choosing the optimal number of rules, since a large number of factors are involved in the decision, e.g. performance of the controller, efficiency of computation, human operator behaviour, the choice of linguistic variables etc.
- Even when human operators exist, their knowledge is often incomplete and episodic, rather than systematic.
- It is not possible to show the stability of the controlled system, since the model is not known.
- It is not guaranteed that rules are coherent. It is possible to have a mismatch between the rules.
- Computing time could be long, because of the complex operations such as fuzzification and particularly defuzzification.

As a result, many researchers have been trying to automate the modelling process of fuzzy systems [Sun91]. The task can be divided in two parts:

- Structure identification of the system (related to finding a suitable number of rules and a proper partition of the feature space).
- Parameter identification (adjustment of the membership functions).

These two reasons have led to the idea of merging the neural network approach and fuzzy technology.

3 KHEPERA MOBILE ROBOT

We think that the only way to check out a control algorithm for a mobile robot is to apply it to the real robot. Simulation is a very nice approach, but the problem is that simulation is often not an approach that represents veraciously the processes from the real world.

Our work is based on experiments with a miniature robot named Khepera (Figure 9) developed in our lab [Mon94]. It has cylindric shape, measuring

Figure 9 Khepera mobile robot

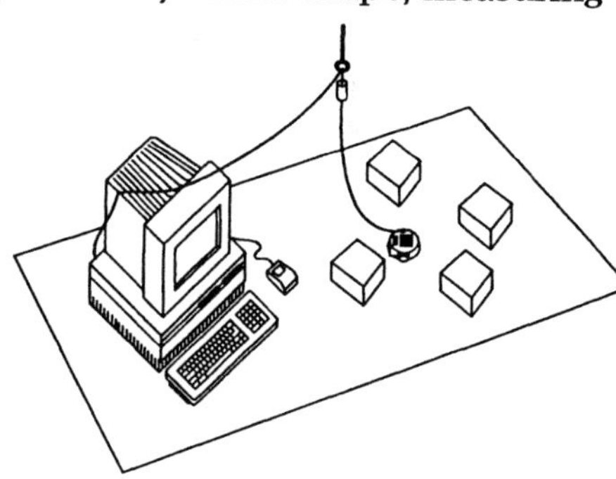

Figure 10 Khepera's working environment

55 mm in diameter and 30 mm in height. Its weight is only 70 g. Its small size allows experiments to be performed in a small work area (Figure 10). In this configuration, Khepera is remote controlled by a work station through a serial link.

The basic configuration of Khepera is composed of the CPU and of the sensory/motor boards. The CPU board is a complete 32 bit machine including a 16 MHz microcontroller, system and user memory, analogue inputs, extension busses and a serial link allowing a connection to different host machines (terminals, visualization software tools, etc.). The microcontroller includes all the features needed for easy interfacing with memories, with I/O ports and with external interruptions.

The sensory/motor board includes two DC motors coupled with incremental sensors, eight analogue infra-red (IR) proximity sensors (S0...S7 on Figure 11) and on-board power supply. IR sensors are composed of an emitter

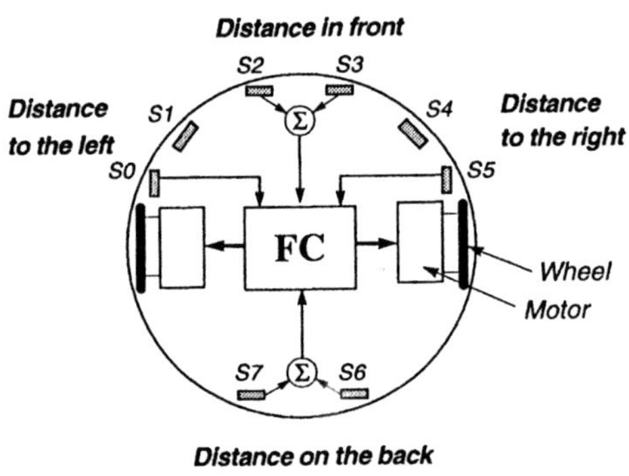

Figure 11 Position of the sensors on the Khepera mobile robot

and of an independent receiver. The dedicated electronic interface is built with multiplexers, sample/hold's and operational amplifiers. This allows the measurement of the absolute ambient light and the estimation, by reflection, of the relative position of an object from the robot. This estimation gives, in fact, an information about the distances between the robot and the obstacles.

What can one do with a robot that has 8 not very accurate sensors? Apply fuzzy logic? It can be one of the ideas. There are some others: genetic algorithms [Flo94], subsumption architecture [Bro86], distributed adaptive control [Ver92], learning with neural networks [Gau93, Zre94], artificial intelligence approaches [Bas93]. In next section, we will describe a very simple fuzzy controller for this robot.

3.1 A simple fuzzy controller for obstacle avoidance

Suppose that the first goal is to do obstacle avoidance with a mobile robot. Whatever the mobile robot has to perform, it has to avoid obstacles. Follow a guide, transport objects, go to a goal... All these actions are supposed not to damage the robot's hardware. Nor its environment!

The aim of our first "fuzzy" experiment was to design a simple controller for Khepera with a minimal number of the linguistic rules. The simple fuzzy controller could be designed in the following way:

Suppose that we take all the available informations. The controller would have 8 inputs and 2 outputs. The inputs are the distances between the robot and the obstacles and the outputs are the motor speeds. For every input we can define 2 membership functions (Small distance, Big distance) and for every output 3 membership functions (Backward, Stop, Forward). It means that there are 2304 ($2^8 * 3^2$) possible rules and a complete fuzzy engine needs 256 (2^8) rules! It is obvious that the design of such a controller without a precise methodology is not possible. In its absence, one is forced to reduce the number of inputs (the number of outputs cannot be reduced) and simplify the fuzzy inference engine.

We have chosen to deal with four inputs: distance to the right, to the left, in front and on the back of the robot (Figure 11). The structure of the implemented controller is shown in Figure 12. For each input we defined three membership functions, therefore classifying the distances as: Big, Average and Small. Each motor speed is represented by seven linguistic variables: Backward Fast, Backward Medium, Backward Slow, Stop, Forward Slow, Forward Medium and Forward Fast.

The linguistic rules can be defined in an almost intuitive way. The first very necessary rule is: ***If*** there are no obstacles around a robot, ***then*** it has to go straightforward. The second might be for example: ***If*** there is an obstacle in front of the robot, ***then*** it has turn to the left.

The minimum number of necessary rules is three but of course, the obtained behaviour is very primitive. These rules are represented in Table 1 and can be read in the following way:

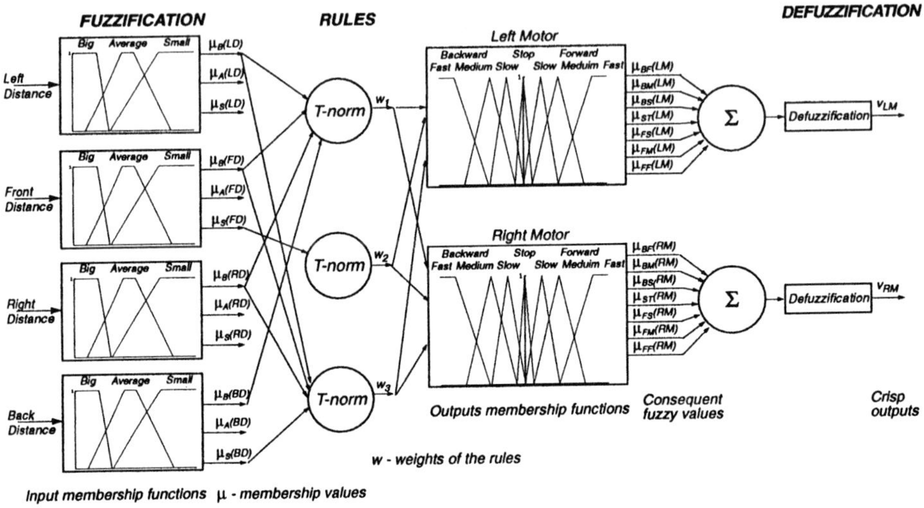

Figure 12 Structure of a fuzzy controller with 3 rules

Rule 1: ***If*** the distance to the left is Big ***and*** the distance in front is Big ***and*** the distance to the right is Big ***and*** the distance on the back is Big ***then*** left motor speed is Forward Medium and right motor speed is Forward Medium.

Rule 2: ***If*** the distance in front is Small (other distances are not considered) ***then*** left motor speed is Forward Medium and right motor speed is Backward Medium.

Rule 3: ***If*** the distance to the left is Big ***and*** distance in front is Big ***and*** distance to the right is Big ***and*** distance on the back is Small ***then*** left motor speed is Forward Fast and right motor speed is Forward Fast.

	Distances				Motor speeds	
	Left	Front	Right	Back	Left Motor	Right Motor
Rule 1:	Big	Big	Big	Big	Forward Medium	Forward Medium
Rule 2:		Small			Forward Medium	Backward Medium
Rule 3:	Big	Big	Big	Small	Forward Fast	Forward Fast

Table 1 Linguistic rules for the fuzzy controller

The robot avoids obstacles turning always in the same way. It turns while the back sensors do not "see" the obstacle. This algorithm allows the robot to move without a specific goal in a simple environment. It means that it cannot get out of a dead end for example.

The next step was the building of a more complicated controller by adding linguistic rules. With 16 rules, Khepera avoids obstacles very successfully turning to the left and to the right. Unfortunately, tuning by hand is very difficult when the controller has more than 16 or 20 rules because of possible mismatching of the rules. An automatic method for the choice of the rules and the parameters is necessary.

The critical problem of this algorithm is that the number of possible rules is very large and it is very difficult to make appropriate definitions of parameters for membership functions. Moreover, a very important factor is the big computational time and this was one of the main reasons that led us to simplify the fuzzy algorithm.

We made some experiments with Takagi and Sugeno's method [Tak83] where output membership functions were defined as singletons (see Section 2.2). The behaviour of the robot was not changed in comparison with the previously described method. The only difference was the computing time which decreased significantly. The centre of gravity method takes around 70% of computational time of the fuzzy controller.

This approach allows a possibility to realize very interesting behaviours of the robot by simply changing linguistic rules. The robot can follow the edge of a wall, approach (or run away from) a light source and so on.

4 NEURAL NETWORKS

The aim of neural networks is to model networks of biological neurons in the brain. From the neuroscientist's point of view, these models are extremely simplified. However, we use them because we hope that they give some insight into principles of biological "computation".

Artificial neural network models have many names in the literature like: connectionist models, parallel distributed processing models, neuromorphic systems, associative networks, etc. Their structure is based on our present understanding of biological nervous systems. In fact, they are parallel structures composed of many computational elements connected by links with variable weights.

4.1 Biological neuron

There are about 10^{11} neurons of different types in a brain. The main parts of a neuron are: *cell body* or *soma*, *dendrites* and *axon* (Figure 13). The

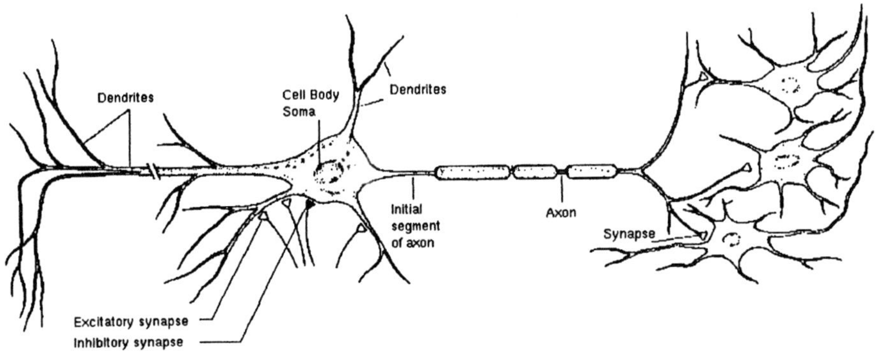

Figure 13 Biological neuron

cell nucleus is located in the soma, while dendrites are fibres connected to it. The axon is one long fibre that is the extension of the cell body. It branches into strands and substrands. At the end of these are *synapses* to other neurons. One axon makes typically a few thousand of *excitatory* or *inhibitory* synapses with other neurons.

The process of transmitting a signal from one neuron to another is chemically complex and is beyond the scope of this paper. Shortly, the receiving cell's *electrical potential* raises or lowers depending on the incoming signal. When this potential reaches a *threshold*, an *action potential* of fixed strength and duration is sent down the axon. The neuron "fires" and the action potential is transmitted through the axon to the synapses with other cells. After a *refractory period*, the cell can fire again [Her91].

4.2 Model of artificial neuron

The first model of an artificial neuron was proposed in 1943 by McCulloch and Pitts [McC43]. Although it is a very simple computation unit, it is the basic element of connectionist neural networks.

In the neuron in Figure 14, x_j $(j = 1, \ldots, n)$ is the *input stimuli* that can

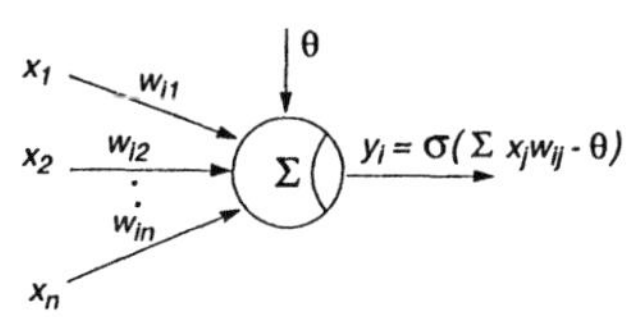

Figure 14 Model of artificial neuron

be either 1 or 0 representing the state of neuron j as firing or nonfiring respectively. The *synaptic weight* w_{ij} can be excitatory or inhibitory and represents the strength of synapse that connects the neuron j to the neuron i. θ is a *threshold*. The neuron potential is:

$$p_i = \sum_{j=1}^{n} x_j w_{ij}$$

This neuron sums n weighted inputs and passes the result through a nonlinearity (or activation function). The node is characterized by an internal threshold or offset θ and by the type of the nonlinearity. Figure 15 illustrates some common types of nonlinearities. The most employed is

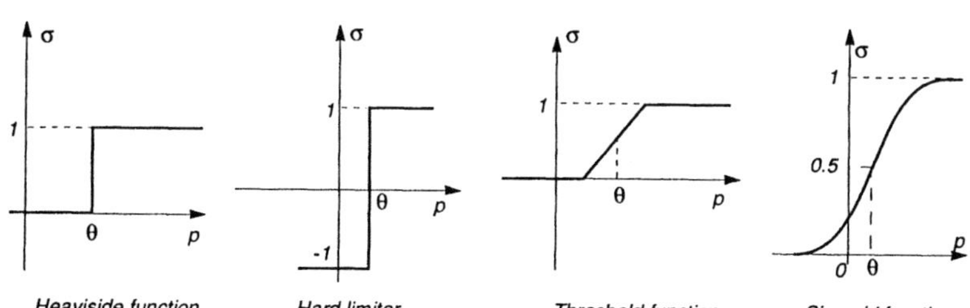

Figure 15 Nonlinearities in the model of artificial neuron

a sigmoid function because it is considered to be very close to the input/output function of a real neuron [Bla90].

It is obvious that this model does not imitate the real neuron. Why then all this excitement about neural networks in the last 40 years? The most important reason is their ability to learn and to generalize to new situations. Once a neural network has been trained for a set of data, it can interpolate and produce answers for the cases not present in the training set.

Neural networks are specified by their *topology* or *connection matrix*, *node characteristics* and *training* or *learning rules*. These rules specify an initial set of weights and indicate how weights should be adapted during learning to improve performance. This can be done in two ways:

- **Supervised learning.** The weights are adjusted on the basis of comparison of the output of the network with the desired (or correct) answer. It means that we "teach" the network to perform a desired computation.
- **Unsupervised learning.** This method is used when the learning goal is not defined at all in terms of specific correct examples. The only available information is in the correlation of the input data. The network has to create categories from these correlations and produce output signals corresponding to the input category [Her91].

4.3 Application to the Khepera mobile robot

A very simple neural network has been implemented on Khepera. It is based on Braitenberg's idea to directly connect sensors with actuators [Bra84] as shown in Figure 16. Every connection has a positive (excitatory) or negative (inhibitory) weight. The vehicle can perform different kinds of behaviours depending on the value of the weight and the kind of sensors. It can be attracted or repulsed from the light source, it can follow another vehicle, avoid obstacles and so on.

To allow a comparison with the fuzzy controller applied to the same robot, we will consider the problem of obstacle avoidance [Mon93]. A learning procedure has not been applied because of the simplicity of the task. Tuning of weights has been done by hand. The only problem was to find the proper weights for all connections.

If one examines the matrix shown in Table 2 and even if the principle of the algorithm is known, it is not possible to define which behaviour the robot will perform. From this point of view the fuzzy controller is superior

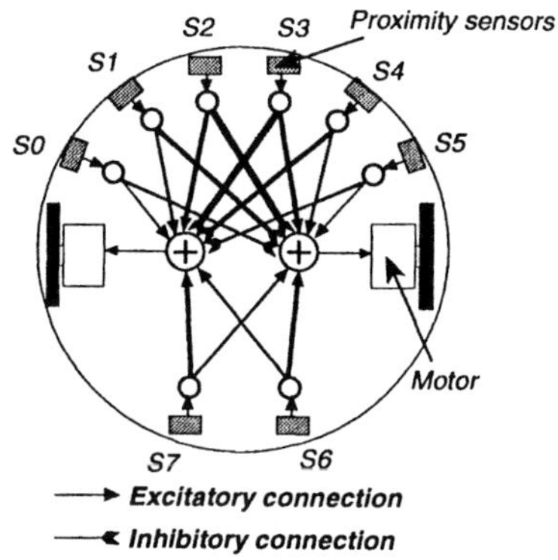

Figure 16 Braitenberg vehicle

Sensor:	S0	S1	S2	S3	S4	S5	S6	S7	Threshold
Motor 1	0.2	0.2	0.3	-0.72	-0.46	-0.2	0.26	0.15	0.30
Motor 2	-0.26	-0.51	-0.77	0.31	0.2	0.2	0.15	0.26	0.30

Table 2 Matrix of weights for Braitenberg's vehicle

because the behaviour is defined by a set of linguistic easily comprehensible rules. However the computational time for neural networks is shorter than that for fuzzy controllers.

4.4 Similarities between neural networks and fuzzy systems

Standard controllers deal with systems that have a precisely defined mathematical model. In the design of a fuzzy controller one does not need to know a precise mathematical model of the plant to be controlled. If the model however exists, it can be used for the simulation and for the test of the control strategy. The main advantages of fuzzy controllers are:

Neural network technology and fuzzy theory were developed at the same time. Their similarities are often discussed in the literature [Kos92, Tak90]. So far, the main direction of the research has been on automatic design and fine-tuning of the membership functions used in fuzzy control through learning by neural networks.

Neural networks and fuzzy systems are highly parametric parallel structures. In neural nets, both knowledge extraction and knowledge representation are difficult. On the other hand, the advantage of fuzzy systems is that knowledge is represented in the form of the comprehensive linguistic rules.

Fuzzy systems are able to treat the uncertain and imprecise informations. The weak points of fuzzy controllers are caused mainly to the difficulty of defining accurate membership functions and lack of the systematic procedure for the transformation of the expert knowledge into the rule base. The shape of the membership function in fuzzy systems and the threshold function in neural networks are similar. ***Multiply-add*** operation of artificial neurons is very close to ***max-min*** operation of approximate reasoning.

These reasons lead to the idea of merging these two approaches. There are two possible methods: a method by which individual merits are combined, and another by which analogies between these are overlapped. One possibility is to endow learning functions to fuzzy logic systems, or to conduct pattern processing before fuzzy logic is applied. The other is to incorporate fuzzy logic and linguistic rules into the structure of neural networks.

5 FUZZY NEURONS

There are a lot of *fuzzy neurons* proposed in the literature. Also, some learning and adaptation mechanisms for the proposed neurons are given [God94, Gup91, Jan92]. We will focus on only three of them.

5.1 Basic structure of fuzzy neuron

The theoretical structure of fuzzy neuron is shown in Figure 17. Since it has to be able to cope with fuzzy informations [Gup91], the inputs are fuzzy sets $x_1, \ldots, x_n$ on the universes of discourse $U_1, \ldots, U_n$, respectively. These fuzzy sets can be labelled by linguistic terms like: warm, high, large... First, they are weighted, but in a different way than in neural networks. After,

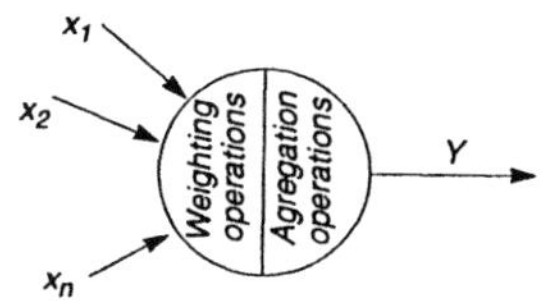

Figure 17 General model of fuzzy neuron

the inputs are aggregated by the fuzzy operation which is, in general the fuzzy implication function.

5.2 Models of fuzzy neurons

The principal motivation for building fuzzy neurons is to have models that behave in much the same way as biological neurons [Gup91]. Even if the motivation is the same as in the field of neural networks, we do not agree that these simplified models imitate biological neurons because natural structure is much more complicated than the models presented below.

The first model is an attempt to design the linguistic if-then rule. The experience of the neuron is stored in the fuzzy operator $\circ$ which is usually a T-norm or a T-conorm [Gup91]. It is composed from the current inputs and the past experiences. There are currently few, if any, learning methods proposed in the literature.

The second model is the fuzzy extension of the non-fuzzy neuron. It has n fuzzy or non-fuzzy inputs and the weighting operations are replaced by membership functions. These operations are *synaptic*. The results are membership values of the corresponding input in the fuzzy set. These values are then aggregated and the aggregation operation is called *somatic*. It means that the output, considered as a level of confidence or degree of truth, is a crisp value from the interval [0,1]. The aggregating operation $\otimes$ may be one of the operators such as ***min*** or ***max***. Learning method for this kind of neuron is proposed in [God94].

The third fuzzy neuron is also an extension of a non-fuzzy neuron. It has fuzzy inputs $X_1, \ldots, X_n$. Each input undergoes a synaptic operation which results in another fuzzy set. All modified fuzzy sets $X'_1, \ldots, X'_n$ are aggregated to produce another n-dimensional fuzzy set Y. The operator

Model 1 Model of the linguistic if-then rule	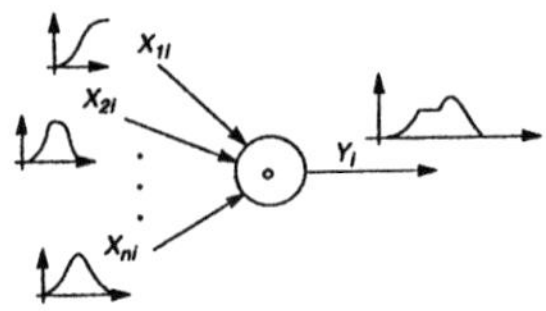	***Rule:*** ***If*** X_{1i} ***and*** X_{2i} ***and*** ... ***and*** X_{ni} **then** Y_i $Y_i = X_{1i} \circ X_{2i} \circ \cdots \circ X_{ni}$
Model 2 Fuzzy extension of the non-fuzzy neuron	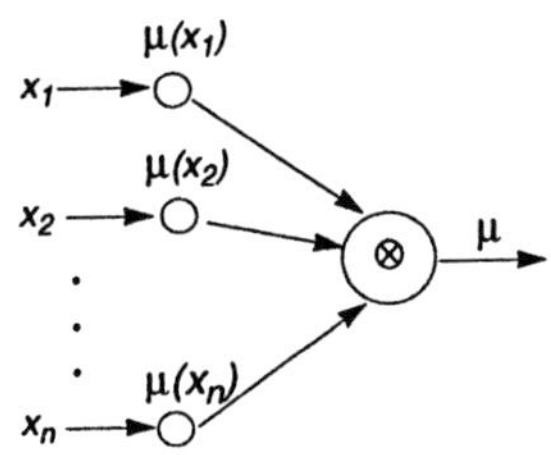	***Fuzzification*** $\mu(x_1, x_2, \ldots, x_n) =$ $\mu_1(x_1) \otimes \mu_2(x_2) \otimes \cdots \otimes \mu_n(x_n)$
Model 3 Extension of the non-fuzzy neuron	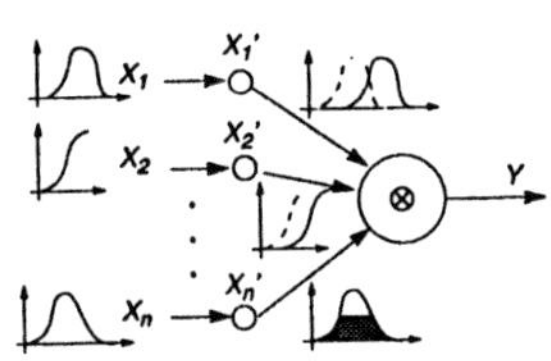	***Fuzzy inference*** $Y = X_1' \otimes X_2' \otimes \cdots \otimes X_n'$ $X_i' = G_i(X_i), i = 1, 2, \ldots, n$

Table 3 Models of fuzzy neurons

$\otimes$ may be one of the T-norms or T-conorms. Similarly to the first model, learning method for this one is not proposed in the literature.

5.3 Learning methods

The aim of a learning method is to train a system to perform the desired computation by iterative adjustment of the synaptic weights. The proposed adaptation mechanisms for neurons described above can be *synaptic* or *somatic* [God93]. They are based on the supervised neural network learning scheme.

Synaptic adaptation means that all the inputs are constantly modified by synaptic operations and then forwarded to the neuron body. These operations are simple in the non-fuzzy case (multiplication), but in fuzzy neurons they are very complex. If the membership functions and inputs are triangular, the proposed modification are shown in Figure 18. The dotted lines

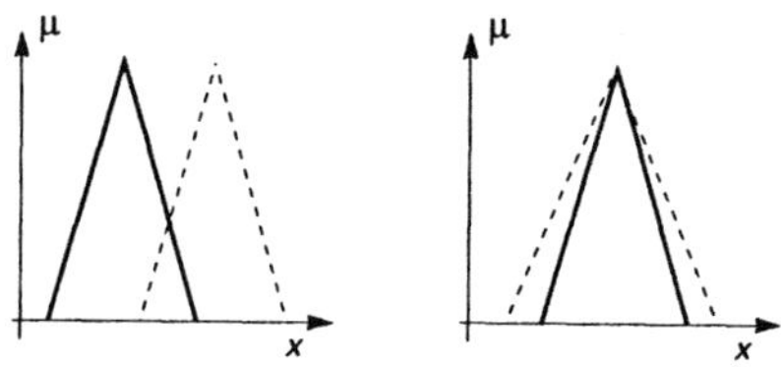

Figure 18 Changes of membership functions during learning phase

represent the fuzzy inputs before modification, and the solid ones after modification.

Somatic adaptation implies that the modifications are applied to the structure of the fuzzy neuron. These modifications include: changing the rule base, changing the membership functions assigned to the fuzzy terms in every rule and changing the aggregation functions.

These types of adaptation can be applied in supervised learning procedures with well known methods like the gradient descent algorithm.

6 NEURO-FUZZY CONTROLLER

Usually, neuro-fuzzy controllers are fuzzy inference systems implemented under the framework of neural networks. The idea of the design of neuro-fuzzy system is to exploit the advantages of fuzzy logic, as an important branch of fuzzy set theory that emulates human reasoning process, and a neural network's ability to learn.

Learning methods for fuzzy controllers are usually the application of conventional adaptive control techniques. Since these techniques are used in neural networks, as for example the least square algorithm, adaptive fuzzy controllers are named neuro-fuzzy networks or neuro-fuzzy controllers. One of the interesting architectures for neuro-fuzzy controller has been proposed by Jang [Jan92].

In this section we will describe one learning method for a fuzzy controller [Nom92]. The starting point is the Takagi-Sugeno method where output membership functions are defined as singletons.

6.1 Linguistic rules

Assume that a fuzzy controller has m inputs $x_1, x_2, \ldots, x_m$ and one output y and that we defined n linguistic rules in the form:

R_i: ***If*** x_1 is A_{i1} ***and*** x_2 is A_{i2} ***and*** ... ***and*** x_m is A_{im} ***then*** y is w_i , $i = 1, \ldots, n$

where i is the index of the rule, A_{ij} is a fuzzy set for i-th rule and j-th linguistic variable defined over the universe of discourse for j-th variable and w_i is a real number that represents a consequent part.

6.2 Membership functions

Every membership function in this controller is defined as symmetric and triangular and has to be partially differentiable (Figure 19). The definition

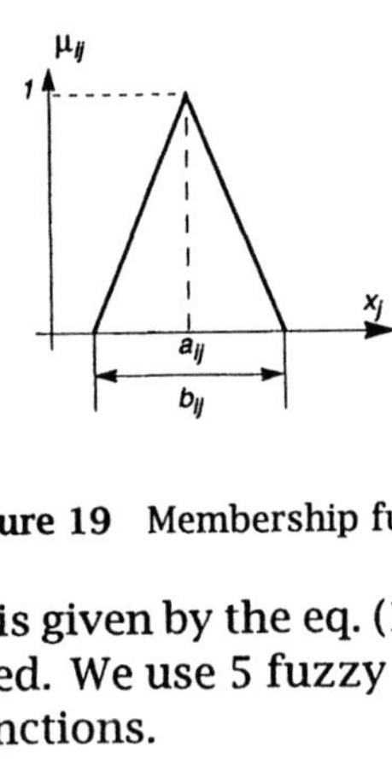

Figure 19 Membership function

of the membership values is given by the eq. (12.3). For every input the universe of discourse is defined. We use 5 fuzzy sets as uniformly distributed triangular membership functions.

$$\mu_{ij} = \begin{cases} 1 - \frac{2|x_j - a_{ij}|}{b_{ij}} & , \text{ for } a_{ij} - \frac{b_{ij}}{2} < x_j < a_{ij} + \frac{b_{ij}}{2} \\ 0 & , \text{ otherwise} \end{cases} \tag{12.3}$$

6.3 Fuzzy inference

For the application of the rules we need to define a fuzzy inference, and in this case, we will take product operator as T-norm. It means that the firing strength of every rule is:

$$\mu_i = \mu_{i1}\mu_{i2}\cdots\mu_{in} \tag{12.4}$$

The consequent part of the rules are crisp values, and the evaluation of a centre of gravity is given by:

$$y = \frac{\sum_{i=1}^{n} u_i w_i}{\sum_{i=1}^{n} u_i} = \frac{u_1}{\sum_{i=1}^{n} u_i} w_1 + \frac{u_2}{\sum_{i=1}^{n} u_i} w_2 + \cdots + \frac{u_n}{\sum_{i=1}^{n} u_i} w_n = \sum_{i=1}^{n} \overline{u_i} w_i \tag{12.5}$$

A general scheme of this system with two inputs and one output is represented in Figure 20. This presentation is analogous to the architecture of a

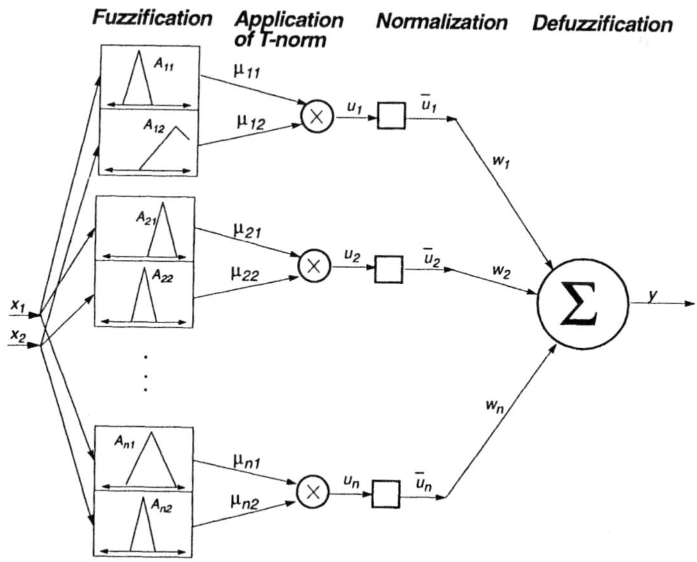

Figure 20 Neuro-fuzzy controller

feed-forward artificial neural network. The first layer performs a fuzzification for each linguistic rule and it is a layer of neurons of model 1 (Table 3). Outputs from this layer are membership values (eq. 12.3) and they are fed into the next layer which performs a T-norm operation -product- (eq. 12.4). The result is a firing strength for each rule and in the next layer, firing strengths are normalized (eq. 12.5). The last layer computes the overall output as the weighted sum of the incoming signals.

6.4 Computational aspects

Suppose that the parametric model of one non-linear system is given in the following form:

$$\frac{dx}{dt} = f(x, u, z, t)$$
$$y = g(x, u, z, t)$$

where x is a state vector, u an input vector, z a parameter vector and y is the output vector. The learning problem is a parameter estimation problem. All solutions to parameter estimation problems consist in finding the extremum of *criterion (loss) function* V considered a function of the parameters of the unknown system. The function to minimize is:

$$V(z) = \frac{1}{2}\left(y(t) - y_d(t)\right)^2 \tag{12.6}$$

where y_d is the desired output provided by an expert. The minimization of this function can be done in several ways [Ast71]. All of them use the scheme suggested by Robbins and Monro in [Rob51]. Here we will use the criterion (eq. 12.6) and apply the *method of stochastic approximation* to identify the parameters of the fuzzy system. It is an iterative procedure given by:

$$z(t+1) = z(t) - \Gamma \nabla_z V\left[z(t)\right]$$

where z is the vector of parameters to adapt, Γ is the predefined constant and $-\nabla_z V$ is the notation for the gradient of V with respect to z:

$$\nabla_z V = \left[\frac{\partial V}{\partial z_1}, \ldots, \frac{\partial V}{\partial z_n}\right]$$

6.5 The learning algorithm

In the fuzzy controller presented above, z is given by:

$$z = (a_{11}, \ldots, a_{nm}, b_{11}, \ldots, b_{nm}, w_1, \ldots, w_n)$$

The number of parameters to estimate is $p = 2nm + n$. The vector which minimizes the loss function is given by:

$$\left(-\frac{\partial V}{\partial z_1}, -\frac{\partial V}{\partial z_2}, \ldots, -\frac{\partial V}{\partial z_p}\right) = 0$$

and the learning rule:

$$z_k(t+1) = z_k(t) - \Gamma \frac{\partial V(z)}{\partial z_k} \quad , \quad k = 1, \dots, p \tag{12.7}$$

It follows from (eq. 12.3 - 12.7) that the equations for the adaptation of the parameters of fuzzy system are:

$$a_{ij}(t+1) = a_{ij}(t) - \Gamma_a \frac{u_i}{\sum_{i=1}^{n} u_i} (y - y_d)(w_i - y) \operatorname{sgn}\left(x_j - a_{ij}(t)\right) \frac{2}{b_{ij}\mu_{ij}(x_j)}$$

$$b_{ij}(t+1) = b_{ij}(t) - \Gamma_b \frac{u_i}{\sum_{i=1}^{n} u_i} (y - y_d)(w_i - y) \frac{1 - \mu_{ij}(x_j)}{b_{ij}\mu_{ij}(x_j)}$$

$$w_i(t+1) = w_i(t) - \Gamma_w \frac{u_i}{\sum_{i=1}^{n} u_i} (y - y_d)$$

The iterative procedure for the adaptation of parameters and for the minimization of the criterion function can be summarized as follows:

1. Initialisation of parameters.
   - Consequent values w_i are random numbers.
   - Choice of antecedent parameters a_{ij} and b_{ij} (all membership functions on the universe of discourse are regularly distributed and have the same width).
2. Data input $(x_1, x_2, \dots, x_m, y_d)$.
3. Fuzzy inference.
4. Adaptation of consequent parameters w_i.
5. Repeat once the step 3 with new values for w_i and step 4.
6. Adaptation of parameters a_{ij} and b_{ij}.
7. Evaluation of the criterion function V.
8. Repeat the steps 3 to 7 until V is smaller than one threshold value.

This method is very similar to the method of gradient descent for the Perceptron [Rum89].

6.6 Application of the neuro-fuzzy controller to the Khepera

We applied this algorithm to the obstacle avoidance problem of the Khepera mobile robot [God95]. The goal was to adjust the parameters of membership functions using a supervised learning procedure. The robot started with 625 basic rules and non-adjusted membership functions. During the learning phase, the adaptation is done each time the robot encounters an unknown situation.

In the first step we have limited the learning method to the adaptation of parameters with a fixed rule base. After learning, we suppressed all rules not used in the supervised learning phase. The problem is that some membership functions "disappeared" from the universe of discourse or overlapped with other functions. It was necessary to make an analysis of the obtained results and to suppress useless functions and rules.

This method is very interesting because it has the ability to express the knowledge acquired from input-output data in the form of linguistic rules.

7 CONCLUSION

In this paper some basic notions of fuzzy controllers and neural networks have been presented. The application of these methods is described in the case of the obstacle avoidance behaviour. As an experimental platform we used the Khepera mobile robot. It is shown that a simple fuzzy controller can be applied for a realization of very simple tasks. Even if the number of rules is small, the robot shows interesting behaviour.

The same problem is treated in the frame of a neural network based approach. We presented a simple neural net which enabled the robot to have different kinds of behaviour. The problem is that the extraction of knowledge from the neural network is impossible.

The advantage of fuzzy systems is that knowledge is represented in the form of the easily comprehensible linguistic rules. On the other hand, neural networks have the ability to learn. This lead to the idea of supplying fuzzy controllers with learning functions. This was implemented on a controller based on the Takagi-Sugeno method. The learning procedure is the stochastic approximation method known in the field of identification of systems which corresponds to supervised learning of neural networks. The results show that fuzzy systems have an ability to learn and that the methods used in the neural network field can be applied. The method is tested on Khepera and the experiments show that the robot can learn by adjusting the parameters of the controller.

Acknowledgements

The author would like to thank André Guignard, Edo Franzi and Francesco Mondada for the design of Khepera and for the help in testing the algorithms presented in this paper. She also gratefully acknowledge the constructive criticism on the manuscript and the constant support of Jean-Daniel Nicoud, Paolo Ienne, Pero Subašić, Laurent Tettoni and Andrea Gees.

REFERENCES

[Ast71] K.J. Astrom and P. Eykhoff. System identification - a survey. *Automatica*, 7:123–162, 1971.

[Bas93] A. Basso, F. Mondada, and C. Castelfranchi. Reactive goal activation in intelligent autonomous agent architectures. In *AAI93 symposium on Abstract Intelligent Agents*, Rome, January 1993.

[Bla90] François Blayo. *Une implantation systolique des algorithmes connexionnistes.* PhD Thesis N° 904, École Polytechnique Fédérale de Lausanne, Lausanne, 1990.

[Bra84] Valentino Braitenberg. *Vehicles.* MIT Press, 1984.

[Bro86] R. Brooks. A robust layered control system for a mobile robot. *IEEE Robotics and automation*, RA-2:14–23, March 1986.

[Flo94] D. Floreano and F. Mondada. Automatic creation of an autonomous agent: Genetic evolution of a neural-network driven robot. In *SAB94*, Brighton, 1994.

[Gau93] Ph. Gaussier and S. Zrehen. Emergence of behaviors on a mobile robot: Learning with neural networks. In *Learning days in Jerusalem*, Jerusalem, 1993.

[God93] Jelena Godjevac. State of the art in the neuro fuzzy field. Technical Report 93/25, École Polytechnique Fédérale de Lausanne - DI, April 1993.

[God94] Jelena Godjevac. Comparison between classical and fuzzy neurons. In *EUFIT*, volume 3, pages 1326–1330, Aachen, Germany, September 1994.

[God95] Jelena Godjevac. A learning procedure for a fuzzy system: application to obstacle avoidance. to be published ISFL '95, Zurich, Switzerland, 1995.

[Gup91] M. M. Gupta and J. Qi. On fuzzy neuron models. In *Int. Joint Conf. on Neural Networks, IEEE+INNS*, volume 2, pages 431 – 435, Seattle, July 1991.

[Her91] John Hertz, Anders Krogh, and Richard G. Palmer. *Introduction to the Theory of Neural Computation.* Santa Fe Institute Studies in Sciences of Complexity. Addison-Wesley, Redwood City, Calif., 1991.

[Jan92] J.-S. Roger Jang. Anfis, adaptive-network-based fuzzy inference systems. *IEEE Trans. on systems, Man and Cybernetics*, 1992.

[Kic78] W. J. M. Kickert and E. H. Mamdani. Analysis of a fuzzy logic controller. In *Fuzzy Sets and Systems 1*, pages 29–44. North Holland Publishing Company, 1978.

[Kos92] Bart Kosko. *Neural networks and fuzzy systems.* Prentice-Hall, 1992.

[Lee90] C. C. Lee. Fuzzy logic in control systems: fuzzy logic controller - part 1 and part 2. *IEEE Trans. on systems, Man and Cybernetics*, 20 (2):404–435, 1990.

[Li88] Y. F. Li and C. C. Lau. Development of fuzzy algorithms for servo systems. In *IEEE International Conference on Robotics and Automation*, Philadelphia, Pensylvania, April 1988.

[Mam75] E. H. Mamdani and S. Assilian. An experiment in linguistic synthesis with a fuzzy logic controller. *Int. J. Man-Machine Studies*, 7:1–13, 1975.

[McC43] McCulloch and Pitts. A logical calculus of the ideas immanent in nervous activity. *Bulletin of Mathematica Biophysics*, 5:115–133, 1943.

[Mon93] Mondada F. and E. Franzi. Biologically inspired mobile robot control algorithms. In *NFP-PNR 23 Symposium*, Zurich, Switzerland, October 1993.

[Mon94] Francesco Mondada, Edoardo Franzi, and Paolo Ienne. Mobile robot miniaturization: A tool for investigation in control algorithms. *Informatik*, pages 17–20, February 1994.

[Nom92] H. Nomura, I. Hayashi, and N. Wakami. A learning method of fuzzy inference rules by descent method. In *Proceedings of IEEE Int. Conf. on Fuzzy Systems*, pages 203–210, San Diego, 1992.

[Rob51] H. Robbins and S. Monro. A stochastic approximation method. *Annals of Mathematical Statistics*, 22:400–407, 1951.

[Rum89] David E. Rumelhart and James L. McClelland. Learning internal representations by error propagation. In *Parallel distributed processing*. MIT Press, Cambridge, Mass., 1989.

[Sun91] C.-T. Sun and J.-S. Jang. Fuzzy modeling based on generalized neural networks and fuzzy clustering objective functions. In *Proc. of the 30th Conference on Decision and Control*, pages 2924–2929, December 1991.

[Tak83] T. Takagi and M. Sugeno. Derivation of fuzzy control rules from human operator's control actions. In *Proc. of the IFAC Symp. on Fuzzy Information, Knowledge Representation and Decision analysis*, pages 55–60, July 1983.

[Tak90] H. Takagi. Fusion technology of fuzzy theory and neural networks, survey and future directions. In *Proc. of the International Conf. on Fuzzy Logic& Neural Networks*, pages 13–26, Iizuka, Japan, July 1990.

[Ver92] P. F. M. J. Verschure, B. J. A. Koese, and R. Pfeifer. Distributed adaptive control: The self-organization of structured behavior. *Robotics and Autonomous Agents*, 9:181–196, 1992.

[Zad65] Lotfi Zadeh. Fuzzy sets. *Information and Control*, 8:338–353, 1965.

[Zad73] Lotfi Zadeh. Outline of a new approach to the analysis of complex systems and decision processes. *IEEE Trans. on systems, Man and Cybernetics*, 3(1):28–44, January 1973.

[Zad94] Lotfi Zadeh. Fuzzy logic, neural networks and soft computing. *Fuzzy Systems, Communications of the ACM*, 37(3):77–84, March 1994.

[Zim90] H.-J. Zimmermann. *Fuzzy Sets Theory - and Its Applications.* Kluwer Academic Publishers, 1990.

[Zre94] S. Zrehen and Ph. Gaussier. Why topological maps are useful for learning in an autonomous agent. In *Proceedings PerAc*, Lausanne, September 1994. IEEE Press.

SUBJECT INDEX

aggregation rules, 175, 184
ample field, 92
andness, 187
anonymity, 182
artificial neuron, 307

binary structure function theory, 201
biological neuron, 306

cartesian product, 4
citizen sovereign, 182
classical possibility, 125
closedness, 35
cognition, 227, 239
complete lattices, 68
component(s), 207, 215
compositional rule of inference, 272, 295
continuity of mappings, 19

decision making, 138, 155, 293
defuzzification, 248
deterministic system, 140, 145
direct closedness, 41
direct image, 38, 98
dispersion, 188, 192, 194
distributive lattices, 68
dynamic programming, 137

extension principle, 97

fuzzification, 4, 21
fuzziness, 3, 18, 138
fuzzy
  algebraic structures, 6
  controllers, 245, 248, 253, 267
  control, 245, 246, 268, 291
  dynamic programming, 137
  graphs, 272
  implication operators, 20, 38, 71, 274
  Lagrangc interpolation, 276
  logic, 225, 230, 291
  multivalued mappings, 37
  neuron, 311
  number, 15, 155
  piecewise interpolation, 276
  preferences, 155, 158, 175
  relational equations, 67
  set, 3, 94
  singleton, 40
  system, 142
  topological space(s), 10, 39
  topology, 7

group decision making, 175

homomorphism, 74

independence of irrelevant alternatives, 181
inference mechanism, 276
infimum-morphism, 73
information aggregation, 175
intelligence, 227

linguistic rule, 294
linguistic variables, 293
local closedness, 41
lower closedness, 42
lower inverse image, 39

maximally surjective triangular norm, 67
mobile robots, 291
multistate structure function theory, 201

neural networks, 232, 291, 305
neural systems, 225, 291
neural-fuzzy logic, 236
neuro-fuzzy controllers, 291, 313
neuro-fuzzy control, 291
neuron, 233
neutrality, 181
non dictatorship, 182
non-void, 38
nonlinear systems, 235
normalized, 38
notion state, 201

ordered hierachies, 189
ordered weighted averaging, 178, 187
orness, 187, 191, 195

positive responsiveness, 182
possibilistic
  extensions, 98
  logic, 89, 115
  truth values, 105, 109, 116
  variable, 95
possibility measure, 94

regularity, 41
relational calculus, 16
reliability theory, 201, 205
robot, 291
root system, 69

soft computing, 292
soma, 234
state space, 212
stochastic system, 140, 146
strongly quasi-separated, 41
supervised learning, 308
supremum-morphism, 73
surjective, 38
synapse, 234
system(s), 207

transitivity, 155
triangular norm (t-norm), 38, 67, 92, 177
truth value, 96
truth-functional approximation, 112

unanimity, 182
uncertainty, 227
unrestricted domain, 181
unsupervised learning, 308
upper closedness, 43
upper inverse image, 39
upper semi-continuity, 35

viability, 271